MRMU
: MID-RISE MIX-USE

PUSAN NATIONAL UNIVERSITY
SYRACUSE UNIVERSITY SCHOOL OF ARCHITECTURE
JOINT RESEARCH PROJECT
FALL 2022

Sponsored by ING Architecture

MRMU

MID-RISE MIX-USE

Pusan National University
Keojin Jin, Assistant Professor

Syracuse University School of Architecture
Bing Bu, Visiting Critic

STUDENTS

Kihyun Kim
Ryulhyeon Kim
Myeungbin Kim
Eunhye Kim
Sangjeen Moon
Changmin Park
Gyusik Son
Juhae Yu
Sangmin Lee
Heonju Lee
Hyeonjin Lim
Seungho Jang
Unghui Jung

Adrienn Virag
Sarah Bannan
Matthew Williams
Aryan Ambani
Michela Brittis-Tannenbaum
Donovan Hernandez
Ruohan Jiang
Tianchonghui Fang
Peter Wang
Yiting Zhong
Kelly Chan
Tony Dai

ACKNOWLEDGEMENT

A sincere thanks to our sponsor and colleagues who made this studio possible.

SPONSOR

ING Architecture
Song Juyeong, CEO

EDITOR

Hyeonjin Lim

GRAPHIC DESIGNER

Leonardo Citterio, INCLS Shanghai

MRMU

2, Busandaehak-ro 63beon-gil, Geumjeong-gu
Busan
Republic of Korea

Photo Credits
Students of PNU Studio
Students of SYR Studio

ING GROUP

SYRACUSE ARCHITECTURE

CONTENTS

INTRODUCTION

In modern cities, where development on a scale that humanity has never experienced is possible, the main focus of architectural planning is how to turn the hybridity of various socio-economic factors, including ownership patterns and actual use of a building (or a collection of buildings) into a catalyst for positive regional development.

The third-year architectural design studio of Pusan National University in Korea (taught by Keojin Jin) and the third-year architectural design studio of Syracuse University in the U.S. (taught by Bing Bu) performed a joint, semester-long research project. The subject of the research was "Mid-rise mix-use (M.R.M.U.)." Through this research, a solution to create a dense, healthy, and active urban typology was explored.

Each student team was given an active urban block in two vibrant cities, each with a distinct style of urban formation: Busan and Los Angeles (LA). Residential blocks in both cities suffer from under-utilized street space and excessive motor vehicle usage. While Busan's uncontrolled crowdedness presents distinct problems (typical in quickly grown Asian cities) relative to LA's problem of inefficient horizontal spread (a typical issue in most cities in the US), both cities share the common issue of creating vehicle-dependent and isolated communities.

This research promotes carefully designed mid-rise mixed-use as a solution for both cities and explores the various possibilities that typology will create for these cities. Based on the research and analysis done at the beginning of the course, students chose small sites in each city for which to design a residential complex.

The overall design research process evolved around shifting complex social issues and interpreting the site's proposed spatial quality from various perspectives. Topological abstraction (diagram), which combines the various socio-economic factors and empirical analysis digested through creative representation (description) served as two primary tools for the design process.

들어가는 말

인류가 역사상 경험하지 못한 규모의 개발이 가능해진 현대 도시에서 건축 계획의 주요 초점은 건물(또는 건물 집합체)의 소유 형태를 포함한 다양한 사회경제적 변수, 그리고 사용자의 물질적 편의성 간의 복잡한 혼성(Hybridity)을 어떻게 지역 개발의 긍정적 촉매로 전환하는 것이다.

한국 국립부산대학교 3학년 건축디자인 스튜디오(진교진 교수)와 미국 시러큐스대학교 3학년 건축디자인 스튜디오(빙부 교수)가 한 학기 동안 이 문제를 다루는 공동 연구 프로젝트를 진행했다. 연구 주제는 "중층 주상 복합 건물 유형(Mid-rise Mix-use: M.R.M.U)"이다. 본 연구를 통해 조밀하고 건강하며 활동적인 도시 유형학을 만들기 위한 다양한 건축적 방안을 탐색하였다.

학생들은 부산과 로스앤젤레스(LA)라는 독특한 도시 형성 스타일을 가진 활기찬 두 도시에서 활동적인 몇몇 도시 블록을 팀별로 선정하였다. 두 도시의 주거 블록은 활용도가 낮은 거리 공간과 과도한 차량 사용으로 인해 도시 환경의 질 저하라는 어려움을 겪고 있다. 부산의 통제되지 않은 난개발로 인한 혼잡(빠르게 성장하는 아시아 도시에서 일반적인)은 LA의 비효율적인 수평 확산 문제(미국 대부분의 도시에서 일반적인)와 대비되어 뚜렷한 도시적 문제의 차이를 제시하지만, 두 도시는 차량에 의존하고 고립된 커뮤니티를 만드는 공통 문제를 공유한다.

본 연구는 두 도시를 위한 솔루션으로 세심하게 설계된 중층 복합 건축의 유형학이 이러한 도시를 위해 만들 수 있는 다양한 도시적 가능성을 탐구한다. 연구 과정 초기에 수행된 조사 및 분석을 기반으로 학생들은 각각 주거 단지를 설계할 각 블록 속의 중대형 규모의 부지를 선택하였다.

전반적인 디자인 연구 프로세스는 복잡한 사회적 문제를 변화시키고 다양한 관점에서 사이트의 제안된 공간 품질을 해석하는 것을 중심으로 발전했다. 다양한 현대사회의 사회 경제적 변수와 지형(Topology)적 조건이 중첩된 추상 다이어그램을 통한 탐색과, 다양한 사료적 분석의 창의적 표현(설명)은 설계 프로세스의 두 가지 주요 도구로써 사용되었다.

MID-RISE MIX-USE

Bing Bu
Principal, INCLS
Visiting Critic, Syracuse Architecture

Mid-Rise Mix-Use buildings could be considered as a most basic module of urban form, a city on a building scale. They have appeared globally along with modern urbanization, in varied forms such as apartments over arcades in Europe, or shop houses in trading towns of southern China. From the cases that we will collect throughout history and around the world, it is hard to generalize and define them into a single building typology. But we may still categorize the programmatic constituents of MRMU into four kinds of spaces, the domestic, the communal, the commercial, and the public. The interlocking yet blurring boundaries/interfaces between the four and the overall topological configuration will be our research and design focus.

MRMU has certain advantages of efficiency in many aspects, such as being easier to build with conventional construction methods, more feasible in the financing, less relying on mechanical means for vertical circulation, and more environment friendly as it provides work-live proximity and reduces car dependence. All such advantages are affected and defined by a broad range of urban contexts, near or far, local or global, in terms of social, technological, ecological, economical, and political. Exploring such contexts and representing them visually will be a key task of our research.

An MRMU needs to be situated on a site to examine its possible responses to the contexts. The situation could be a seamless infill to a densely built area, or a lone castle in an open field of suburb low-rise, or anywhere in-between, either heralding the coming intensification of an urban core or mediating with the fragments of constant up-and-down.

The overall design research process of this project will be evolving around the translation of complex social issues into formal operations, and the interpretation of proposed spatial quality under multiple perspectives. Topological abstraction (diagram) and experiential representation (narratives) will be the two primary tools.

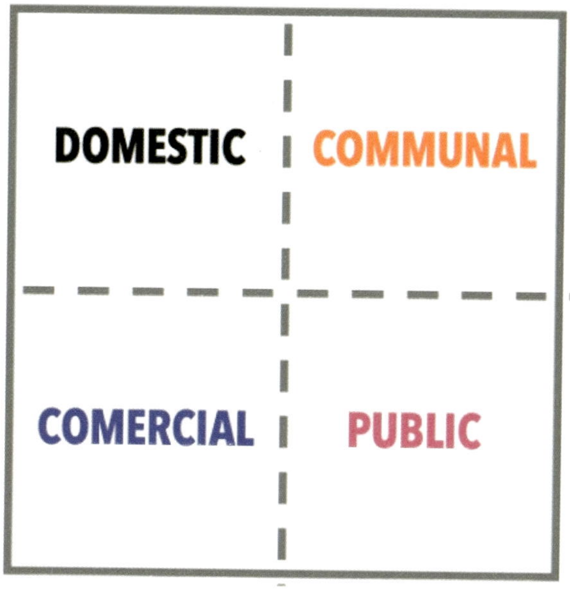

중층 주상복합 건물(이하 MRMU)는 도시 형태의 가장 기본적인 모듈, 즉 건물 규모의 도시로 간주될 수 있다. 그들은 유럽의 아케이드에 있는 아파트나 중국 남부의 무역 도시에 있는 상점과 같은 다양한 형태를 가지고 현대 도시화와 함께 세계적으로 나타났다. 이제 우리가 역사를 통해, 그리고 전세계에서 수집해 올 사례들을 하나의 건물 유형학으로 일반화하고 정의하기는 아직 어렵다. 그러나 우리는 여전히 MRMU 사례를 구성하는 프로그램 요소를 주거(domestic), 공동(communal), 상업(commercial), 공공(public)의 네 가지 공간으로 분류할 수 있다. 분류한 네 가지지 공간들과 전체적인 위상학적 구성 사이의 상호 연결되지만 흐릿한 경계/인터페이스가 우리의 연구와 설계의 초점이 될 것이다.

MRMU는 기존 공법을 사용하되 시공하기가 더 쉽고, 자금 조달이 더 가능하며, 수직적 순환을 위한 기계적 수단에 덜 의존하며, 일과 생활 사이의 상호 근접성을 제공하며 자동차의 의존도를 줄여 더 환경친화적인 등 여러 측면에서 효율적으로 확실한 이점을 가지고 있다. 이러한 모든 이점은 사회적, 기술적, 생태학적, 경제적 및 정치적 측면에서 가깝거나 먼 지역, 혹은 세계적인 광범위한 도시 상황에 의해 영향받고 정의된다. 그러한 맥락을 탐구하고 시각적으로 표현하는 것이 이 연구의 핵심 과제이다.

MRMU는 맥락에 대해 가능한 응답을 검사하기 위해 대지에 위치해야 한다. 여기서 조밀하게 지어진 지역을 빈틈없이 매끄럽게 채울 수도 있고, 교외의 저층 공터에 자리를 잡은 외딴 성일 수도 있고, 그 사이 어디든 도시 중심부의 강화를 예고하거나 또 끊임없이 오르내리는 작은 파편들로 중재할 수도 있다.

이 프로젝트에서 전반적인 디자인 연구 과정은 복잡한 사회적 이슈를 공식적 작업으로 전환하고 제안된 공간 품질에 대한 다양한 관점에서의 해석을 중심으로 점차 발전했으며, 위상적 추상화(다이어그램)와 경험적 표현(서술)이 두 가지 주요 도구가 되었다.

MID-RISE MIX-USE
as Korean Residential Typology

한국의 주거 유형의 대안으로서의 중층 주상복합 건축유형

Keojin Jin
Assistant Professor,
Pusan National University

진교진
조교수
부산대학교 건축학과

Ordinary Residential Typology and the Role of the Architect

In Korea, more than 80 percent of the homes built since the 1980s are large, high-rise concrete apartments. The statistic that 60 percent of the Korean population currently resides in high-rise concrete apartments reveals how much this typology dominates the lives of Koreans. Numerous commentaries and discussions have reflected from an architectural point of view on the typology of the apartments covering Korea.

A representative point of view of the public is a critical attitude toward the "one-size-fits-all" design of these buildings. Architects especially lament the scenery of Korean cities, where high-rise concrete apartments fill entire vistas. However, these commonly held points of view are unexpectedly myopic; criticism of the repetition of design is unfairly exaggerated, covering over the actual, major weaknesses that this kind of typology holds.

In fact, cities typically renowned for their beauty–Paris, Barcelona, Vienna or Melbourne, for example–consist of repetitive apartment buildings as well. Rather than condemning these similarities between most of their buildings as "one-size-fits-all design," we often praise the unified look of such cities. Since ancient times, humans have been building repetitive structures and slowly developing their know-how to take control of the complex variables involved in building structures. Just as humans living in caves in the primitive ages had to use fire to warm themselves and then dig the ground around the fire to make a windbreak, humans building structures in subsequent

Figure 1. Concrete Apartment Complex near Gwanghyo, Gyeonggi Province
경기도 수원 부근의 아파트들. Photo by Keojin Jin

일반 다수의 타이폴로지와 건축가의역할

1980년대 이후로 건설된 우리나라 주택의 80퍼센트 이상이 대형 고층 콘크리트 아파트이며, 현재 60퍼센트의 대한민국 인구가 거주하고 있다는 통계는 실로 이 하나의 건축유형(Typology)가 대한민국 국민 모두의 삶을 얼마만큼 지배하고 있는지를 보여준다. 한국을 뒤덮고 있는 이러한 콘크리트 아파트의 건축유형(Typology)에 관해 건축적 시선으로 되돌아보는 수많은 분석과 토론이 이어져 왔다.

이 주제에 관해 토론할 때 대중의 입장에서 종종 지적하는 대표적인 점은 이 건물들의 '천편일률적인 설계와 디자인'에 관한 점이다. 물론, 곳곳에 눈만 돌리면 가득 들어서 있는 고층 콘크리트 아파트들을 보며 각 건축물들의 디자인에 남다른 관심을 가진 건축가들이 한탄할만하기도 하다. 그러나 일반적으로 가지고 있는 이러한 관점은 의외로 근시안적이다. 디자인의 반복에 대한 비판은 부당하게 과장되어, 이러한 종류의 유형학이 지닌 실질적인 주요 약점을 은폐한다.

사실 우리가 아름다운 도시환경이라고 생각하는 프

Figure 2. Streetscape in Vienna.
비엔나의 거리 풍경. Photo by Keojin Jin

eras organized and shared design forms as part of the process of solving numerous problems.

Individuals who build houses based on these repetitive forms of architecture decide on the structure and size of each room in order to maintain the maximum body temperature in cold winters with minimum fuel consumption. The know-how of building these buildings has been handed down and stored in the form of a type; in this way, buildings are no different from various objects such as baskets and other household items used by humans. In the end, this type of form was organized into an architectural typology that can be repeatedly reproduced and developed through the overlapping of human wisdom over time. This typology is expressed in various "styles," which according to a divisionary theory of forms are coagulations of various cultural ideas, symbols, and practical considerations found within the vernacular architectural practice.

Spanish architect Rafael Moneo, in his 1978 article "On Typology," published in the architectural magazine Oppositions, defined "type"–the central term of architectural typology–as a means of thinking in common. (The word type here is a more comprehensive concept different from the type that distinguishes usage in architectural legal system.) Since architecture is complex and cannot be explained only by form, there are always opportunities to overcome the rigidity of such forms for each project. Therefore, most architects initiate a design project by examining the

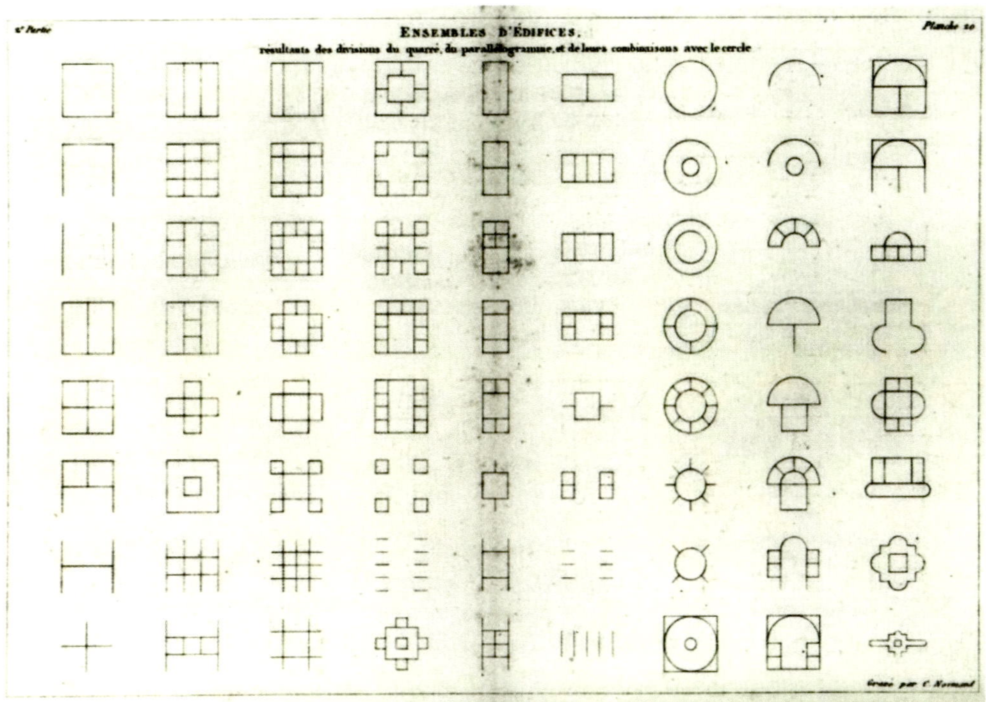

Figure 3. J.N.L. Durand, Plate 21 of Précis des leçons d'architecture donnés à l'École polytechnique, 1802/5.

랑스 파리, 비엔나, 바르셀로나 또는 멜버른에 가더라도 거의 비슷한 모양의 건물들로 도시의 대부분이 가득 차 있다. 이것들을 놓고 우리는 '천편일률적인 설계와 디자인'이라고 욕하기보다는 통일성 있는 도시의 모습으로 추켜세우곤 한다.

인간은 고대에서 부터 반복적인 구조물들을 세우며 그 노하우들을 서서히 발전시켜 왔다. 원시시대 동굴에서 살던 인간이 불을 이용해 몸을 따뜻하게 하다가 그 불을 중심으로 땅을 파 바람막이를 만들어야 했듯, 인간은 수많은 문제 해결의 과정의 일환으로서 디자인의 형식을 정리하여 상호 공유하였다. 이러한 건축의 반복되는 형식들을 기반으로 집을 짓는 개개인들은 최소한의 연료를 가지고 추운 겨울 최대한의 체온 유지 효과를 내기 위해 방의 구조와 크기가 결정하였으며, 오랜 기간 발전을 거듭하여 온돌시스템과 같은 전통 건축 기술로 발전되기도 하였다. 결국 이러한 건물을 짓는 노하우는 형식(Type)의 형태로 전수 및 저장되어왔으며, 이는 기타 인간이 사용하는 바구니나 집기류와 같은 다양한 물건과 다를 바 없었다. 이러한 유형학은 다양한 "스타일"로 표현되며, 형태 다양한 묶음들이 분할 및 정리된다. 이로써 나타나는 각각의 토착 건축(vernacular architecture)의 관행들은 그 위치에서 존재하는 다양한 문화적 아이디어, 상징 및 실용적인 고려 사항이 응고된 결과이다.

스페인 건축가 라파엘 모네오(Rafael Moneo)는 1978년 Oppisitions라는 건축잡지에서 발 표한 'On Typology'라는 글에서 건축 타이폴로지의 포괄적인 의미인 형식(Type)이란 공동으로 생각하는 수단이라고 정의한다. (여기에서의 Type이라는 단어는 건축법적인 의미에서의 용도를 구분하는 타입과는 다른 좀 더 포괄적인 개념이다.) 그러나 건축이란 형식만으로 설명되지 않으며 수없이 많은 변수들이 복합적으로 작용되어 만들어 지는 과정에서 각 프로젝트별로 그러한 형식의 경직성을 극복할 기회들이 항상 존재하기 마련이다. 이는 대부분 건축가

elements of the types and references, but they complete their work by integrating individual decisions about the specific sites, programs, and contexts of the project.

Moneo also said that architectural typology should not be treated as a "frozen device" but instead as a "frame within which change operates" that evolves step-by-step by absorbing the creative ideas of each architect. According to the article, the word "typology" is a combination of the word "type" and the Greek suffix "-ology," where the added "-ology" means dialogue, treaty, theory, or science. Therefore, it also means that the existence of type should be actively used as a subject with a purpose in scientific discussion. Moreover, when we look back at history we realize that the most exciting architectural moments were when new types of architecture emerged. An example is how Brunelleschi created a new type of dome that spread throughout Europe, giving rise to new, scientific ways of thinking about architecture.

However, these typologies only evolve to become efficient when proper care and thought are put into their development. There are many cases of human-made systems developing in an irrational direction due to political or social factors, and architectural typology is no different. For example, modern architects focused on creating new design languages by combining past proportions and modern technologies. Ironically, as architects broke away from formalism and adopted functionalism, they created modern architectural types. Furthermore, functionalism and idiosyncratic solutions were preferred regardless of form throughout post-modernism and the subsequent era. However, creating these individual, experimental buildings was the only interest of modern architects.

The architects advocated escaping this formalism and did not actively consider a typology that governs general residential buildings. They pursued individual, experimental buildings to the detriment of their responsibility for developing mundane residential architecture. This general lack of concern about the practical world of repeated and "ordinary" typologies, which have been recycled with many new variables in the modern era, led to a relinquishing of control over most of the world's design quality.

One of the exciting aspects of Le Corbusier's Ville Radieuse (The Radiant City) proposal, announced in 1924, is that it has historically blurred the boundary between the upper limits of architectural ambition and the majority of ordinary residential typologies that architects often ignore. The proposal was, at first, simply perceived as a way to solve the urban problems of the time. The logic of constructing high-rise towers to provide abundant green space and sunlight to the ground below was a conceptually desirable and logical solution. The perspective drawings such as figure 4 of the Ville Radieuse iconically show the residential concrete building blocks standing as monumental and simple sculptures among vast green lawns within the complicated and dense city. The images powerfully convey life inside the apartment complex, where residents can enjoy a private, quiet life and an unobstructed view away from the hustle and bustle of the city. This plan showed a unique way of creating architecture, where lifestyle, symbolism, and ambition be-

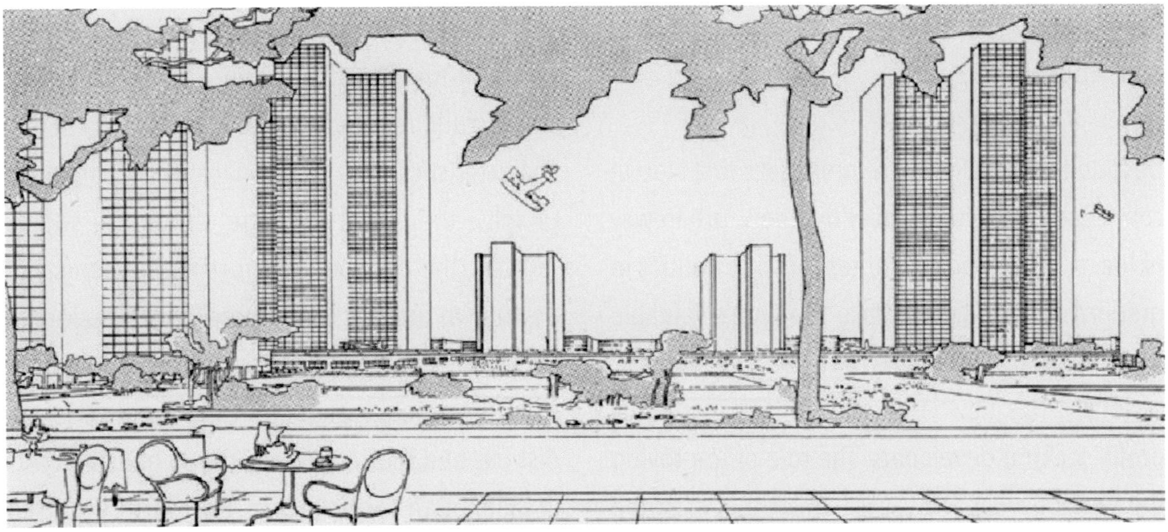

Figure 4. Le Corbusier Ville Radieuse (The Radiant City)

가 프로젝트를 진행하는 방식이 형식(Type)의 요소를 검토함으로써 시작해 구체적인 계획의 개별적 결정들을 통합시켜 하나의 작품을 완성하는 것으로 나타나는 방향성을 보아 알 수 있다.

모네오는 또한 건축의 타이폴로지가 '얼어버린 장치'가 되는 것을 경계해야 하며, 오히려 각 건축가들의 창의적 아이디어들을 흡수하여 차곡차곡 발전하는 '변화를 가동시키는 프레임'(frame whithin change operates)으로서 작동해야만 한다고 주장한다. 건축의 타이폴로지(Typology)는 'Type'이라는 단어와 '-ology'라는 그리스어 접미사의 조합인데 여기서 추가된 '-ology'는 대화, 조약, 이론 또는 과학을 의미한다. 따라서 형식(Type)이라는 존재가 과학적인 토론의 목적성을 띠는 주체로서 적극적으로 이용되어야 한다는 명제를 의미하고 있기도 하다. 이는 건축 역사의 가장 흥미로운 순간들은 새로운 건축의 타입들이 생겨나는 순간들임을 보면 알 수 있다. 예를 들면 부루넬레스키(Brunelleschi)가 새로운 형식의 돔을 만들고 그것이 이후 유럽 전역의 건축에 대한 철학에 과학적 진보를 불러일으켰던 것과 같은 방식이 그 예이다.

그러나 이러한 타이폴로지는 올바른 방향의 고민과 대화의 노력이 있을때만 효율적인 것으로 발전되어 가게 된다. 인간이 만든 많은 시스템이 정치적, 또는 사회적 요소에 의해 때로는 비이성적인 방향으로 굳어져 가는 경우들이 많은데, 건축 타이폴로지 또한 그러하다. 특히 근대 건축가들은 사회에 새로운 이미지들을 주기 위하여 과거의 비례들과 현대의 기술들을 기술적으로 조합하여 새로운 언어들을 만들어 내는 데 집중하였는데 여기서 르 꼬르뷔제(Le Corbusier)는 당시 부상한 대량생산과 산업물들의 '반복되어짐' 과 형식주의의 탈피, 기능주의를 적극 사용하여 새로운 현대 건축의 형식을 (아이러니하게도) 만들어 내었다. 그리고 포스트 모더니즘과 그 이후의 시대를 거치면서 기능주의와 특이한 건축적 형태들을 형식에 무관하게 설계하는 방식들이 선호되며 발전되어 왔다.

그러나 이러한 개별적이고 실험적인 건물을 만드는 것만이 대부분의 현대 건축가의 관심사였다. 건축가들은 이러한 형식주의의 탈피를 주장하며 일반 주거용 건물을 지배하는 유형학을 적극적으로 고려하지 않았다. 그들은 일상적인 주거 건축 개발에 대한 책임을 희생하면서 개별적이고 실험적인 건물들의 탐구만을 추구했으며 심지어 이러한 건물들을 적대시하기도 하였다. 결국 이것들을 적극적으로 고민하지 않은 건축가 집단은 결과적으로 전 세계 대부분의 건축물에 대한 통제 수단을 스스로 놓아버린 셈

come the core value of the architecture, instead of its appearance.

Despite the Ville Radieuse proposal's first step in connecting functionalism with a new, urban way of life, architects and other responsible entities in modern society did not take the type's valuable function of "frame within which change operates" seriously. This new vision was only exploited by profit-seeking developers. The role of improving the typology was left vacant as architects, or the system that runs the field of architecture, voluntarily gave up improving these common typologies. The architectural possibilities in the embryonic vision of Corbusier's Ville Radieuse needed to be established as a typology. Instead, mechanical design techniques with no qualitative spatial consideration were repeated, focusing on the characteristics of economic efficiency and ease of investment. As a result, most buildings were created with typologies eroded by urban economic considerations other than architectural variables, and these typologies ate away at the city.

Urban Density vs. Height of Buildings

The price of physical energy declined as oil dependency increased in the expanding modern economy of the 20th century. At the same time, the development of mechanical construction methods, building material production methods, and architectural and structural methods promoted the construction of high-rise residential buildings. Due to this, the building energy, structure, and material used per unit area of the final building had to increase exponentially to overcome the structural requirements involved with the height.

As pointed out in Edward Glaeser's book Triumph of the City, the denser a city is, the more successfully realized the unique developmental characteristics of its urban space are. In particular, to solve the recent shortage of housing supply in cities, the method of increasing the density of downtown areas is being practiced in major cities worldwide. Take the case of New York City. To solve the problem of gentrification in commercial districts and the continuous rise in housing costs in Manhattan's Soho district, New York City raised the floor-area ratio of the neighborhood by 500% through a project called Soho Rezoning. It used existing construction methods to realize this. This project aims to safeguard the history of the existing city by actively preserving and vertically expanding the city's historic buildings and realizing a desirable, walkable city through an appropriate mix of residential and commercial use.

On the other hand, Korea has ardently utilized the reinforced concrete housing construction method, in the form of high-rise apartments, as an essential and valuable tool for rapid urban improvement. As a result, the technology and planning methods for it have also made rapid progress. However, due to the relatively fast aging rate inherent in the reinforced concrete method, the large amount of carbon emissions generated during the construction process, and the absence of architectural value apart from the surrounding context, the vicious cycle of redevelopment after consumption in a short time is repeated.

The negative characteristics of the current Korean complex-type apartment typology are extensive. Due to its disconnectedness from its setting, this type of development creates a gated community.

이 된 것이다.

1924년에 발표된 르 꼬르뷔제(Le Corbusier)의 Ville Radieuse(The Radiant City) 제안의 흥미로운 측면 중 하나는 건축가들이 종종 간과하는 대부분의 일반 주거 유형, 그리고 건축적 원대한 비전, 이 둘 사이의 경계를 역사적으로 모호하게 만들었다는 것이다. 다시말해 건축가의 일상적 건축유형을 통제하는 방식으로서의 고민의 방향이었던 것이다. 그러나 이 제안은 단순히 당시의 도시 문제를 해결하는 하나의 방법으로 인식되었다. 사실, 아래 지상에 풍부한 녹지 공간과 일조량을 제공하기 위해 고층 타워를 건설하는 논리는 개념적으로 바람직한 논리적인 솔루션으로 보였다. Figure 4에 나타나는 건축가의 개념 조감도에는 일반 시민들의 주거건물이 공원 속 거대하고도 단순한 조형물처럼 강렬하게 도열해 있다. 이 계획은 건축의 외형이 아닌 거주민의 새로운 생활방식, 상징성, 야망이라는 무형의 가치들이 건축의 핵심 가치가 되는 독특한 건축 창조 방식을 보여주었다. 하지만 대중은 이 계획안을 보며, 도시의 번잡함을 떠나 개인적이고 조용한, 그리고 탁 트인 조망을 가질 수 있는 삶을 저렴한 가격에 누릴 기회를 선택적으로 포착하였다.

이렇게 Ville Radieus 제안은 기능주의를 새로운 도시적 삶의 방식과 연결하여 상상과 디자인의 첫 발을 내디뎠으나, 이후, 앞서 언급한 형식(Type)의 순기능인 변화를 가동하는 프레임'(frame within change operates)으로서의 건축유형(Typology)의 작용을 통제할 주체는, 건축가가 그 자리를 스스로 포기함으로써 공석이었다. 그리하여 이러한 건축적 가능성이 그 타이폴로지로서 올바르게 자리 잡지는 못하였으며 그 속의 경제적 효율성과 투자의 용이성 등의 특성이 중심이 되어 공간적 고찰이 전무한 기계적 설계가 반복되어 졌다. 이로써 대부분 건축물들은 건축적 변수 이외의 도시 건축적 요소들에 잠식당한 수동적 타이폴로지들이 만들어지고, 이러한 건축유형(Typology)들이 도시를 잠식해 버리게 된 것이다.

도시의 밀도 vs. 건물의 높이

20세기의 팽창하는 경제사회시스템 안에서 사용되는 물리학적 에너지의 가격은 급격히 낮아졌다. 그리고 이를 이룩할 기계 방식과 건축 자재의 생산 방식, 건축 구조방식이 개발됨으로 인하여 이전에는 소수의 막강한 권력자만이 행할 수 있었던 거대 구조물의 건축행위의 특권이 많은 이들에게 나누어졌다. 이로 인한 인근 도시와의 분절, 그리고 고층 건물을 지을 때 중력을 극복하기 위해 최종 만들어진 단위 면적당 사용되는 건축 에너지와 구조체 그리고 건축 물질의 사용이 기하급수적으로 증가하여야 하였다.

에드워드 글레이져 (Edward Glaeser)가 그의 책, 도시의 승리에서 지적한바, 도시는 고밀화 될수록 도시 특유의 발전적인 특징이 강하게 나타난다. 특히 최근 지속되는 도시의 주거 공급 부족의 해결을 위하여 도심지 밀도를 높여 나가는 방식은 전 세계 주요 도시들에서 행해지고 있다. 예를 들면 뉴욕 맨해튼 소호지구의 지속적인 주거비 상승과 상업지구의 젠트리피케이션 문제를 해결하기 위해 뉴욕시는 Soho Rezoning이라는 프로젝트를 통해 해당 지역의 용적률을 500% 상향 조정하고, 이를 실현하기 위한 건설방식에 기존의 역사적 건물들을 적극 보존, 수직 증축하여 기존 도시의 역사를 보전하고, 주거와 상업 용도의 적절한 혼합을 통한 바람직한 걷기 좋은 도시를 실현하고자 하는 프로젝트를 진행하고 있다.

반면, 우리나라는 아파트로 대변되는 철근 콘크리트 주거 건축방식을 신속한 도시환경 개선에 중요하고 유용한 도구로서 적극 활용하여 왔고, 또 그를 위한 기술과 계획방식 또한 비약적 발전을 이루었다. 그러나 철근 콘크리트 방식에 필연적으로 내재 된 비교적 빠른 노후화 속도, 그리고 건축 과정에서 발생하여지는 많은 양의 탄소 배출, 그리고 주변 맥락과의 단절된 건축적 가치의 부재로 인하여 짧은 시간 소비되고 재개발 되는 악순환이 반복되어 왔다.

우리나라의 고층 아파트 주거단지 유형의 부정적 특

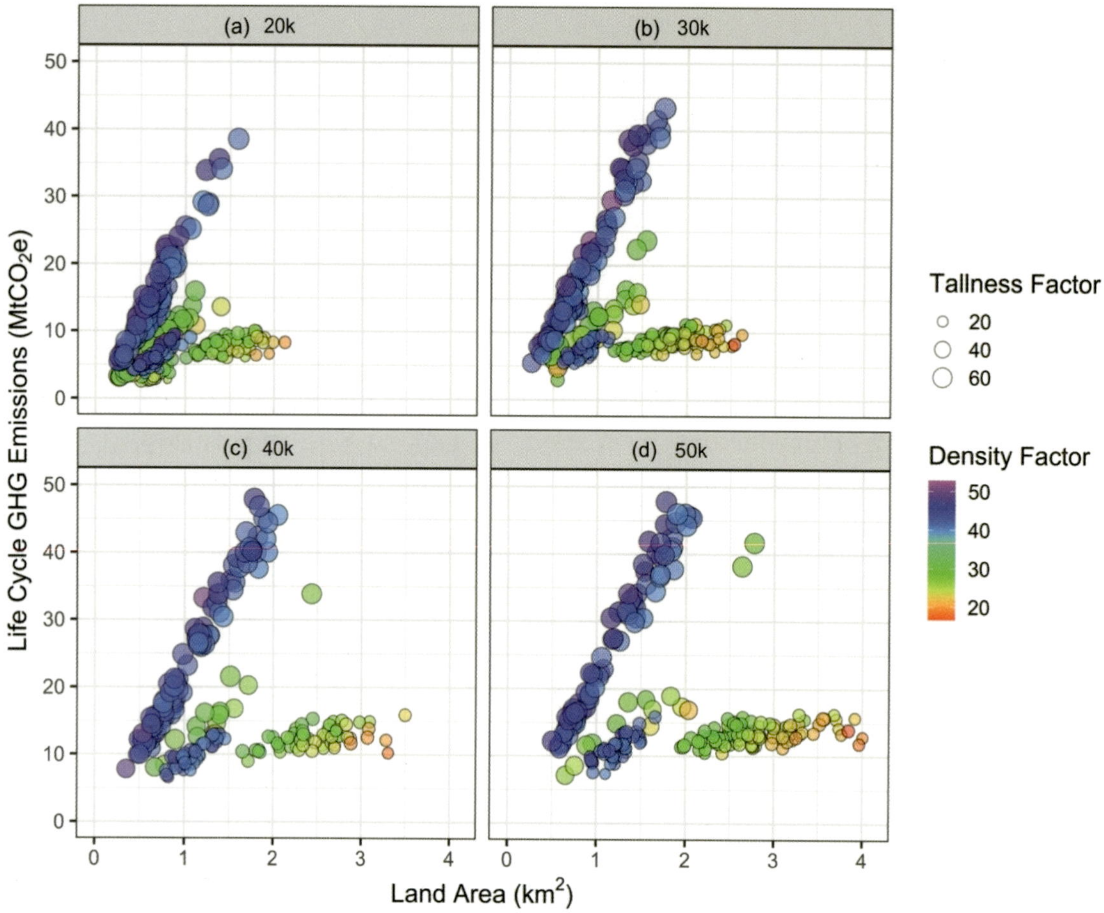

Figure 5. LCGE versus built land area for fixed populations. Results presented for 20 (a), 30 (b), 40 (c), and 50 (d) thousand people. Pomponi et al. npj Urban Sustainability

It also de-activates the surrounding streets due to its unfamiliar scale relative to its context. The disadvantages of non-circulating, monolithic megastructures such as Korea's high-rise residential apartment complexes are extensive enough to be judged as fatally unsuitable for forming a healthy city.

According to UNEP's 2021 Global Status Report for Buildings and Constructions, carbon emissions from building construction account for 10% of the world's total carbon emissions. Such carbon emissions from building construction are even higher in Korea, where the new housing supply is growing more rapidly than in other developed countries.

More than 50% of the buildings that make up our built environment are reinforced concrete and general steel structures. Consequently, a large amount of carbon dioxide is generated at each stage of manufacturing, transportation, construction, disposal, and recycling of materials. Reinforced concrete structures, in particular, have fatal disadvantages in that they cannot compensate for the corrosion of buried rebars. Moreover, since the entire building is integrally connected, one part's aging leads to the entire building's aging. However, the fatal disadvantage of the short aging cycle of these buildings has become

Figure 6. Cover Image of Stewart Brand's book: How Buildings Learn.

성은 광범위하다. 일단 주변 대지와의 연결 관계가 고려되지 않아 그 경계 부분을 단절시키며 담으로 둘러싼 고립된 섬들을 만들어 낸다. 또, 순환 불가한 일체형 거대 콘크리트 구조물에 관한 단점은 건강한 도시를 이루는 데 치명적 부적합 요소로 판단이 될 만하다. UNEP의 2021 Global Status Report for Buildings and Constructions에 따르면 건축물을 짓는 데서 기인한 탄소 배출량은 세계 전체 탄소 배출량의 10%에 달한다. 건축 시공 분야에서의 이러한 탄소 배출은 신규 주택 공급이 활발한 우리나라에서는 그 비중이 더욱더 높다.

우리의 건축 환경을 이루는 건축물들의 50% 이상은 철근 콘크리트 구조와 일반 철골 구조로 이루어졌으며, 이를 건축하기 위한 제조, 자재의 운반, 시공, 폐기 및 재활용의 각 단계에서 대량의 이산화탄소가 발생하게 되어 환경 보존에 부정적인 영향을 끼치고 있다. 특히 철근 콘크리트 구조물은 매립된 철근이 부식되어가면 이를 보완할 수 없고, 건물 전체가 일체형으로 연결되어 있어 일부분의 노화가 건물 전체의 노화로 이어진다는 치명적인 단점이 있다. 그러나 이러한 건물의 짧은 노화 주기라는 치명적 단점은 아파트 건축의 관념적이고 관습적이며 거래와 투자에 유용한 건축유형(Typology)으로 반복적으로 선호되면서 우리의 환경에 대한 장기간의 부정적 영향이 우려된다.

한국 주거 구조의 대안적 유형에 대한 근본적인 물음은 아직 적극적으로 탐구되지 않았다. 고층 아파트의 수명을 보완하기 위해 정부 연구기관에서 많은 연구가 진행되고 있다. 그러나 이러한 연구의 대부분은 평면구성의 가변성 증진을 위한 구조 및 기술적 전략과 시멘트 배합 기술 강화, 철근 도장 개선 등 기술 개선의 방향으로 진행되었다. 그러나 이와 같은 제한적이고, 국지적인 대응으로는 우리나라에서 이러한 대규모 콘크리트 아파트 단지 유형으로 야기되는 다양한 문제들을 해결할 수 있을 것으로 보이지 않는다.

높은 도시 인구 밀도는 고층 건물 유형이 있어야 달성할 수 있는 것으로 생각 될 수 있지만, 의외로 고층 건물의 확산은 도시 밀도의 증가와 직접적인 관련이 없다. 이는 고층 건물이나 아파트가 즐비한 서울이나 뉴욕의 인구밀도가 5층 이하의 건물로 가득 찬 바르셀로나의 인구밀도보다 낮은 사실에서 찾아볼 수 있으며 이는 총 연면적을 따져 보았을 때도 동일하다. 그리고, 고층건물 은 생애주기 탄소 배출면(Life Cycle GHG or LGHG) 에서 저층 타이폴로지보다 훨씬 많은 전생에 주기 탄소 배출을 보인다. CU Boulder and Edinburgh Napier University customary and ideologically acceptable in apartment architecture. As it is repeatedly favored as a useful typology for trading and investment, the long-term negative impact on our environment is a pressing concern.

The fundamental question of an alternative typology for Korean residential structures has yet to be probed. In order to supplement the longevity of high-rise apartment buildings, many studies have been conducted by government research institutions. However, this research was done in the direction of technological improvement, such as increasing the flexibility of plan composition, strengthening cement mixing technology, and improving the reinforcing bar coating through chemical innovation. This kind of limited, local response has seldom shown signs of solving the durability problems for these massive concrete structures in Korea.

Although a high urban population density might seem to require a high-rise building typology, the proliferation of high-rise building is not directly correlated with the increase in urban density. For example, the population density of Seoul or New York, where high-rise buildings and apartments line up in rows, is lower than that of Barcelona, which is filled with buildings with five or fewer floors. This observation shows that, contrary to the general stereotype of the city packed with high-rise buildings, it is sufficiently possible to achieve the same density with low-rise buildings. In addition, the high-rise building typology results in much higher life cycle carbon emissions (Life Cycle GHG or LGHG) than does the low-rise typology. According to a study by CU Boulder and Edinburgh Napier University (figure 5), cities with high-rise apartment blocks reported 150% more LGHG emissions than cities with the same population density and low-rise buildings.

Rediscovery of Low to Mid Rise Urban Residential Typology

In the end, it is necessary to utilize low- or mid-rise high-density typologies, such as emerged in European cities in the past, to achieve an urban design with tangible and intangible advantages such as the following characteristics. In high-density, low-rise residential complexes, natural diversification of building design may be achieved over a long period. The image on the figure 6 is the cover of a book authored by Stewart Brand, titled How Buildings Learn. The buildings in this image show changes in their appearance arising over a long period. The left image depicts the buildings' elevation at their time of completion in 1857, when several identical buildings were built and sold by a commercial real estate construction company. The photo on the right shows the changed appearance of the same buildings 140 years later. Over time, different owners have occupied the buildings or units for years or decades, extending the buildings or changing their elevations to suit their tastes and uses. This manner of change to the personality of a building to suit the tastes and requirements of individual owners cannot be achieved in a high-rise apartment building.

In addition, to attain the various advantages that a city made up of low-rise, high-density housing can provide, urban designers, transportation system designers, and citizens need to form a con-

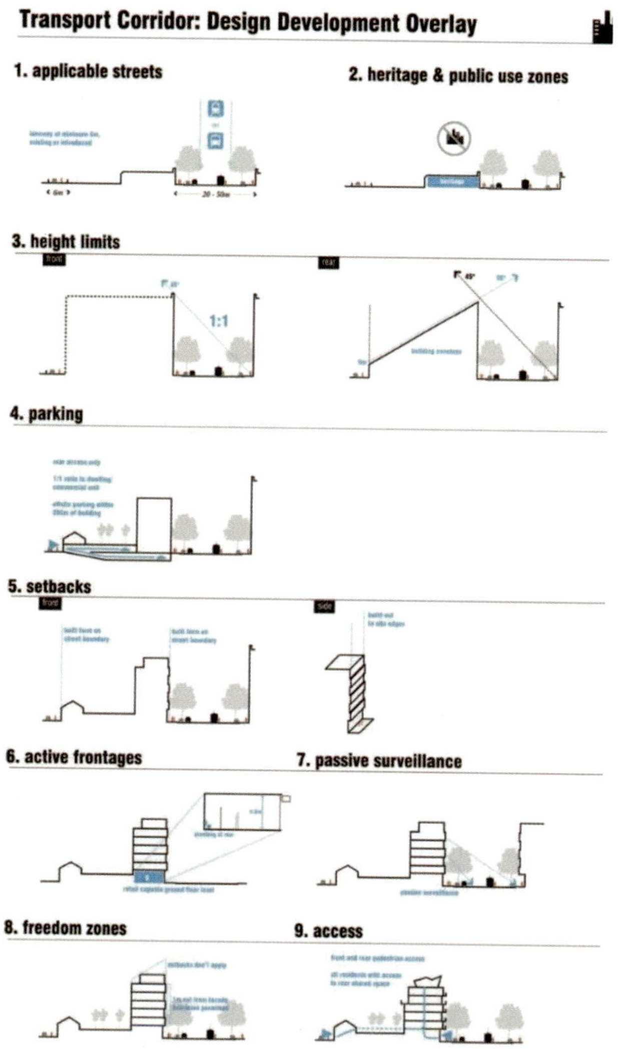

Figure 7. Planning requirements explained on one piece of paper, with simple diagrams.
Illustration by: Steve Thorn, Ralph Webster, Simon Goddard, City of Melbourne

가 진행한 연구에 따르면 (Figure 5) 고층 건물로 이루어진 도시 블록은 동일한 인구밀도의 저층 건물로 이루어진 도시에 비교 시 150%이상의 LGHG배출을 보인다고 보고하였다. 이는 도시는 고층 건물로 이루어져야 한다는 일반적인 고정관념과는 다르게, 저층 건물로 동일한 밀도를 이루는 것이 충분히 가능하며 그것이 친환경적인 수단이라는 점을 보여준다.

중저층 도시주거의 재발견

결국 과거 유럽 도시와 같은 중저층 건축유형(Typology)을 적극 활용하여 아래의 특성들과 같은 유/무형의 장점들을 활용한 도시 디자인을 이룰 필요가 있다. 고밀도 저층 주거단지는 자연스러운 건물 디자인의 다양화가 장시간에 걸쳐서 이루어질 수도 있다. 아래의 Figure 6의 이미지는 Stewart Brand가 1990년대 영국의 다큐멘터리를 통해 방영된 건축의 장시간에 걸친 모습의 변화와 이를 고려한 도시

sensus to create walkable streets and convenient transportation systems. As shown in Figure 7, the city of Melbourne produced guidelines for creating walkable streets and achieved successful downtown regeneration in the 1990s. Through this process, the population of the city center increases, commerce prospers, and tax revenue increases, which is used to rescue various urban infrastructures, including transportation, creating a virtuous cycle.

Diversity of the building's internal programs should also be achieved. For example, the illustrated cross-section in Figure 8 shows various activities inside a building in downtown Paris, where new cultural creations exploded in the mid-19th century. In this high-density city, for various reasons, people from different classes and occupations worked or lived on each floor of a building, interacting frequently. Through these interactions, people influenced each other and made creative actions and discoveries. These mutually intersecting rhythms of life and work must be achieved to take advantage of low-rise, high-density urban housing.

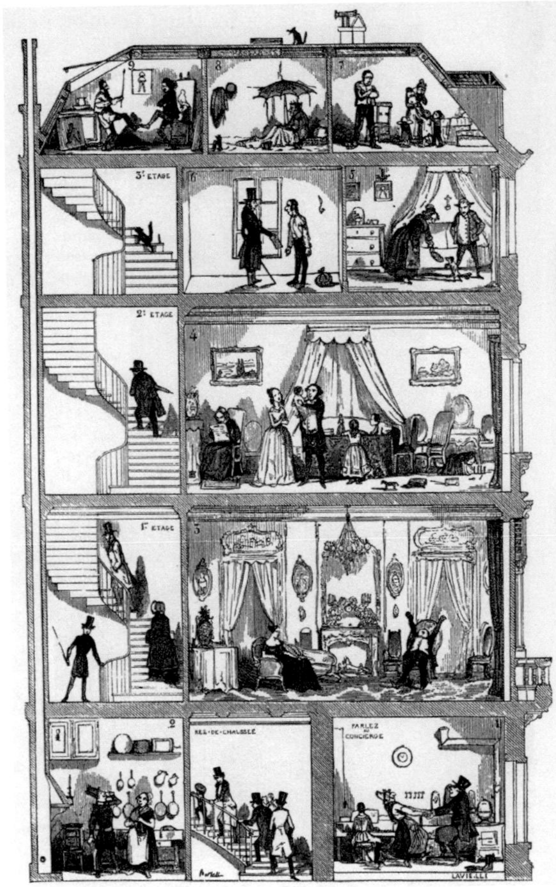

Figure 8. Cross-section of a Parisian house around 1850 showing the economic status of tenants varying by foors. (Edmund Texier, Tableau de Paris, Paris, 1852)

Reference

1) J.N.L. Durand, Plate 21 of Précis des leçons d'architecture donnés à l'École polytechnique

2) Moneo, R. (1978). On typology. Oppositions 13.

3) 황두진 (2016) 무지개떡 건축, 메디치 미디어

4) Pomponi, F., Saint, R., Arehart, J.H. et al. (2021) "Decoupling density from tallness in analysing the life cycle greenhouse gas emissions of cities." npj Urban Sustain 1, 33

5) Edmund Texier, Tableau de Paris, Paris, 1852

6) Illustration by: Steve Thorn, Ralph Webster, Simon Goddard, City of Melbourne

와 건축의 설계에서의 고려를 활자와 이미지로 출간한 책의 표지에 실린 이미지이다. 이를 보면 왼쪽에는 상업적 부동산 건설회사에서 다수의 건물을 똑같이 지어서 분양한 1857년 준공 당시의 입면도를 볼 수 있고 오른쪽의 사진은 동일한 건물의 140년 후의 변화된 모습을 보여준다. 오랜 기간을 거치며 각기 다른 주인들이 건물, 또는 유닛들을 수십 년 점유하면서 자신의 취향과 용도에 맞는 모습으로 건물을 증축하거나 입면을 변화시켰다. 책에서는 이 과정에 관한 구체적인 개인적 사료들과 법령의 변화 과정을 예로 들어 해당 건물, 그리고 비슷한 다양한 건물의 변화과정을 흥미롭게 보여준다. 이러한 개별 공간 소유자들의 취향과 상황에 맞는 건물의 개성의 변천사는 고층 아파트 건물에서는 이루어질 수 없다.

해당 건물들은 가변성을 허용하는 단순한 구조로 이루어져서 이러한 변형과 증축이 손쉽게 가능하였다. 그러나 이러한 변화를 수용해온 건물의 상당수는 우리 아파트와 같은 벽식구조로 건축된 것으로 보아, 현재 우리나라 아파트의 장수명성의 대안으로 제시되는 핵심 아이디어들이 하나인 벽식구조에서 라멘조로의 변형은 이러한 건물의 가변적 장수명성에 그다지 큰 요소가 아닐 수도 있다. 다양성을 원한다면 처음 디자인할 때부터 각기 다른 건축가에게 설계를 의뢰하는 것도 좋겠지만 현대 주택 공급방식의 단기적 효율성은 이를 허용하지 않을 때가 많다. 그러나, 이렇게 가변적 건물을 놓고 시간을 놓고 자연스레 변해가는 과정을 보는 것 또한 매우 흥미로울 것이다.

이외에도 저층 고밀 주거로 이루어진 도시가 이룩할 수 있는 다양한 장점을 이루기 위해서는 도시설계자, 교통 시스템 설계자, 그리고 시민 모두가 공감대를 이룩하여 걷기 좋은 거리와 편리한 교통 시스템을 만들어야 한다. 멜버른시는 Figure 7.에서 보이는 것처럼 걷기 좋은 거리를 만들기 위한 가이드라인을 제작하여 90년대부터 성공적인 도심 재생을 이룩하였다. 이를 통해 도심의 인구가 늘어나고 상업이 번창하여 세수가 늘어나고 이를 이용해 교통을 포함한 다양한 도심 인프라를 구축하는 데 투입하는 선순환이 이루어지고 있다.

건물의 내부적인 프로그램의 다양성 또한 이루어야 한다. Fig 8. 의 삽화 단면도는 19세기 중반 문화적으로 새로운 창조물들이 폭발하던 프랑스 파리 시내의 건물 안에서 벌어지는 다양한 활동들에 관하여 보여준다. 고밀도 도심 속에서 다양한 자연스러운 이유로 건물의 각 층마다 다른 계층 다양한 직업의 사람들이 일을 하거나 주거하며 이것이 도시 속에서 또 건물과 건물들의 다양한 프로그램들이 서로 얽혀가며 마치 숲의 자연 생태계같은 복잡성을 띠며 새로운 일들이 벌어졌다. 그 안에서 사람들은 서로에게 영향을 주며 창조적 행위와 발견을 하였는데 이는 저층 고밀 도심 주거의 장점을 활용하기 위해 반드시 이루어야 할 삶과 행위의 상호 교차 방식이다.

TYPOLOGY OR TOPOLOGY

This section is dedicated to the collection of research on the mid-rise and/or mix-use precedents. The precedents were explored through massing analysis, topological diagrams, and diagrammatic drawings depicting a sequential experience as one moves from the most private part of the domestic space to the public domain, such as the bathroom to the street.

이 섹션은 '중층' 그리고/또는 '주상복합' 건축물의 선례에 대한 연구이다. 본 연구의 대상이 된 실존 건물들의 예제들은 건물의 기하학적 메스와 프로그램의 관계를 분석하는 매스+위상 다이어그램을 중심으로 탐구되었다. 이에 더해, 가정 공간의 가장 사적인 부분에서 욕실에서 거리로 이동하는 것과 같이 공공 영역으로 이동하면서 순차적인 경험을 묘사한 도식과 스케치들을 통해 공간의 경험을 탐색하였다.

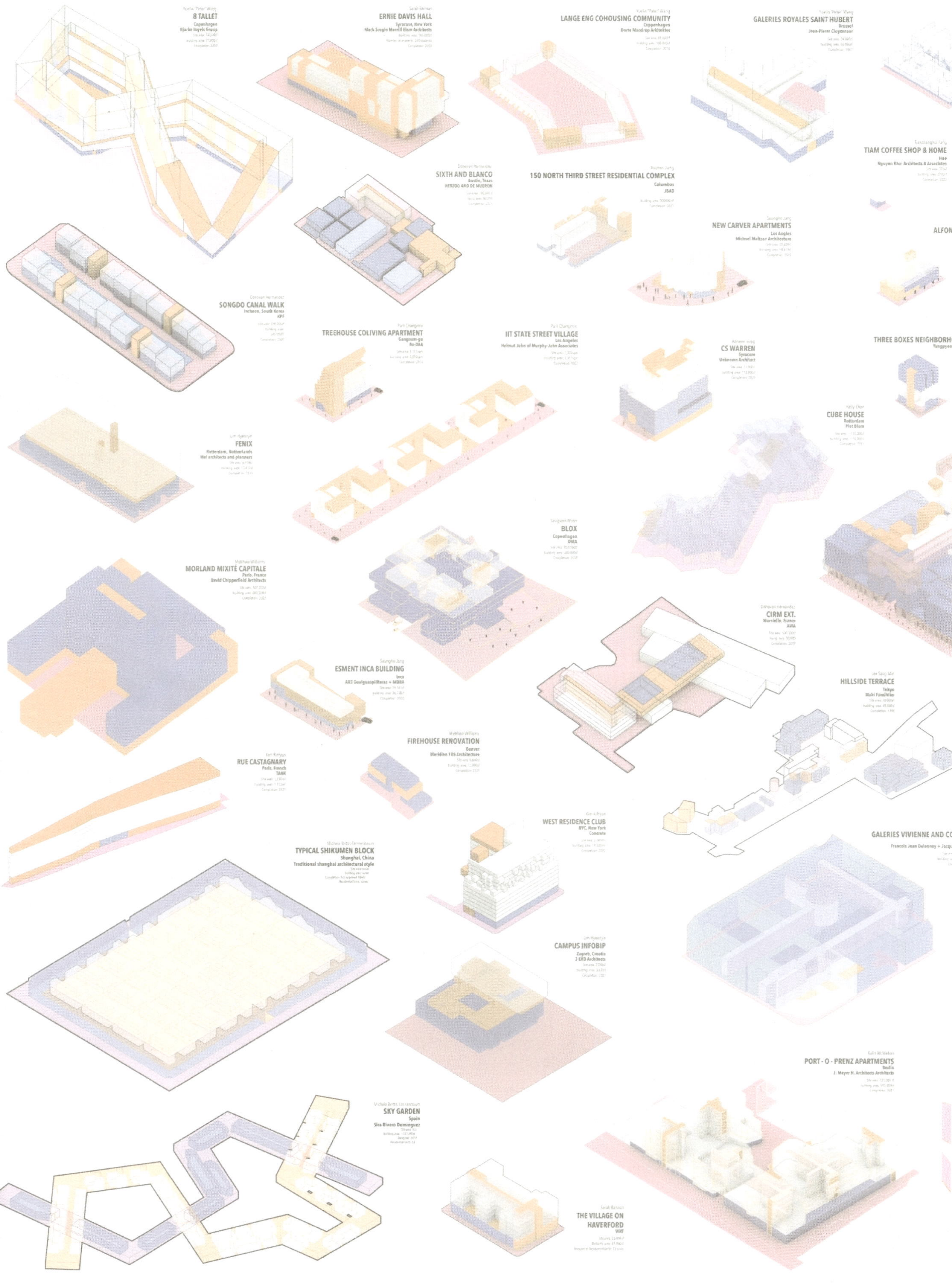

8 TALLET
ERNIE DAVIS HALL
LANGE ENG COHOUSING COMMUNITY
GALERIES ROYALES SAINT HUBERT
TIAM COFFEE SHOP & HOME
SIXTH AND BLANCO
150 NORTH THIRD STREET RESIDENTIAL COMPLEX
NEW CARVER APARTMENTS
SONGDO CANAL WALK
TREEHOUSE COLIVING APARTMENT
IIT STATE STREET VILLAGE
CS WARREN
CUBE HOUSE
FENIX
BLOX
MORLAND MIXITÉ CAPITALE
CIRM EXT.
ESMENT INCA BUILDING
HILLSIDE TERRACE
FIREHOUSE RENOVATION
RUE CASTAGNARY
WEST RESIDENCE CLUB
TYPICAL SHIKUMEN BLOCK
CAMPUS INFOBIP
PORT - O - PRENZ APARTMENTS
SKY GARDEN
THE VILLAGE ON HAVERFORD

MRMU

ARC045 PNU SECTION FALL 2022 | Kim KiHyun

RUE CASTAGNARY

Paris, French
TANK

Site area: 1,230m²
building area: 7,553m²
Completion: 2021

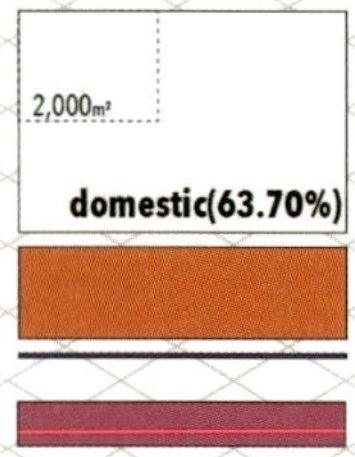

Domestic

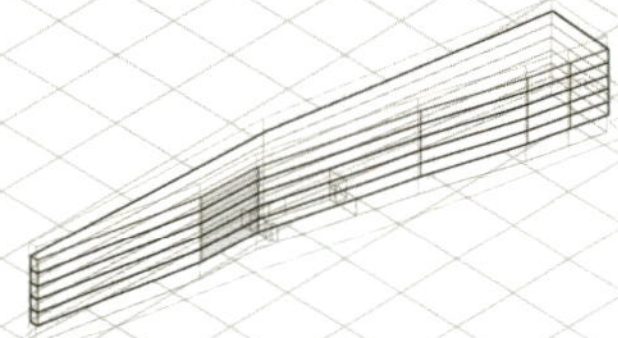

Communal

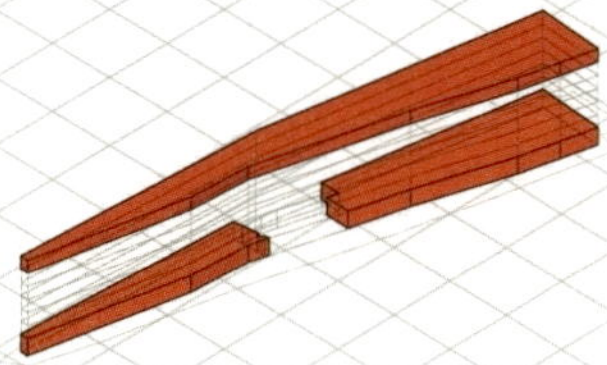

Commercial

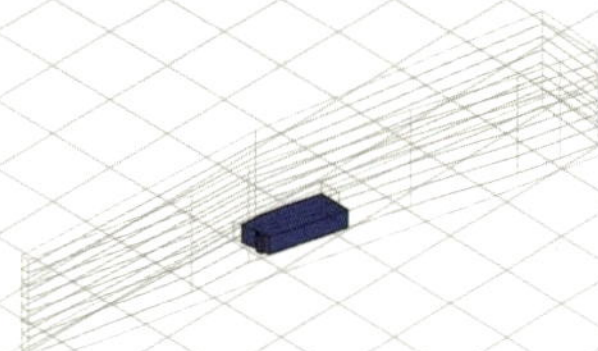

Public

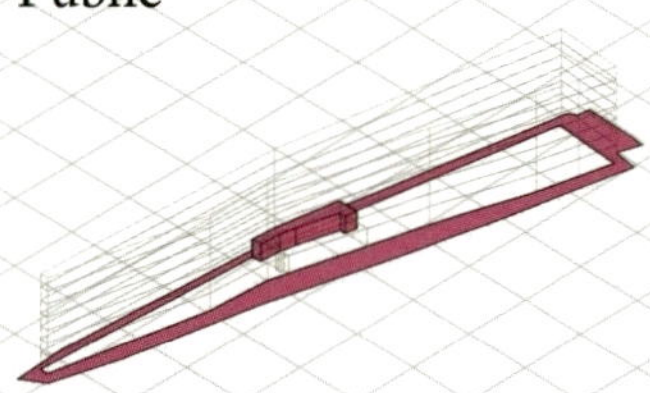

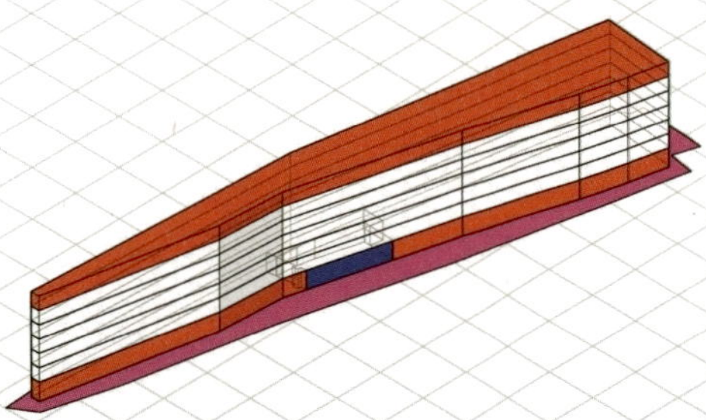

10m² cube

MRMU

ARC045 PNU SECTION FALL 2022 | Kim KiHyun

WEST RESIDENCE CLUB

NYC, New York

Concrete

Site area: 2,760m²

building area: 19,500m²

Completion: 2022

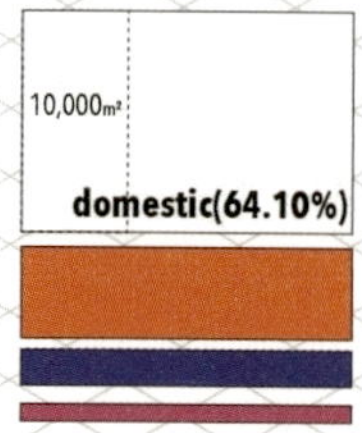

Domestic

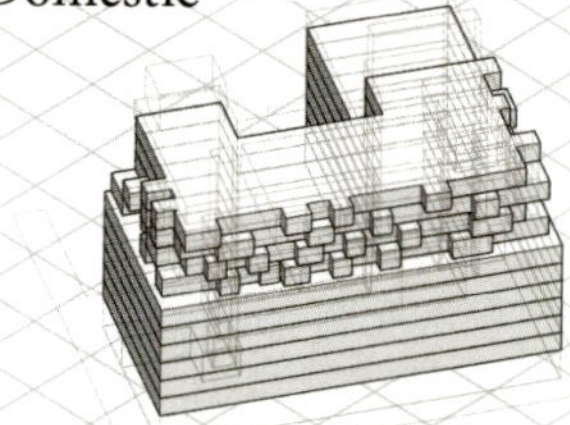

Communal

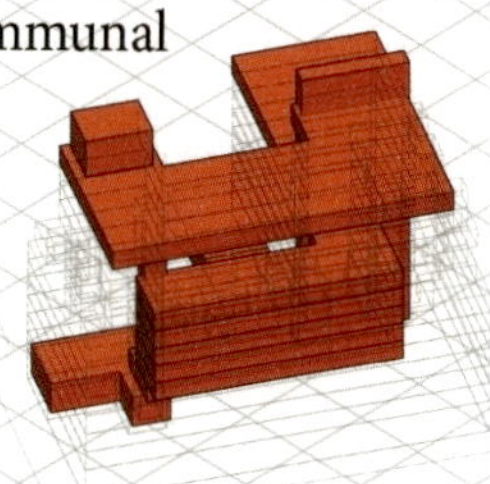

Commercial

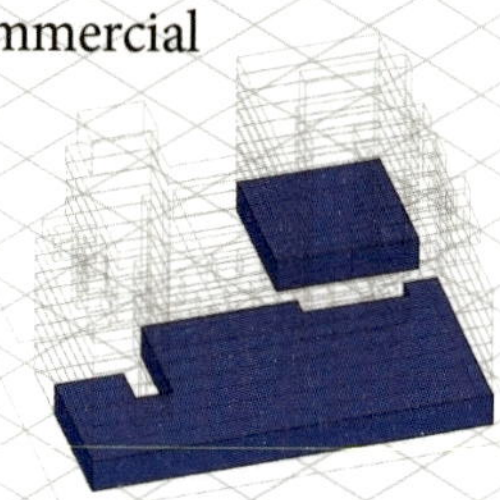

Public

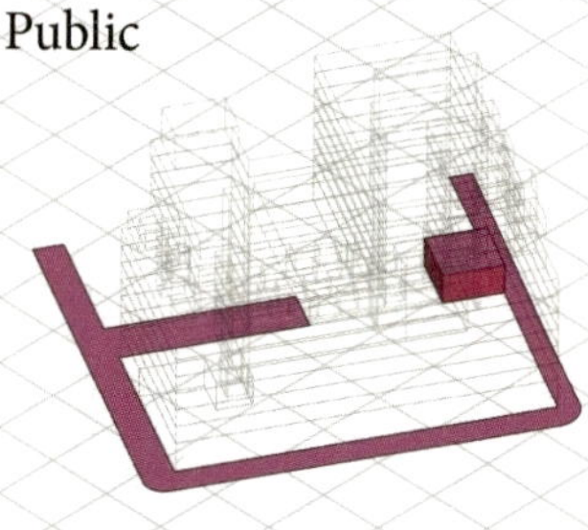

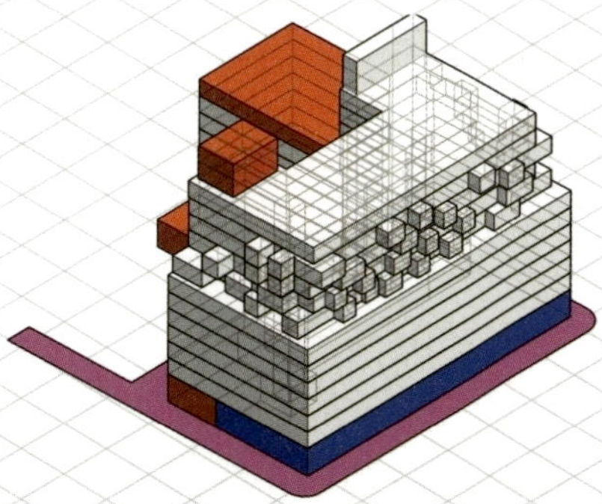

10m² cube

ARC307 BU SECTION FALL 2022 | Kim Ryulhyeon

NEXUS WORLD HOUSING

Fukuoka, Japan
OMA

Site area: 37,362sqft
building area: 62,054sqft
Completion: 1991

5,000 sf
domestic
communal
commercial
public

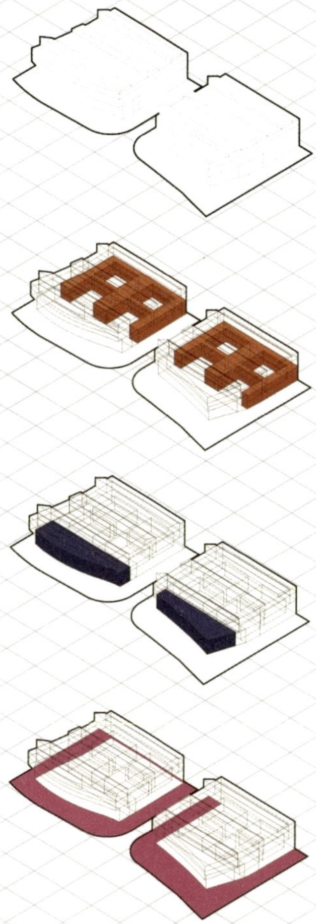

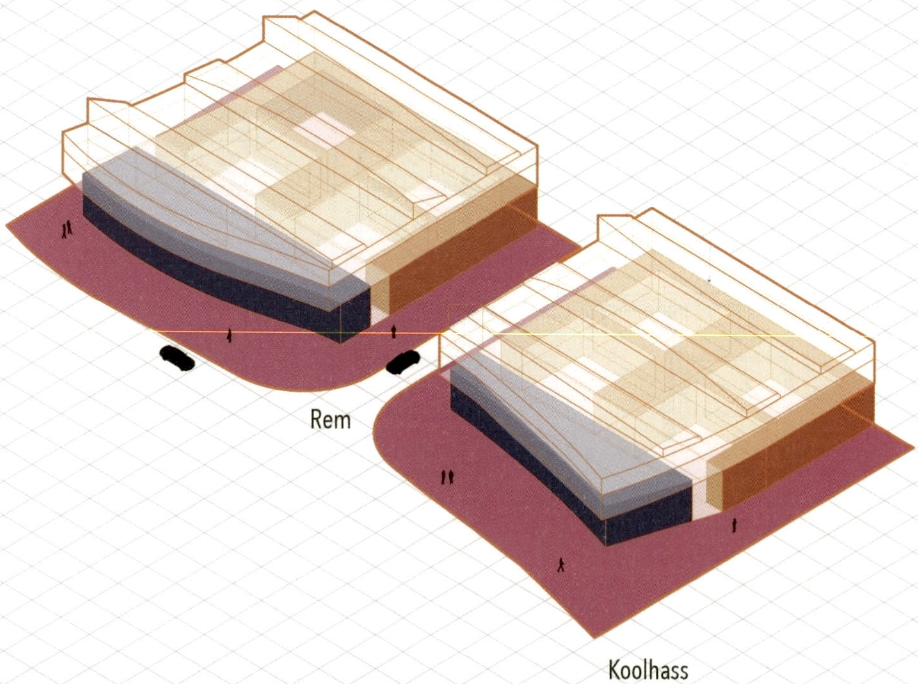

10ft cube

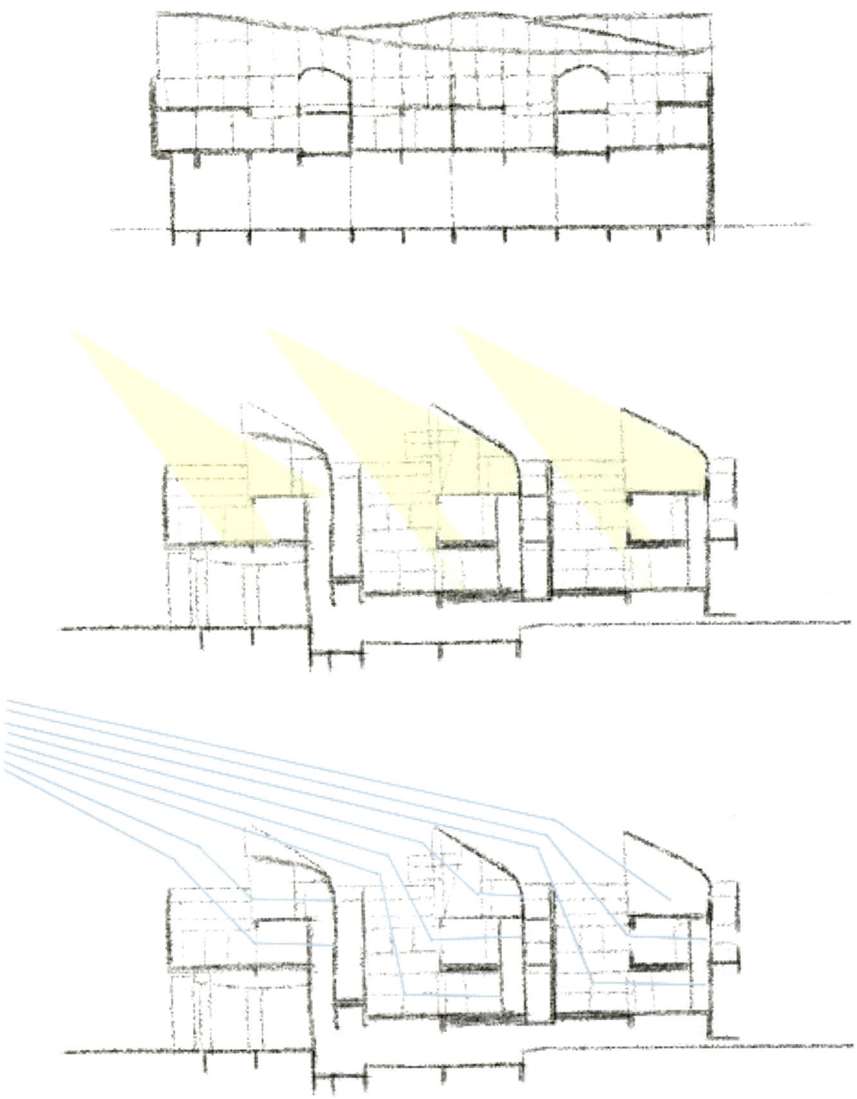

MRMU

ARC307 BU SECTION FALL 2022 | Kim Ryulhyeon

HABOUR CLUB AMSTERDAM

Amsterdam, The Netherlands
LEVS architecten

Site area: 109,792sqft
building area: 70,016sqft
Completion: 2021

10,000 sf **domestic**
communal
commercial
public

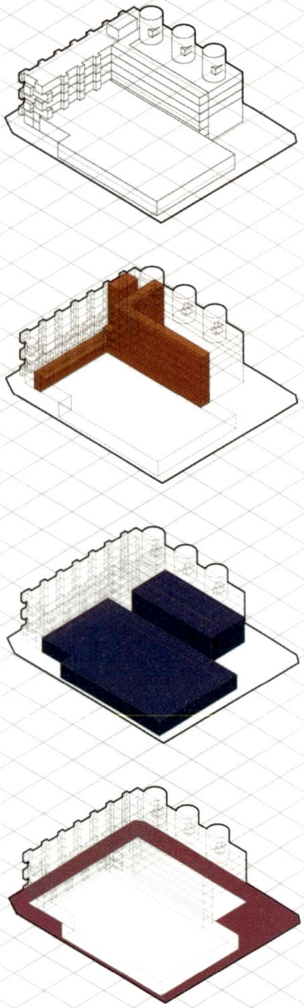

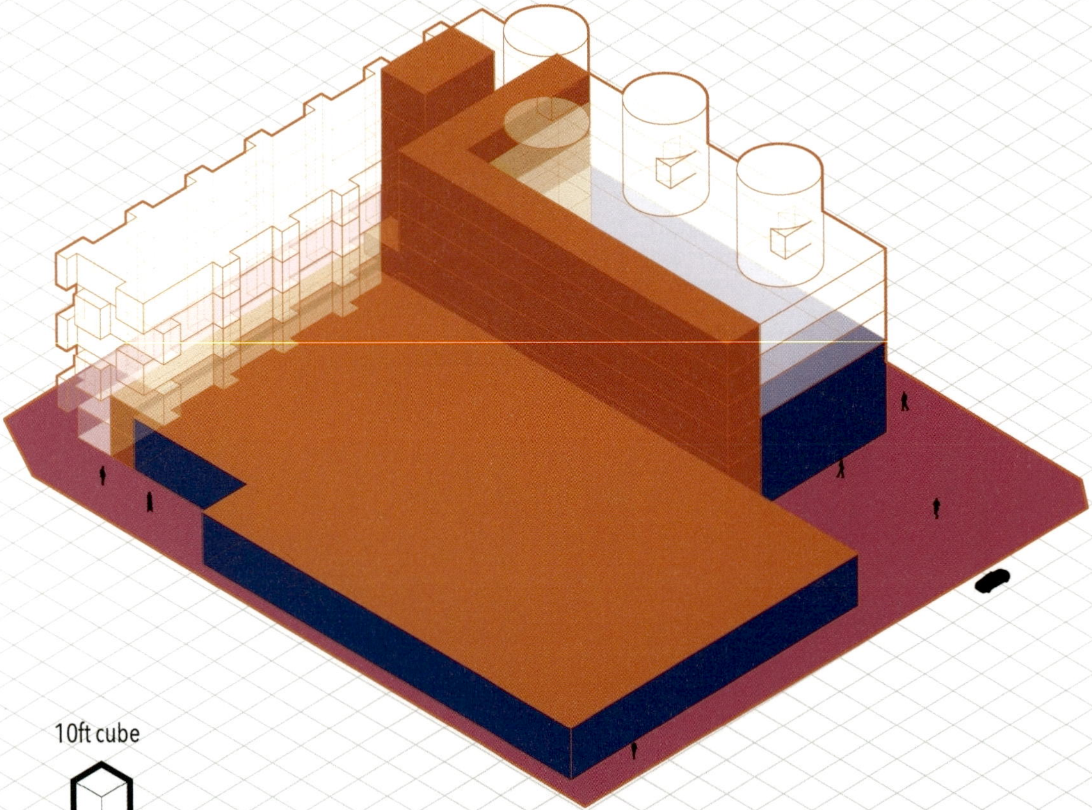

10ft cube

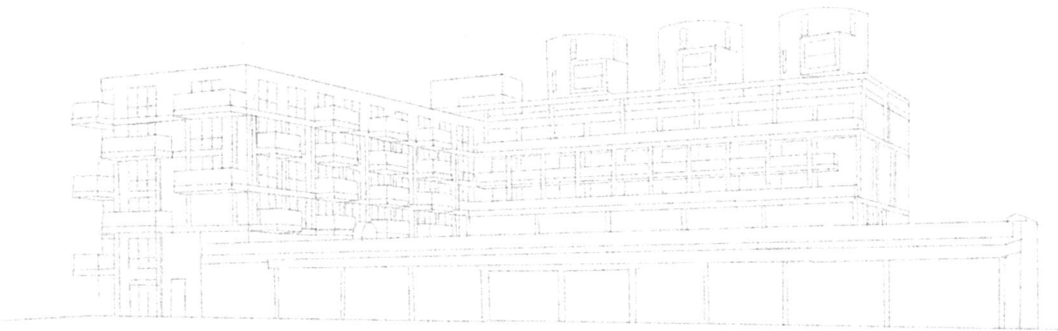

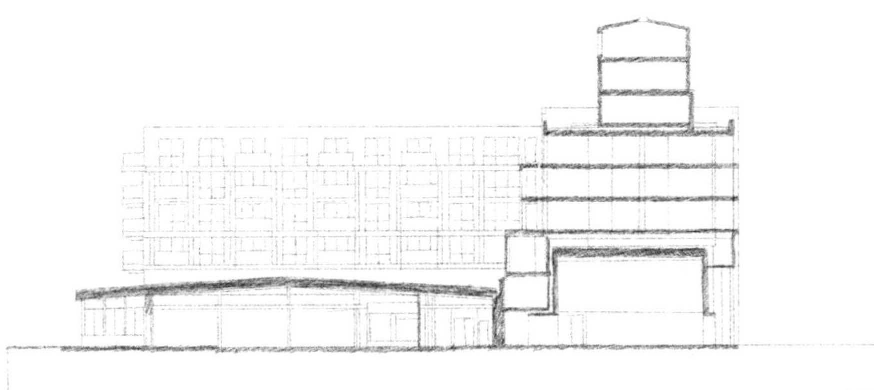

MRMU

ARC307 BU SECTION FALL 2022 | KIM MYEUNG BIN

MISS SARGFABRIK

Vienna
BKK-3

Site area: 10,600sf
building area: 63,200sf
Completion: 2000

10,000 sf

domestic

communal

commercial

public

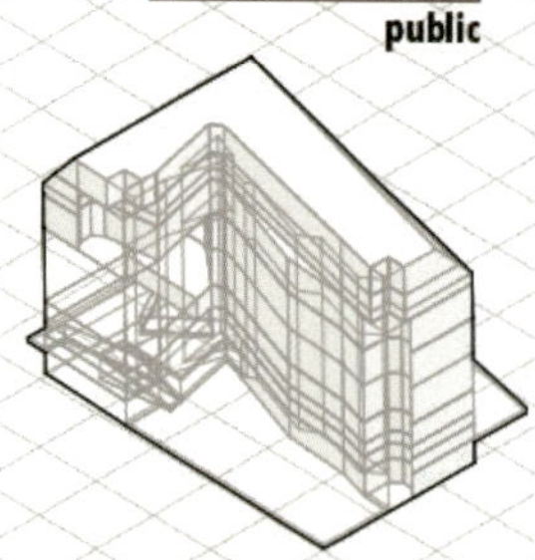

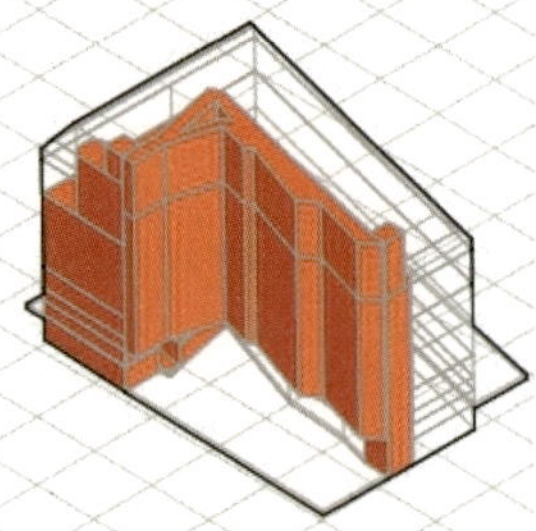

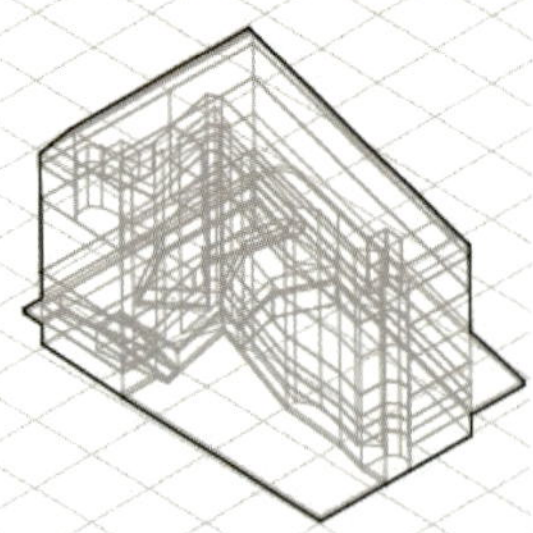

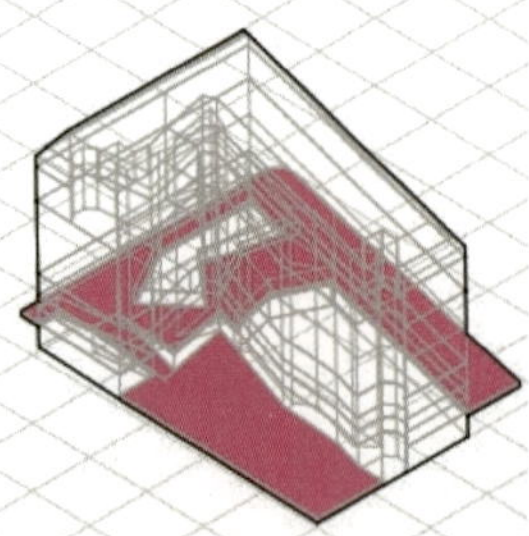

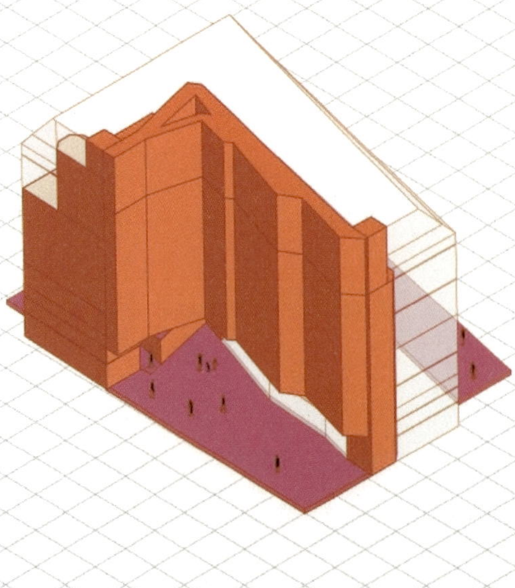

10ft cube

MRMU

ARC307 BU SECTION FALL 2022 | KIM MYEUNG BIN

BELAROÏA RESIDENTIAL BUILDING

FRANCE

Manuelle Gautrand Architecture

Site area: 37,800sf

building area: 255,700sf

Completion: 2019

10,000 sf

domestic

communal

commercial

public

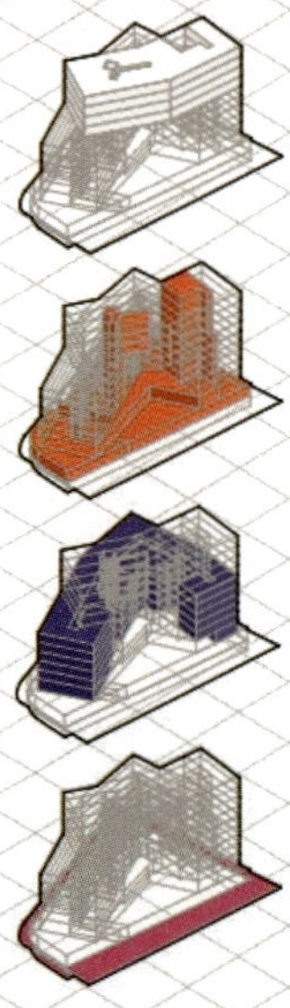

10ft cube

BELAROIA

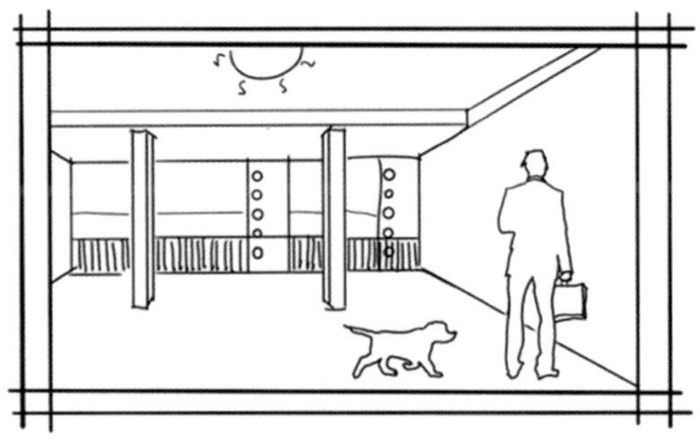

MRMU

ADS045 PNU FALL 2022 | Eunhye Kim

PORT-O-PRENZ APRATMENTS

Berlin, Germany
J. Mayer H. Architects
Site area: 103,655sf
Building area: 301,389sf
Completion: 2021

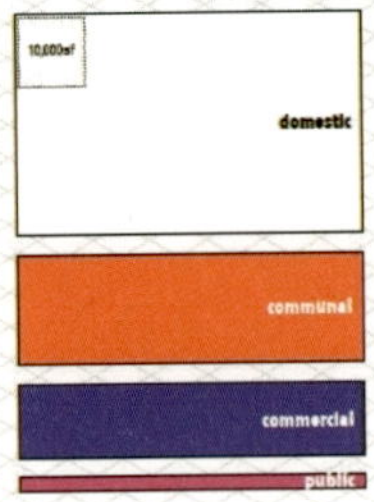

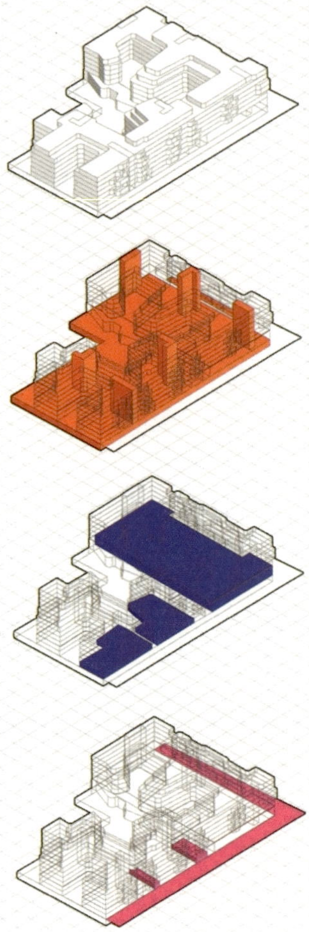

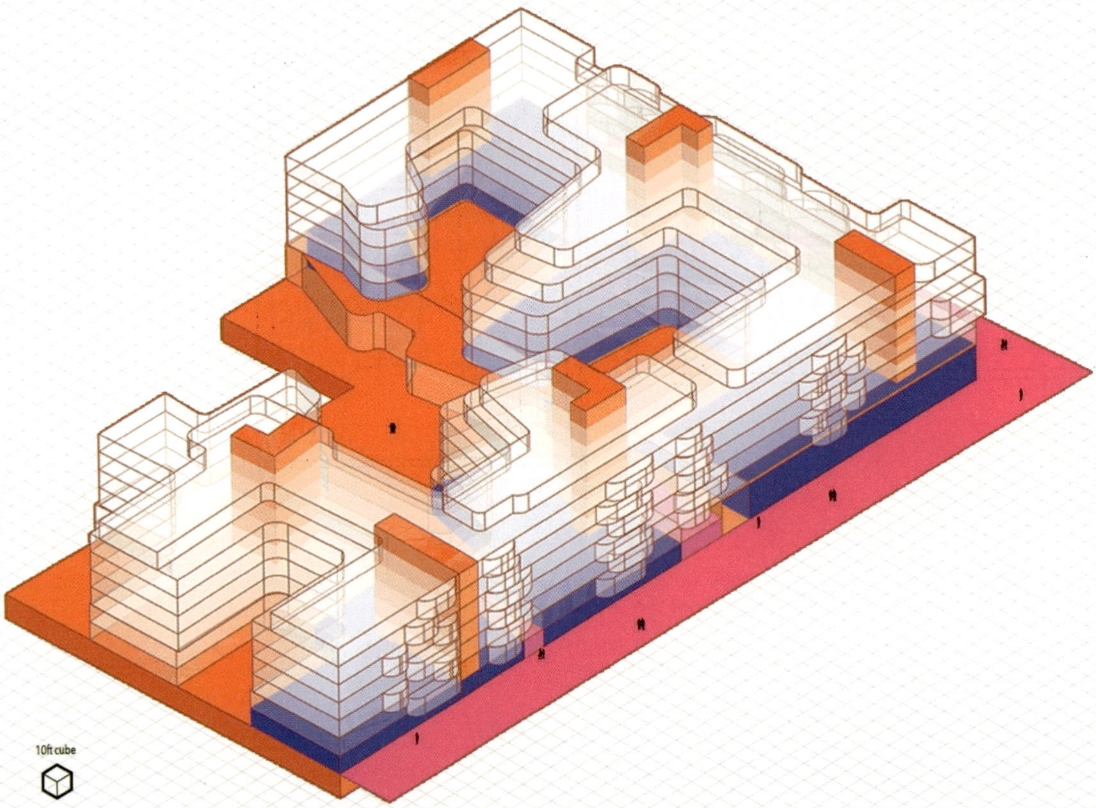

10ft cube

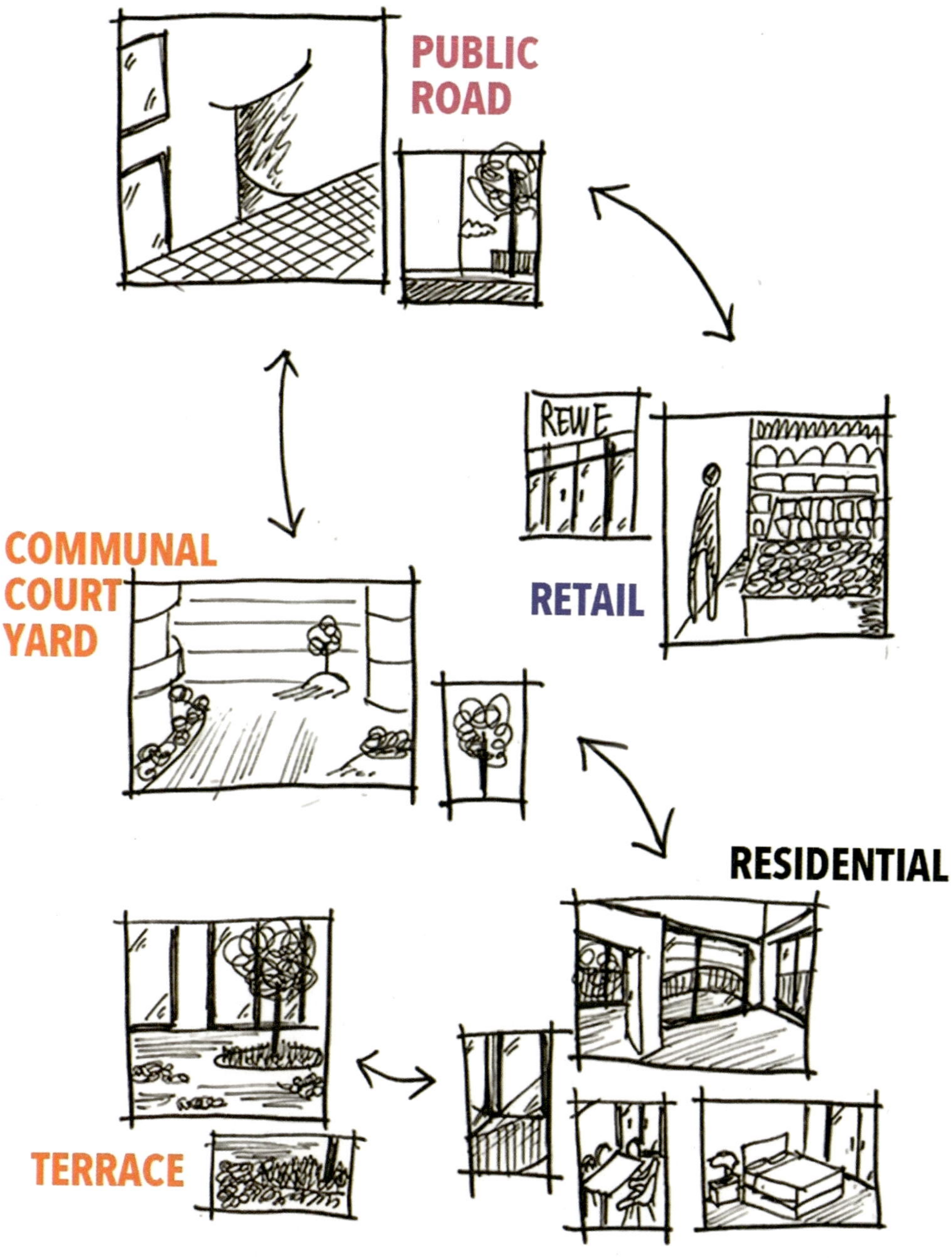
PUBLIC ROAD
REWE
RETAIL
COMMUNAL COURT YARD
RESIDENTIAL
TERRACE

MRMU

ADS045 PUN FALL 2022 | Eunhye Kim

SCHOTS 1&2

Groningen, The Netherlands
Studio Woodroffe Papa

Site area: 113,537sf
Building area: 185,871sf
Completion: 2003

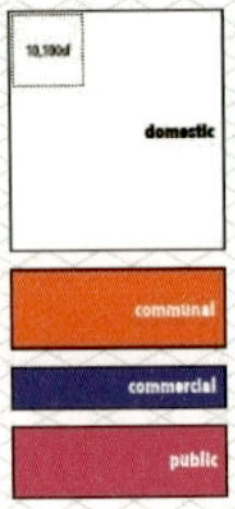

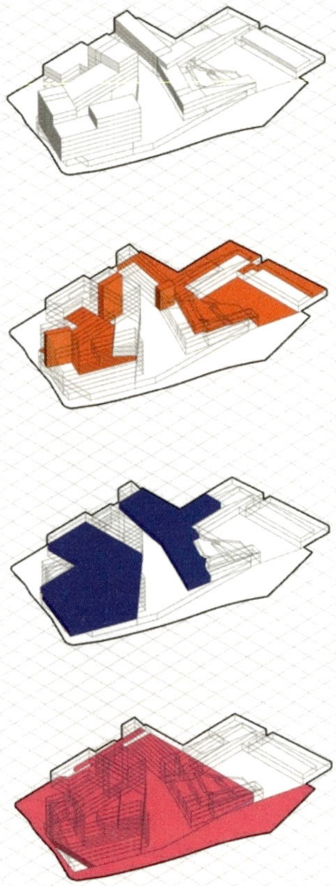

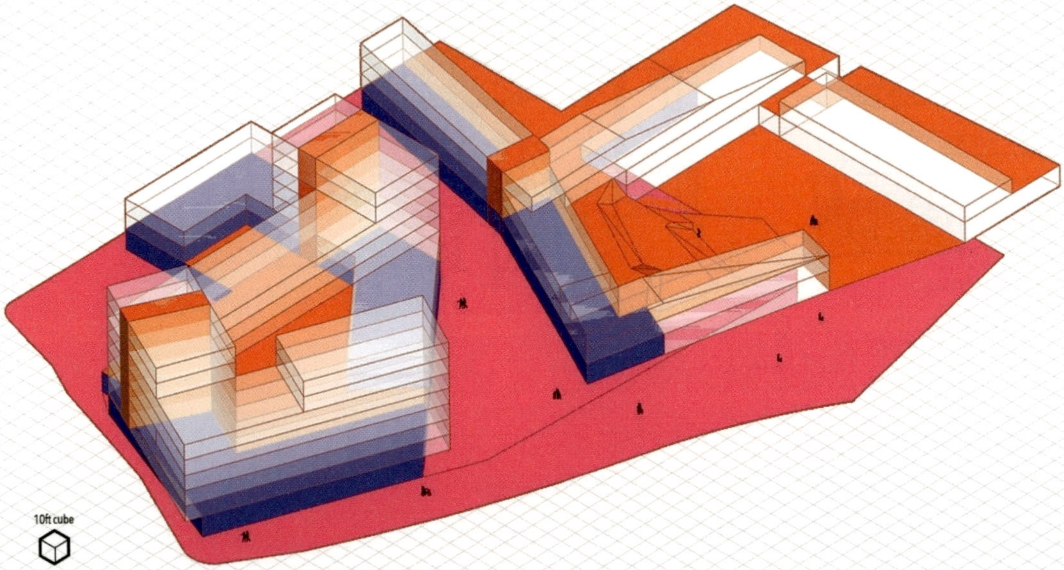

10ft cube

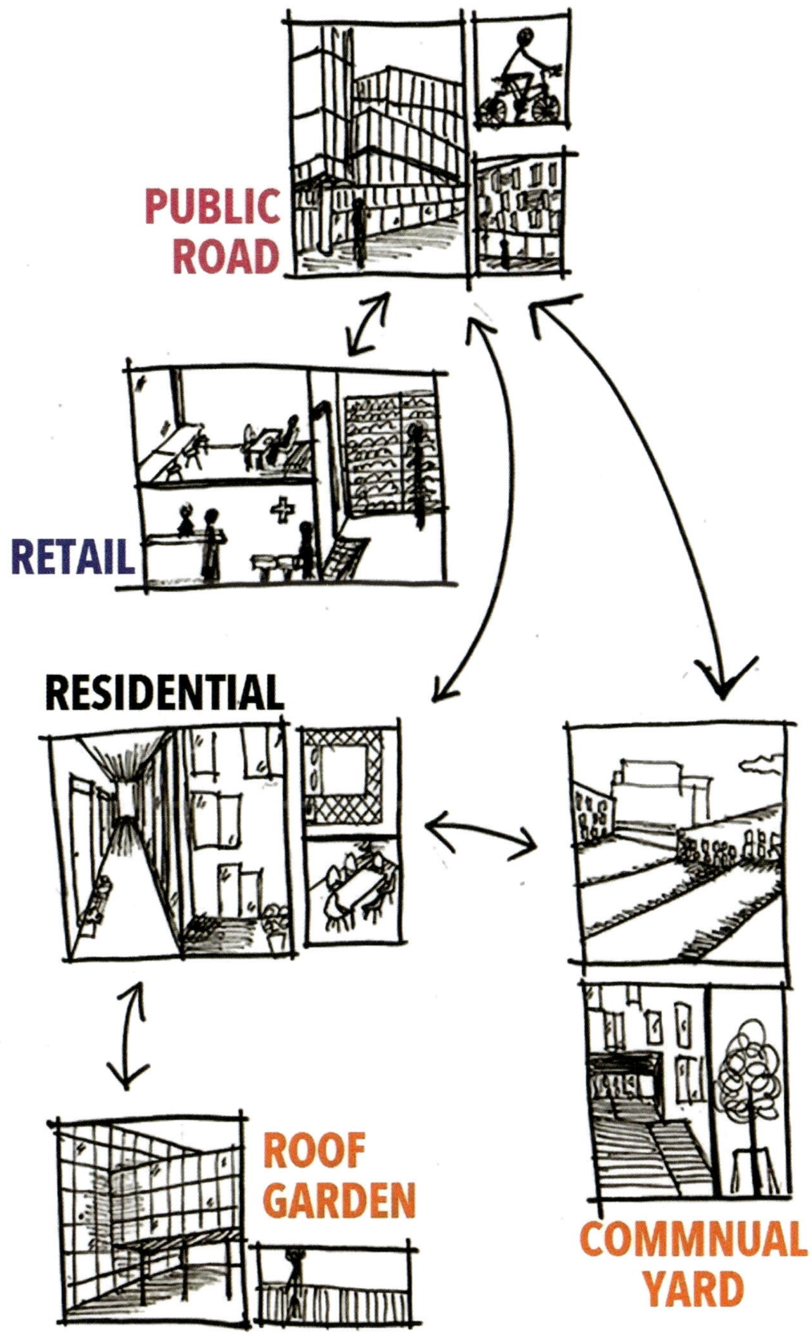
PUBLIC
ROAD
RETAIL
RESIDENTIAL
ROOF
GARDEN
COMMNUAL
YARD

MRMU

ARC307 BU SECTION FALL 2022 | Sangjeen Moon

ALFONSO REYES 58

Mexico city
Derek Dellekamp

Site area: 3,444sf
building area: 18,684sf
Completion: 2003

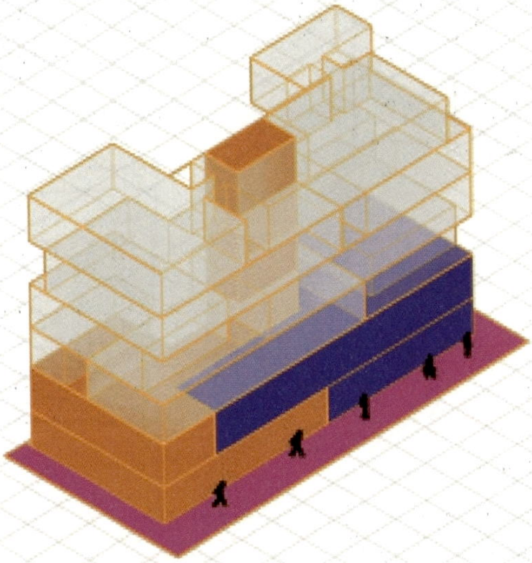

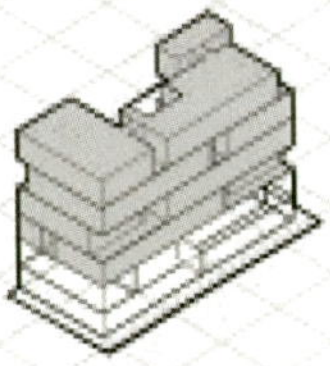

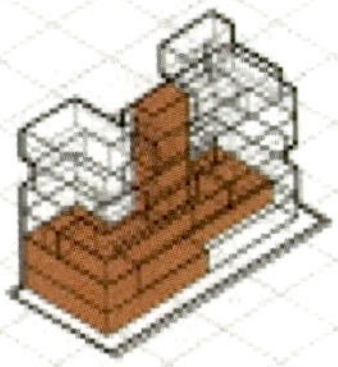

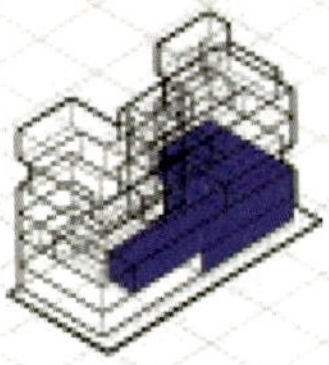

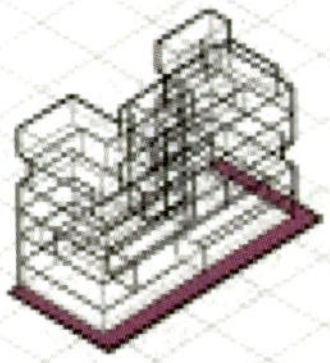

10ft cube

ALFONSO REYES 58

MRMU

ARC307 BU SECTION FALL 2022 | Sangjeen Moon

BLOX

Copenhagen
OMA

Site area: 103,000 sf

building area: 300,000 sf

Completion: 2018

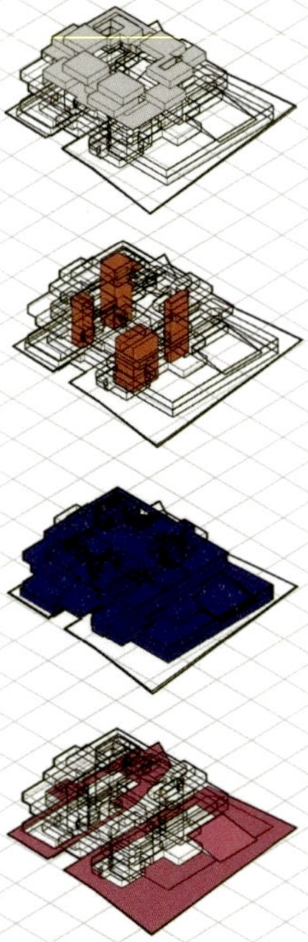

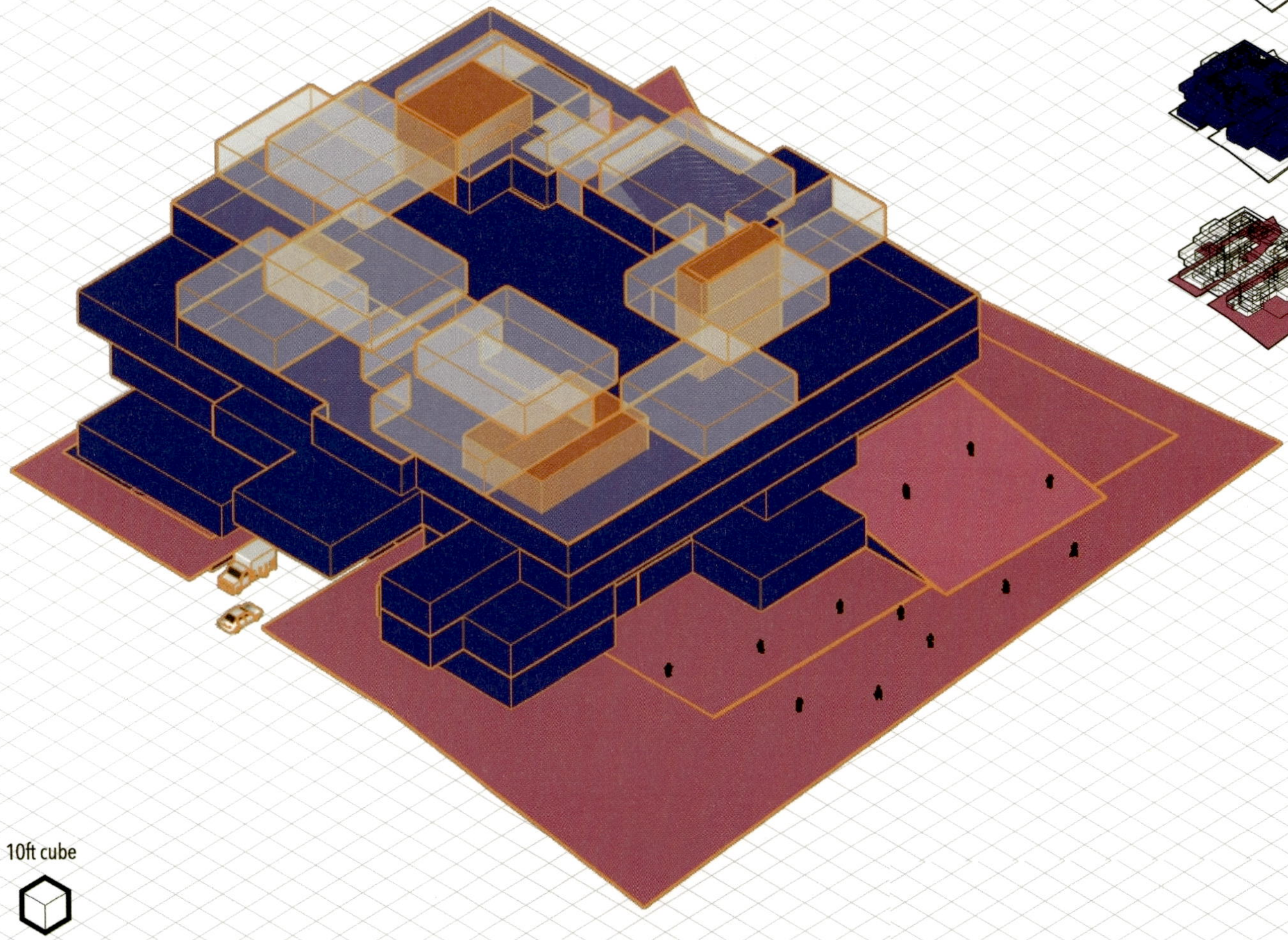

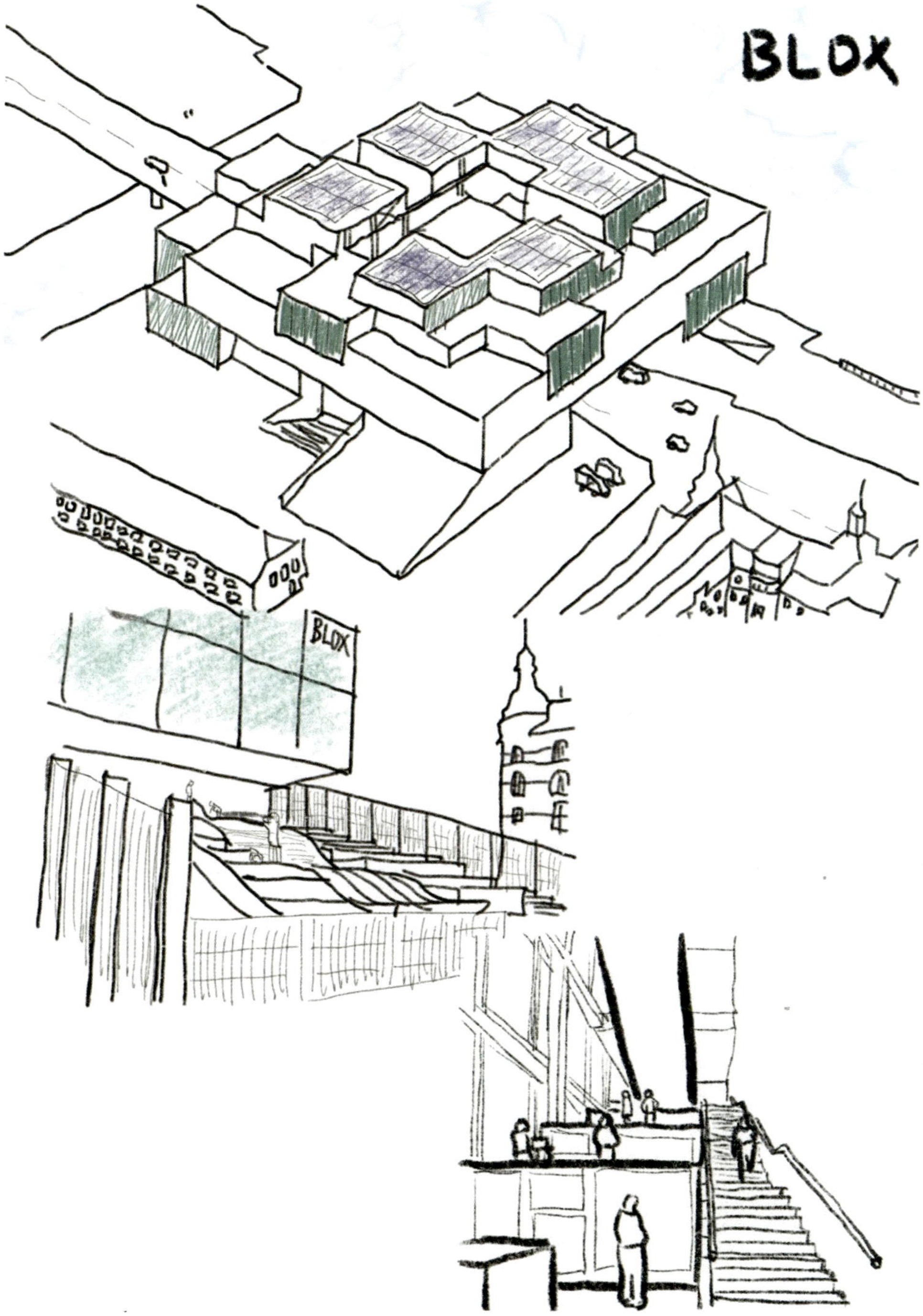
BLOX
BLOX

MRMU

ARC307 BU SECTION FALL 2022 | Park Changmin

IIT STATE STREET VILLAGE

Los Angeles

Helmut Jahn of Murphy-Jahn Associates

Site area: 7,335sqm

building area: 9,387sqm

Completion: 2003

1,000 sqm

domestic

communal

public

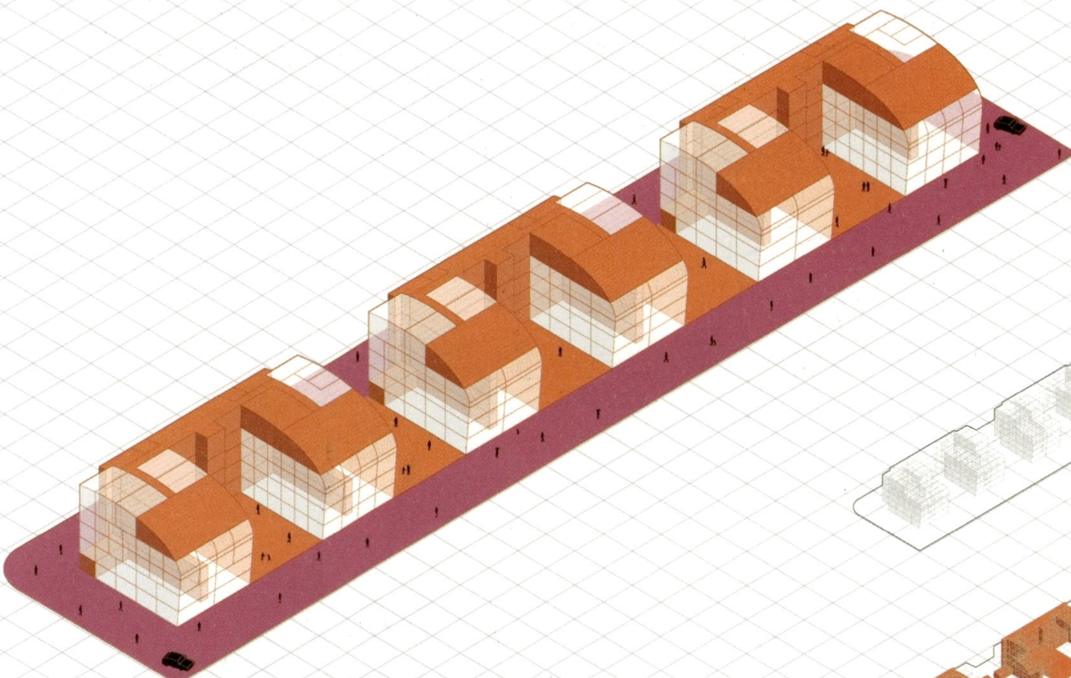

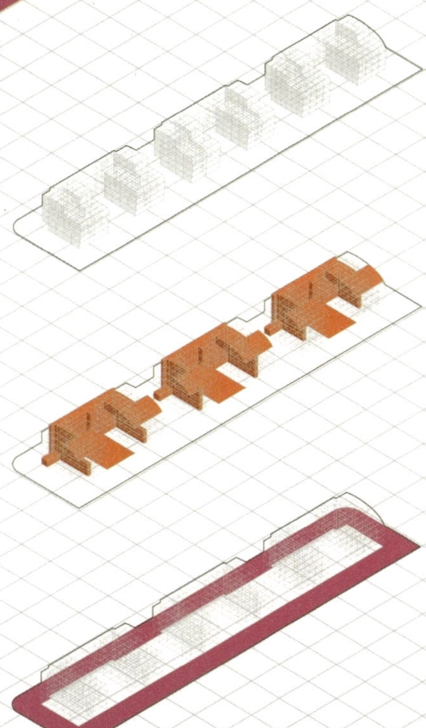

6m cube

MRMU

ARC307 BU SECTION FALL 2022 | Park Changmin

TREEHOUSE COLIVING APARTMENT

Gangnam-gu
Bo-DAA

Site area: 1,171sqm
building area: 4,810sqm
Completion: 2018

1,000sqm
domestic
communal
commercial
public

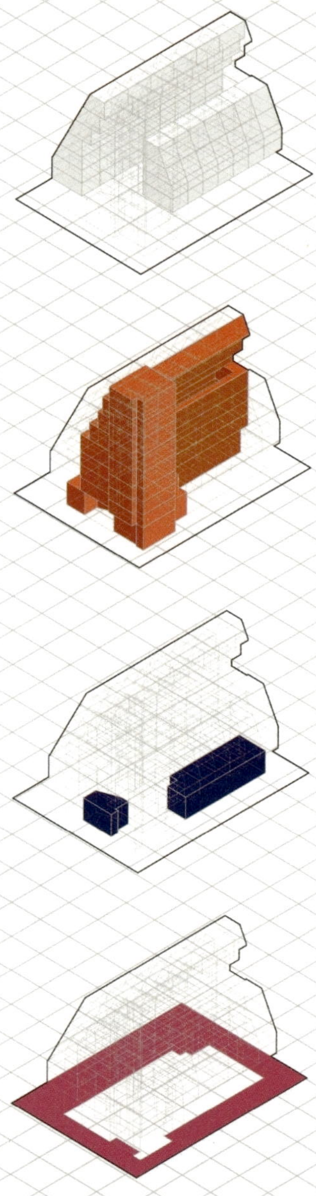

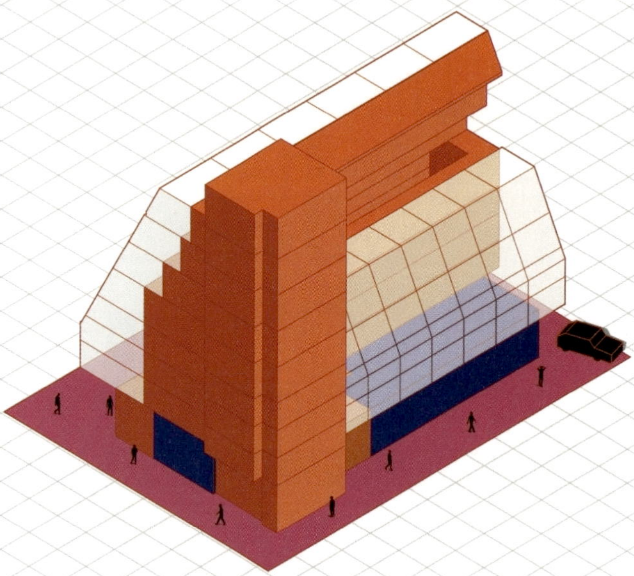

3m cube

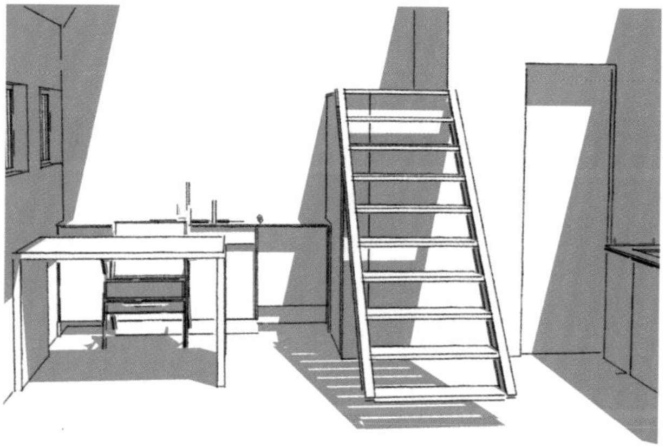

MRMU

ARC307 BU SECTION FALL 2022 | Son Gyusik

MOOKAS HQ

Gyeonggi-do Province
DJH Architects

Site area: 15,000sf
building area: 44,470sf
Completion: 2016

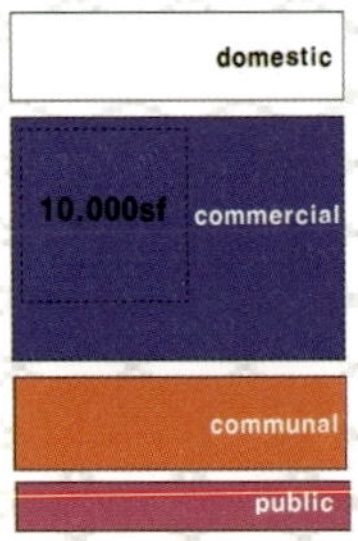

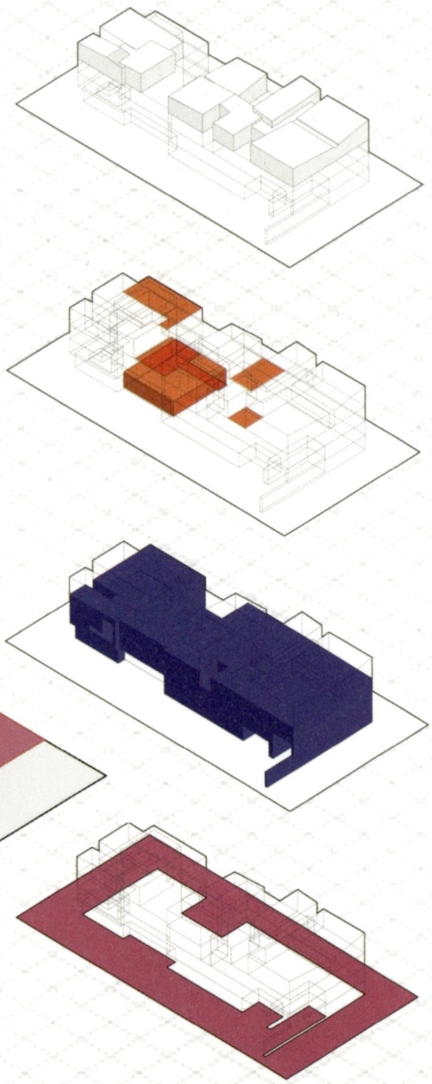

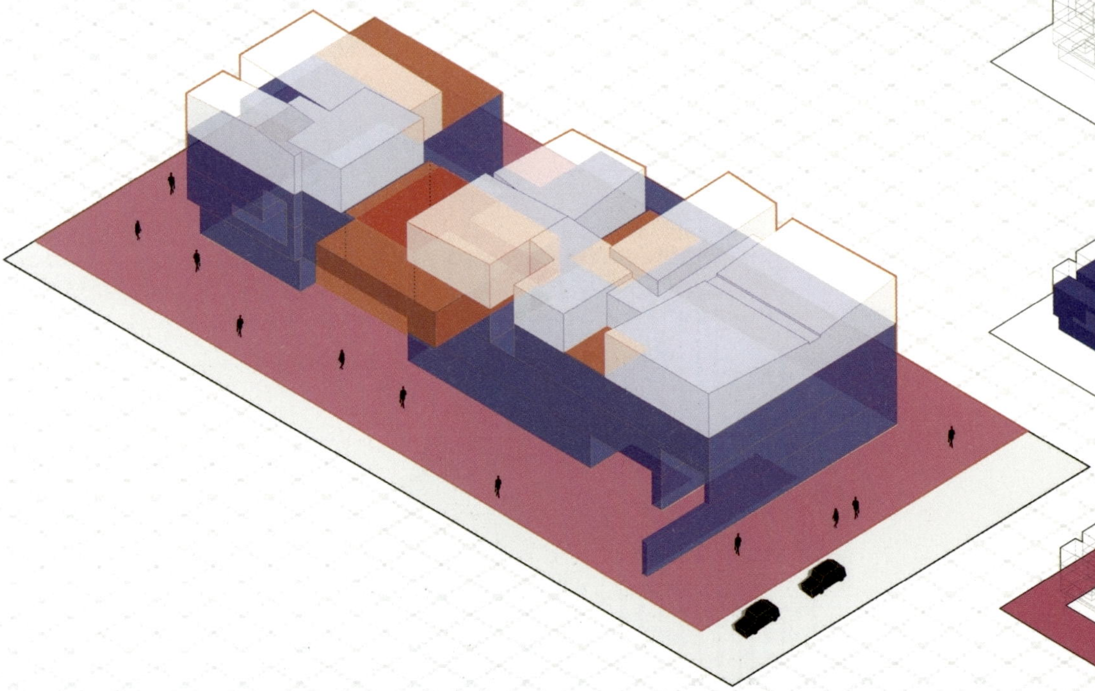

10ft cube

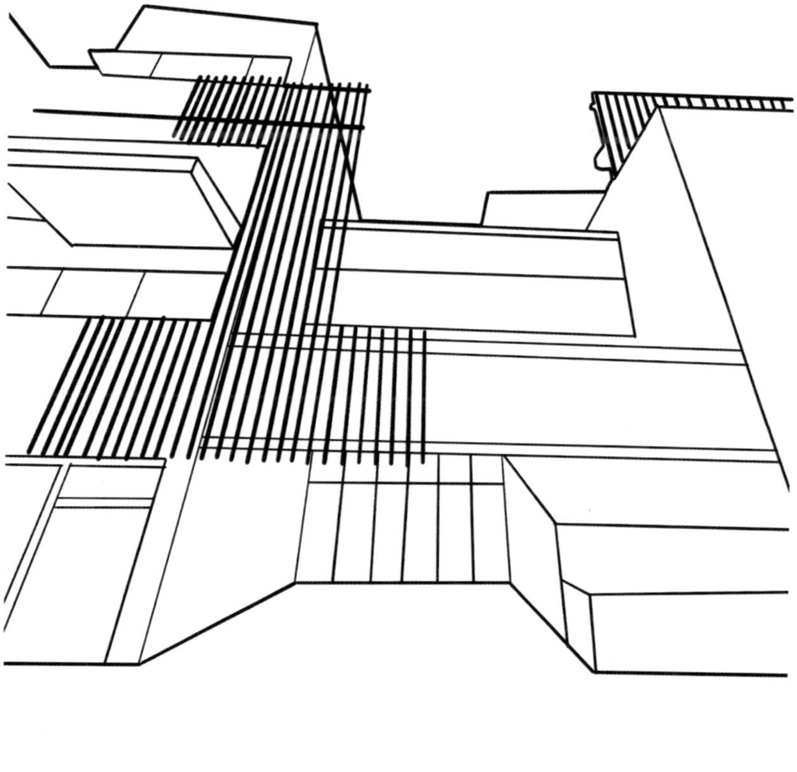

MRMU

ARC307 BU SECTION FALL 2022 | Son Gyusik

LOOSHAUS

Wien, Austria
Adolf Loos

Site area: 15,000sf
building area: 58,000sf
Completion: 1911

10,000sf domestic
communal
commercial
public

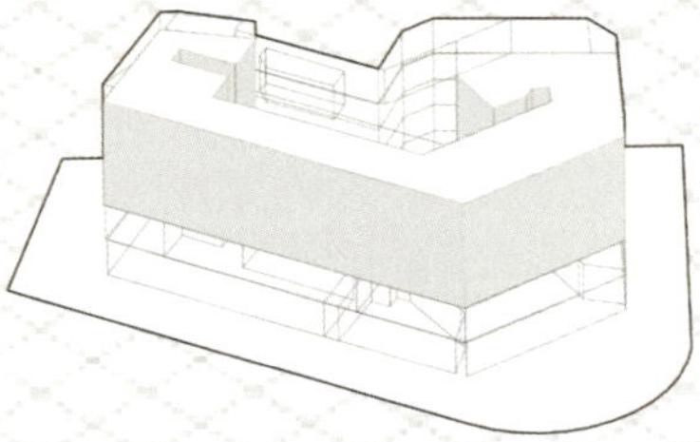

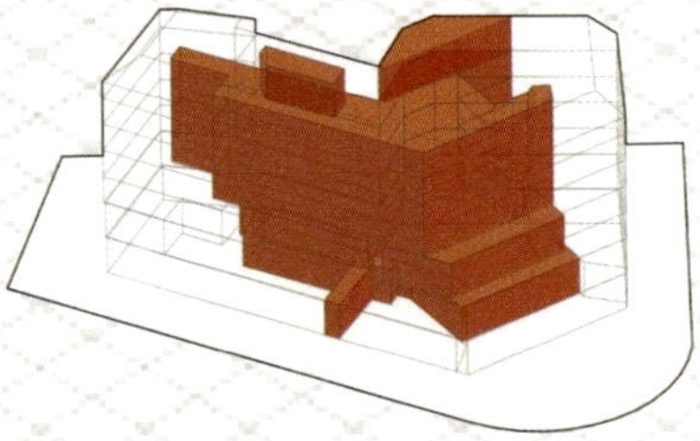

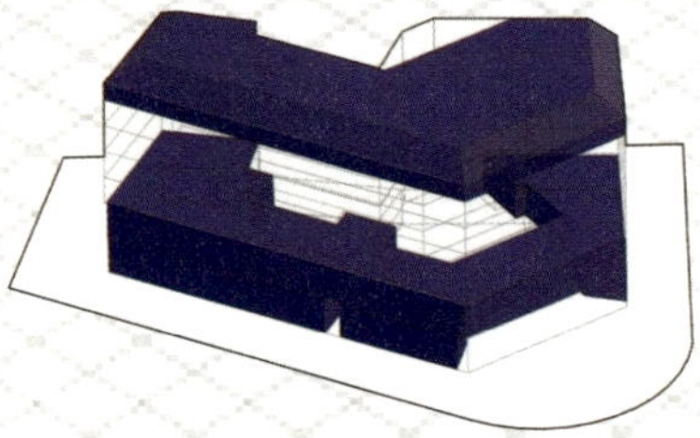

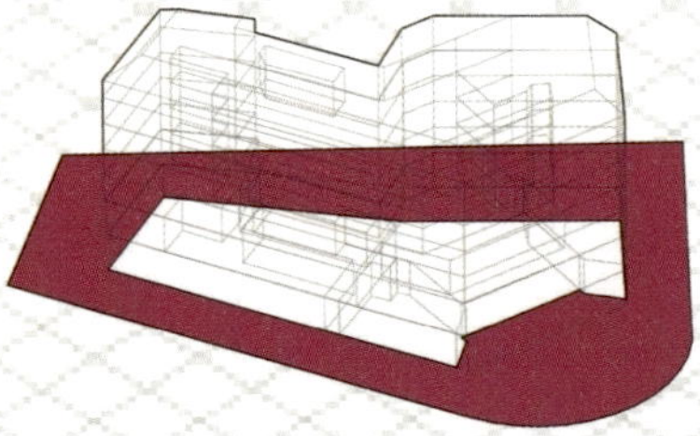

10ft cube

MRMU

ARC307 BU SECTION FALL 2022 | You Juhae

MIXED USE TOWER WINNING PROPOSAL

Beijing, China
Dos Architects

Site area: 24,200 sqm
building area: 219,200 sqm
Completion: 2017

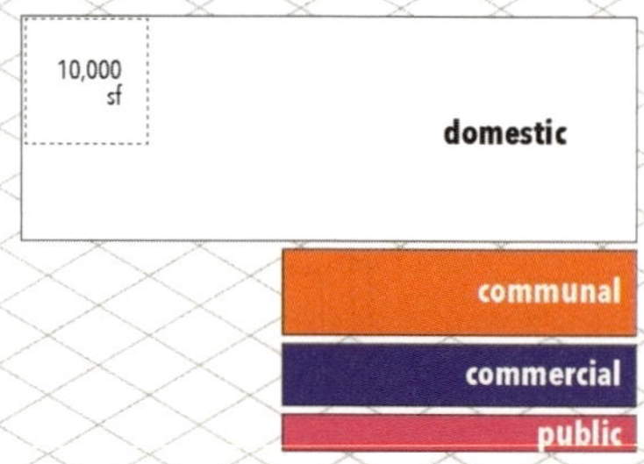

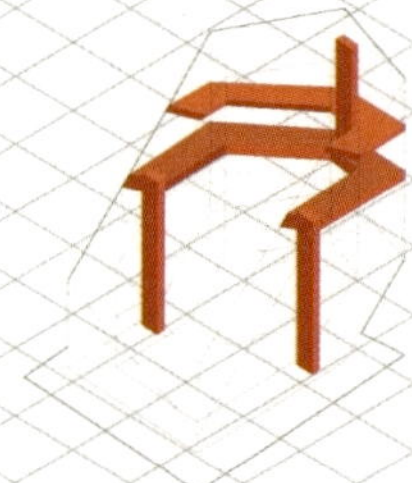

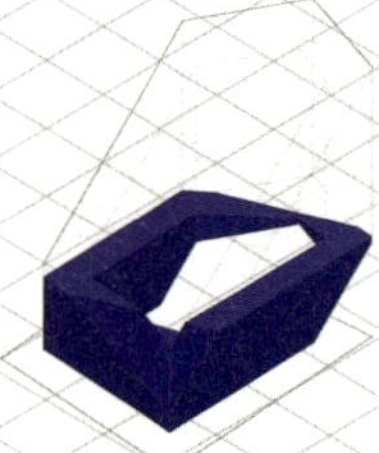

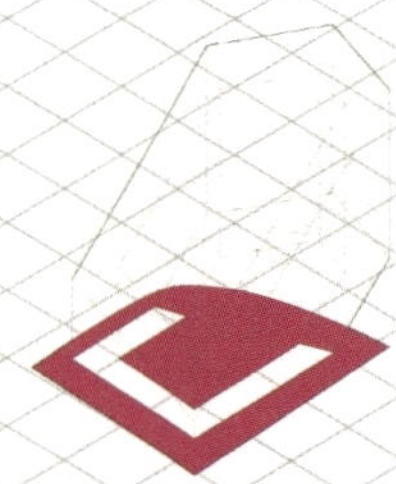

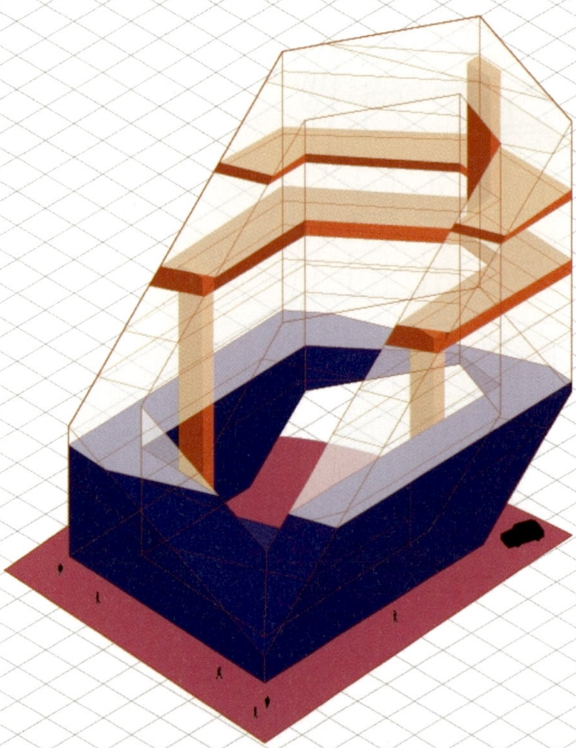

10ft cube

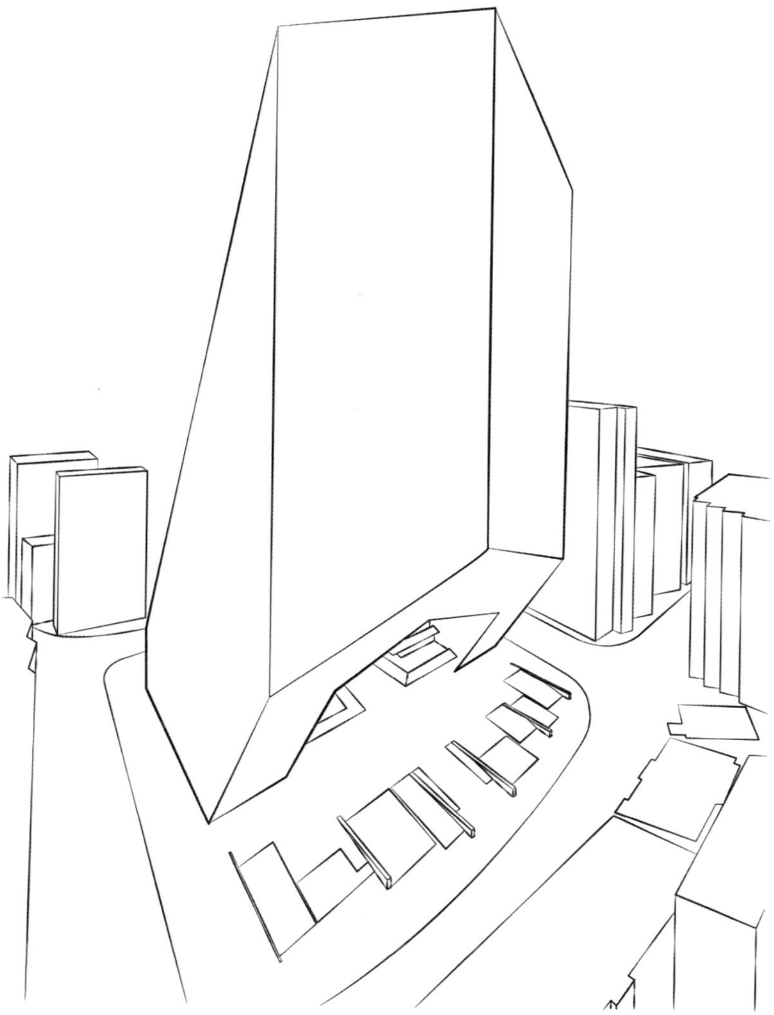

MRMU

ARC307 BU SECTION FALL 2022 | You Juhae

VIA VERDE
New York, USA
Grimshaw + Dattner Architects

Site area: 24,000sf

building area: 290, 000 sq ft

Completion: 2012

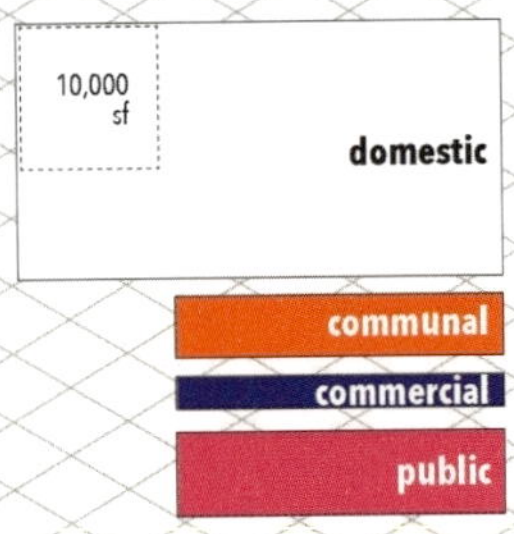

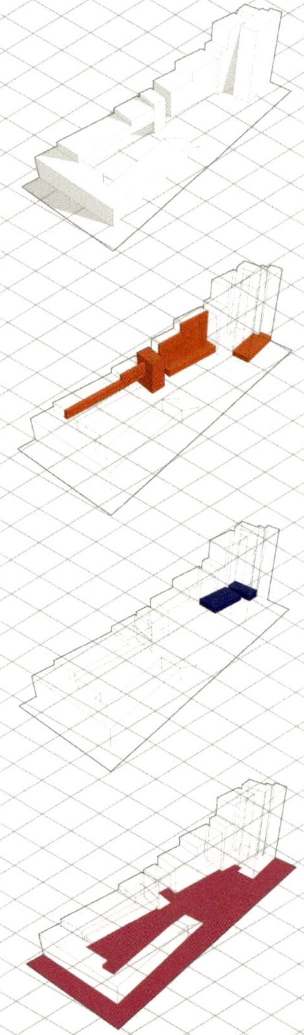

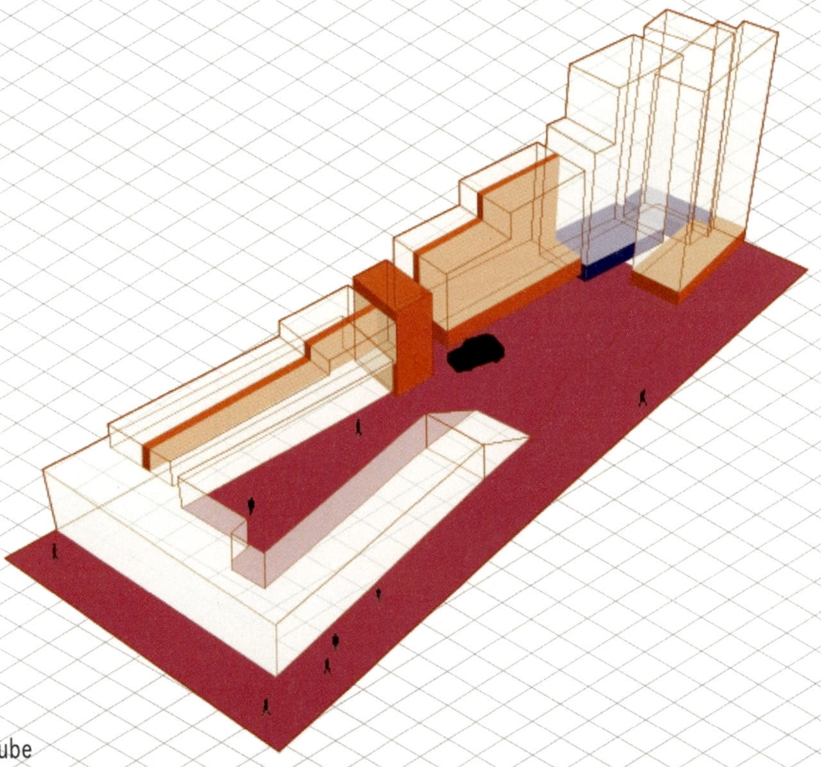

10ft cube

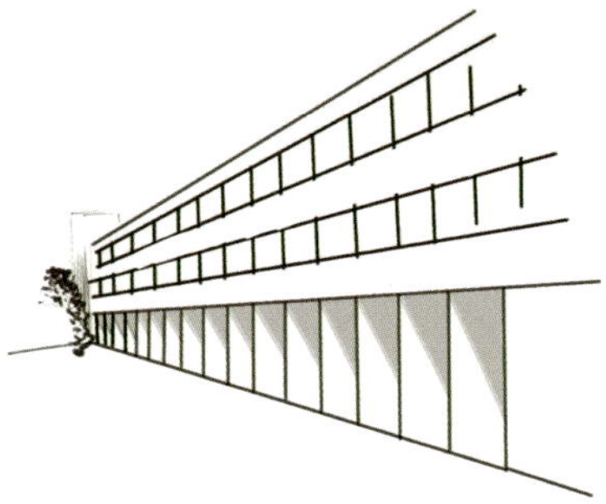

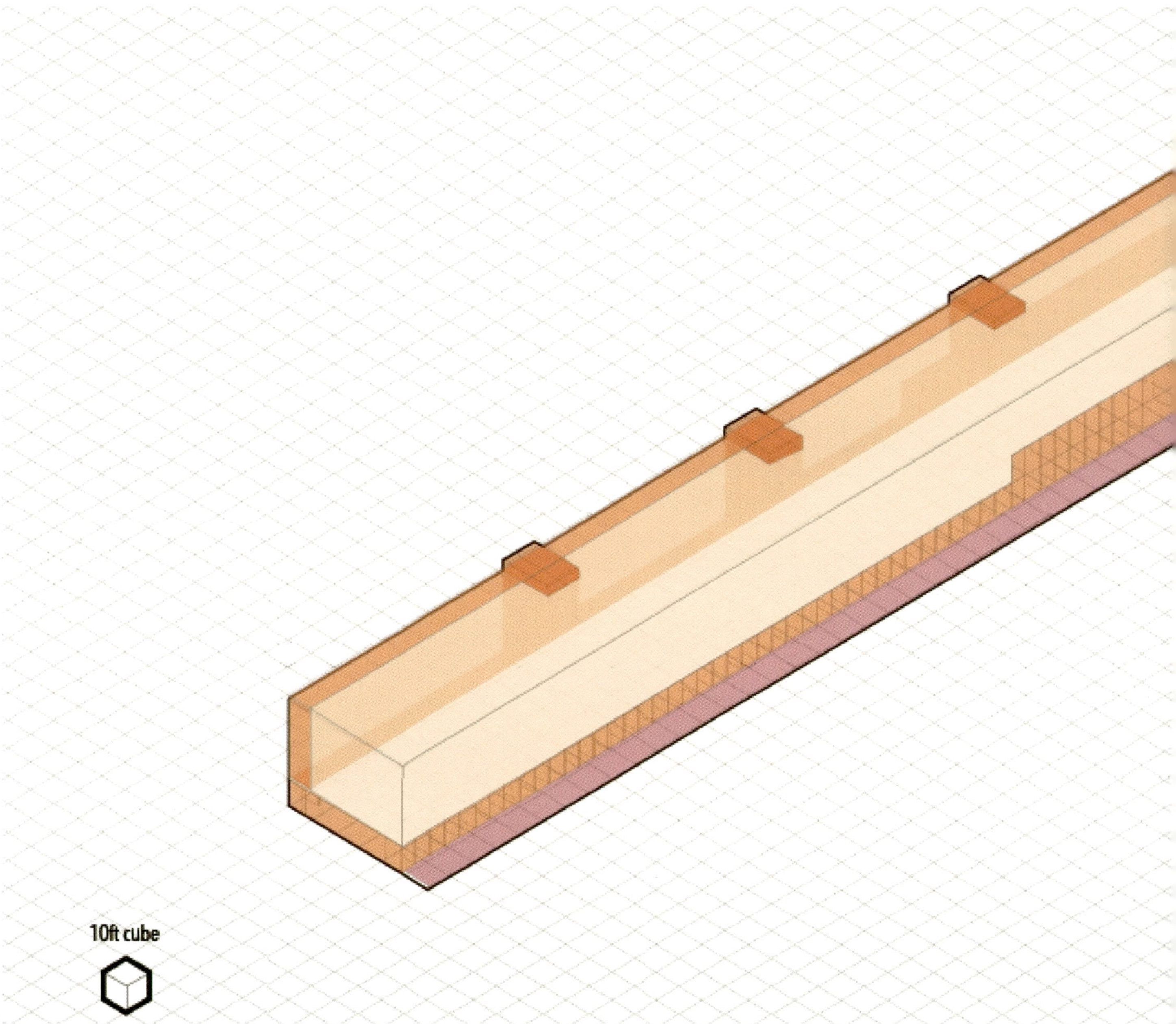
10ft cube

MRMU

ARC307 BU SECTION FALL 2022 | Lee Sang Min

Gallaratese Housing

Milan, Italy
Aldo Rossi

Site area: 38,000sf

building area: 35,000sf

Completion: 1972

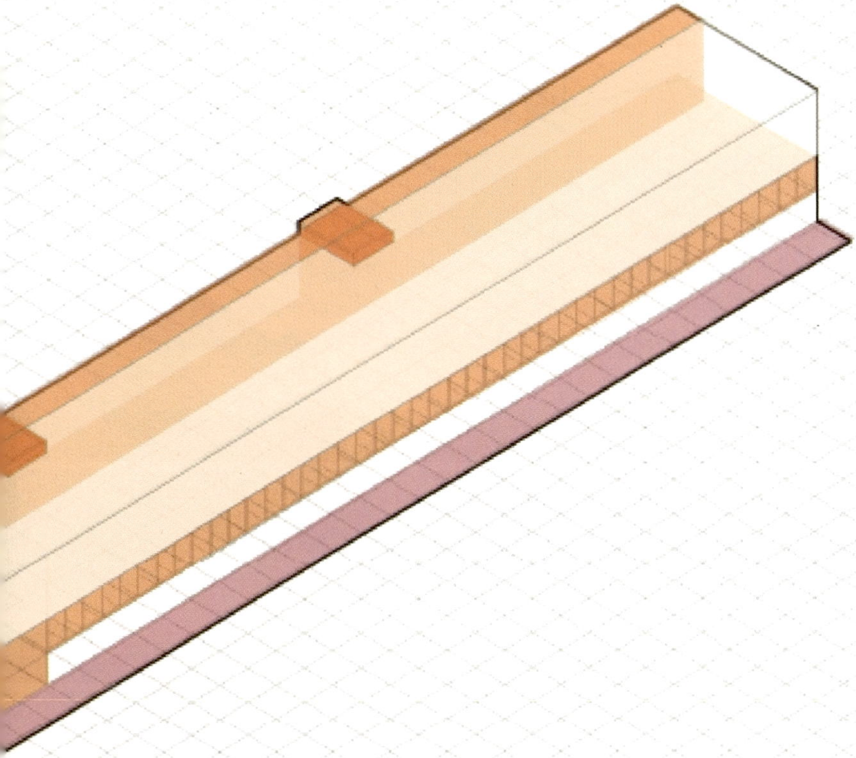

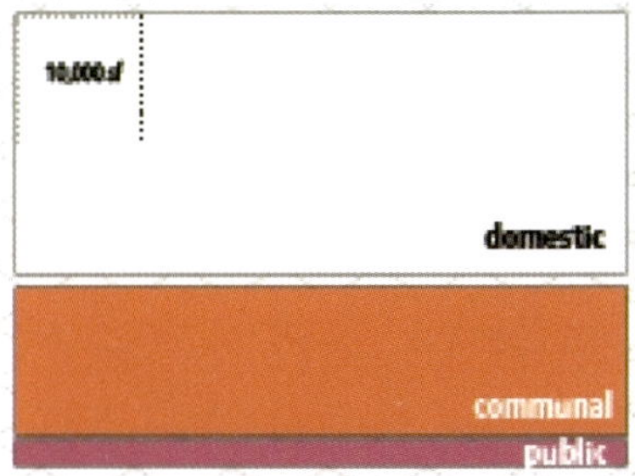

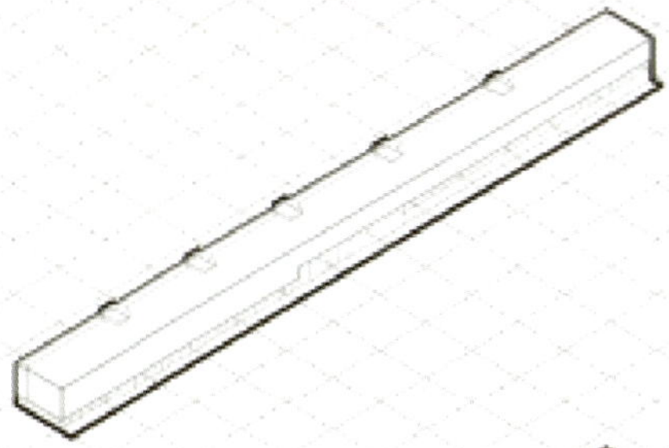

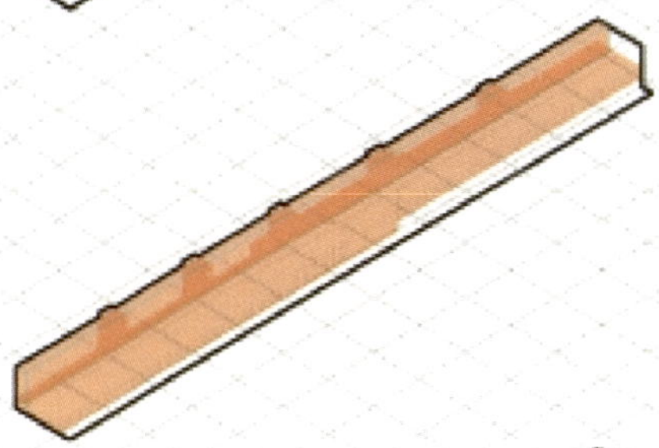

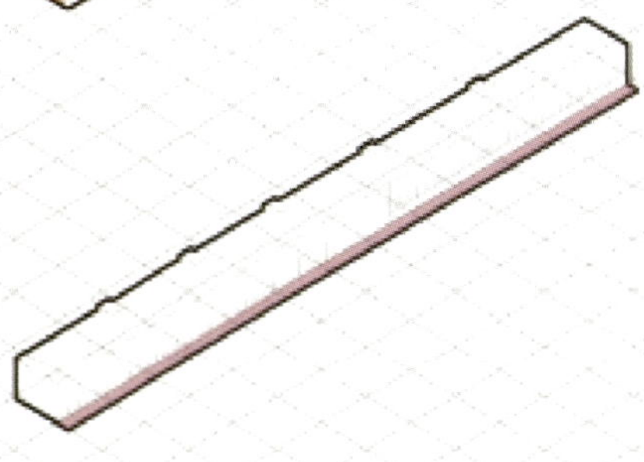

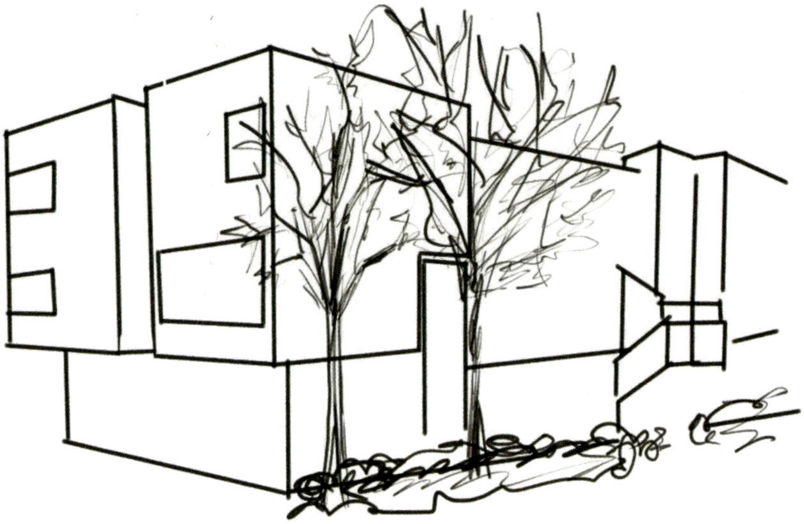

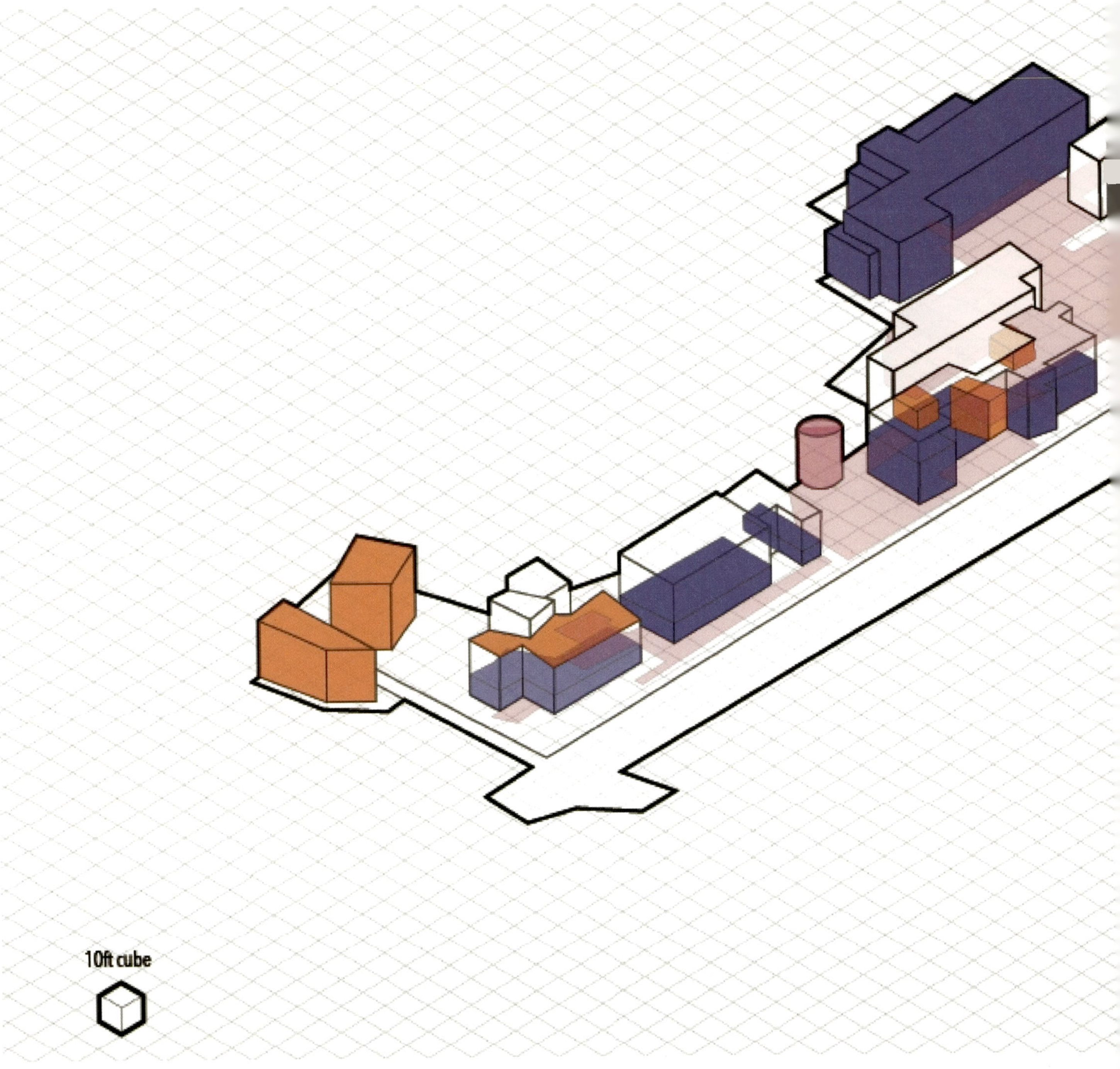
10ft cube

MRMU

ARC307 BU SECTION FALL 2022 | Lee Sang Min

Hillside Terrace

Tokyo
Maki Fumihiko

Site area: 78,000sf
building area: 48,000sf
Completion: 1998

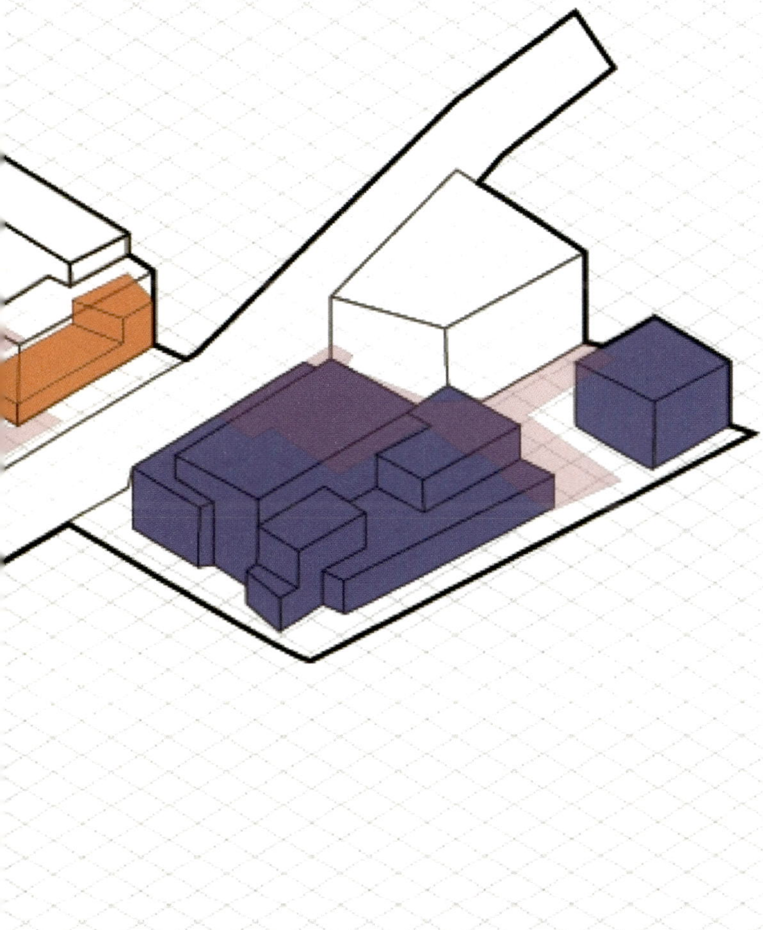

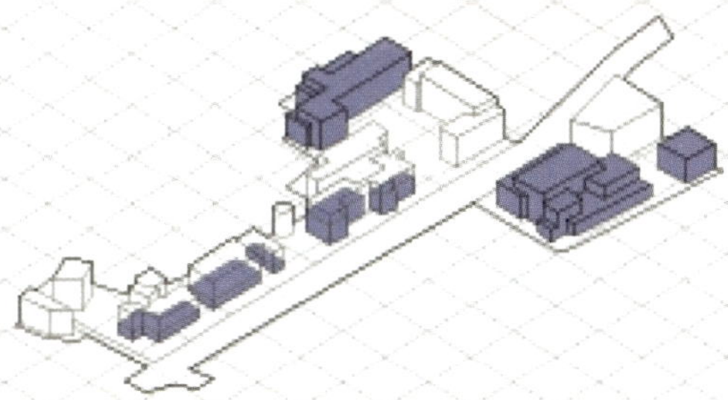

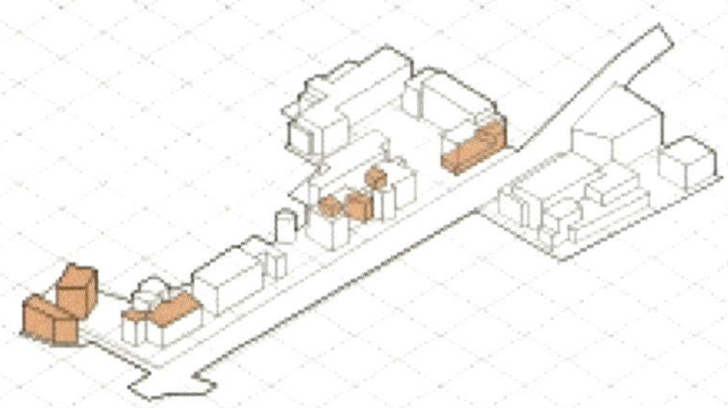

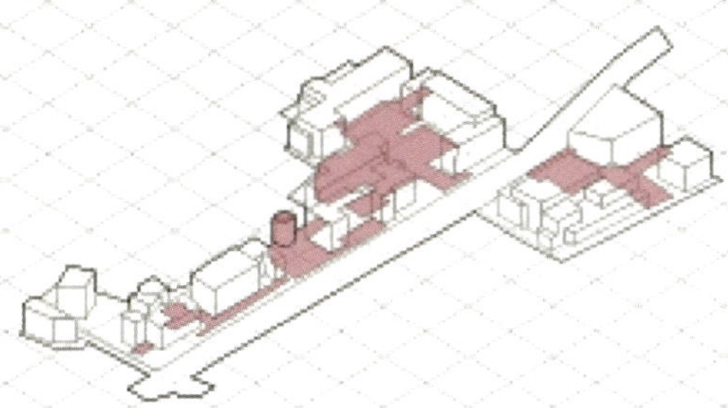

MRMU

ARC307 BU SECTION FALL 2022 | LEE HEONJU

ASSEMBLY 118
SEATTLE
Caron Architecture

Site Area: 70,120 sf
Building Area: 30,398 sf
Completion: 2018

10,000 sf — Domestic

Communal

Commercial

Public

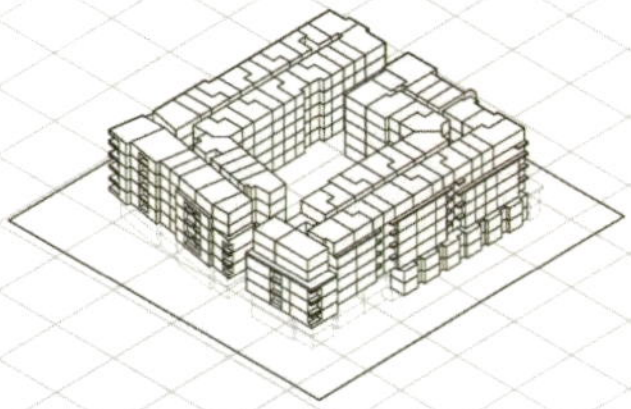

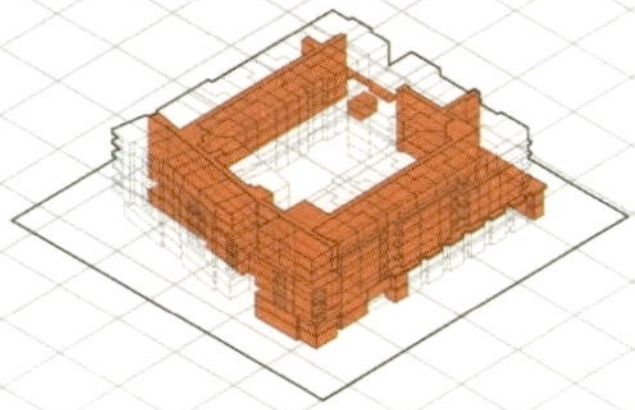

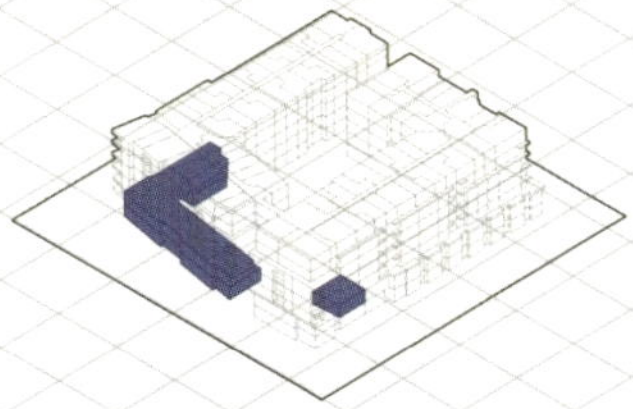

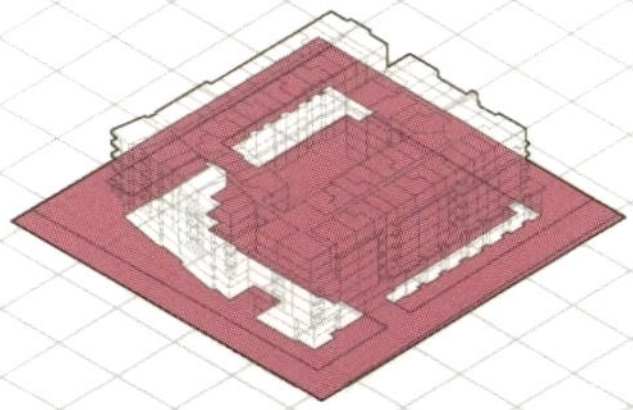

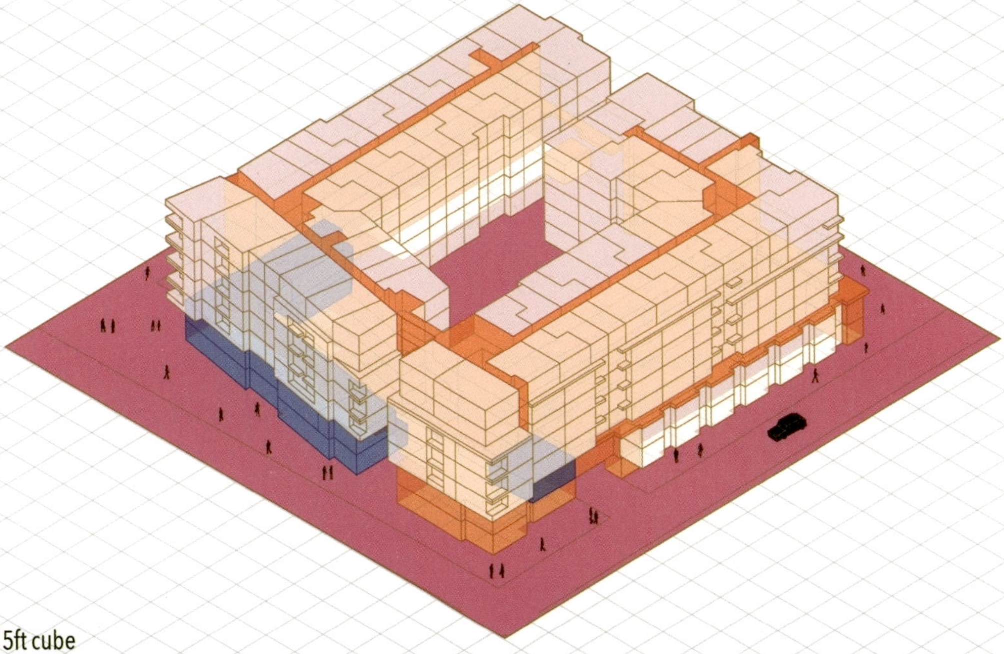

15ft cube

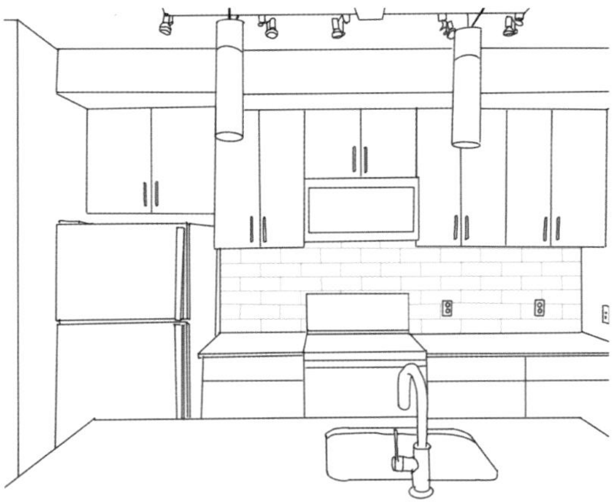

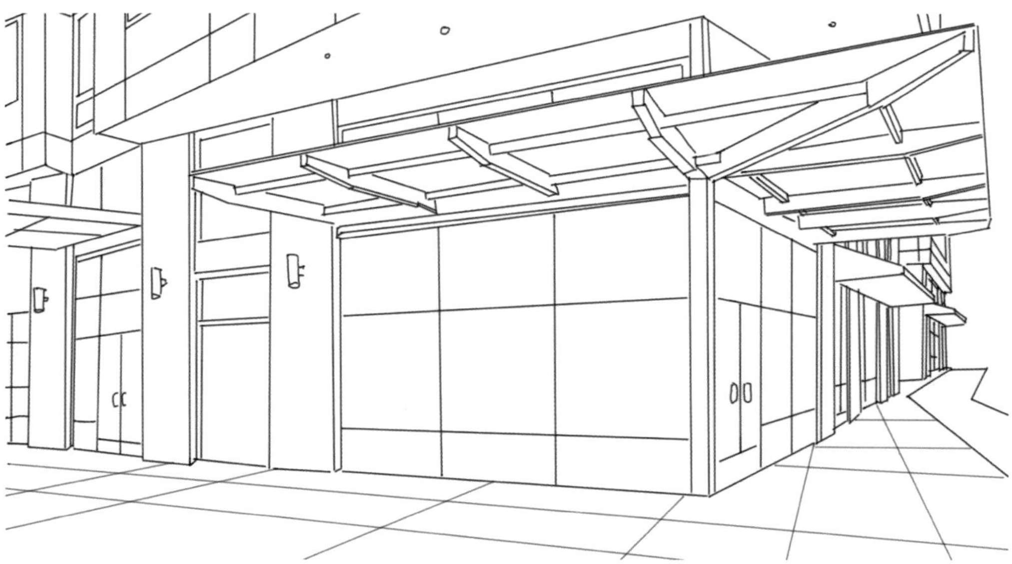

MRMU

ARC307 BU SECTION FALL 2022 | LEE HEONJU

NAKWON SANGA

SEOUL
UNKNOWN

Site Area: 82,409 sf

Building Area: 141,222 sf

Completion: 1969

10,000 sf	Domestic

Communal

Commercial

Public

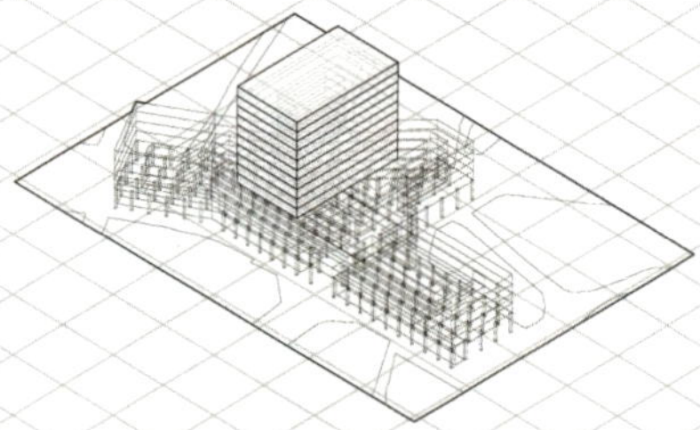

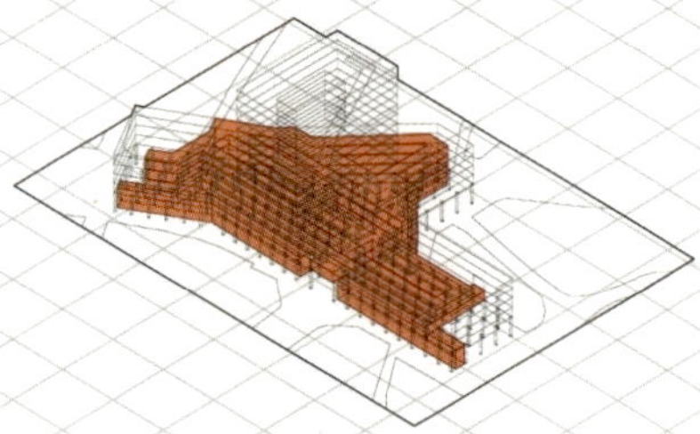

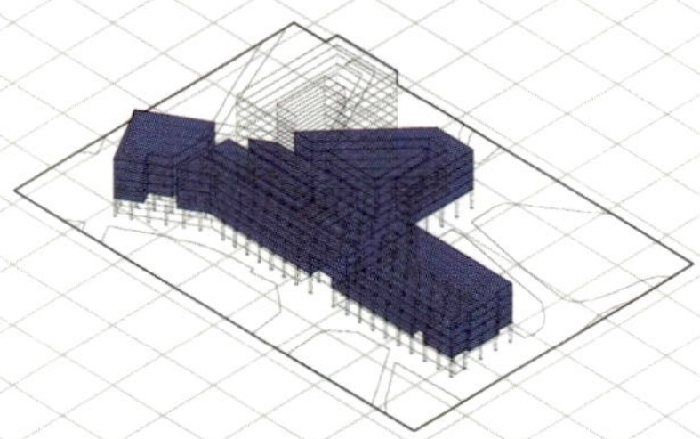

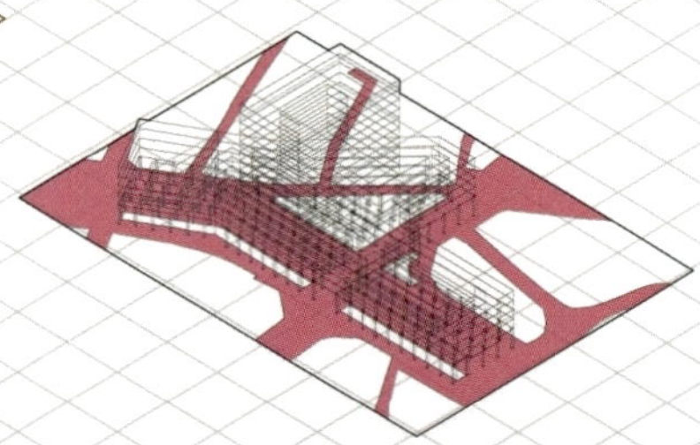

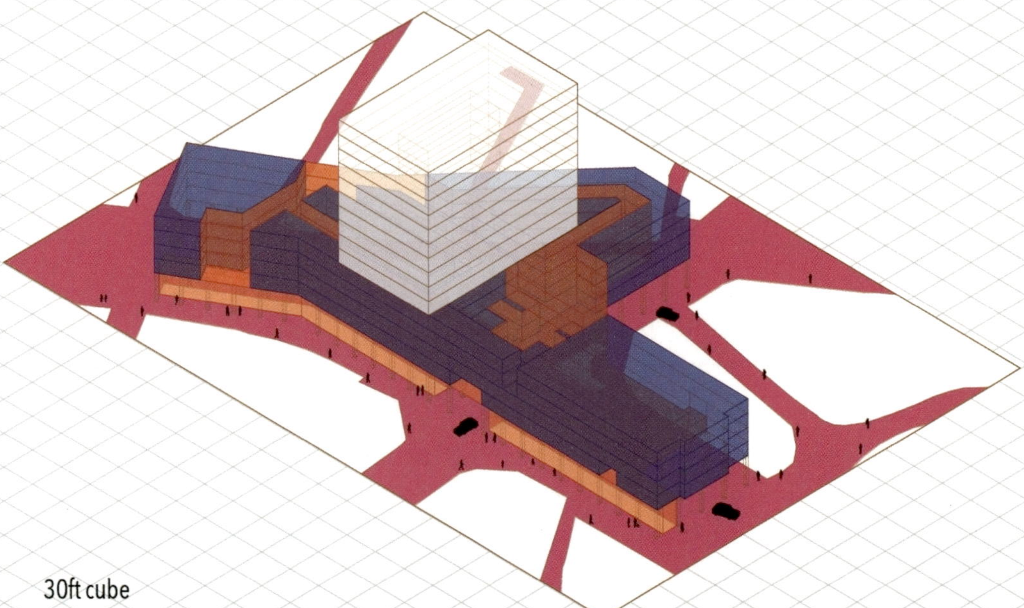

30ft cube

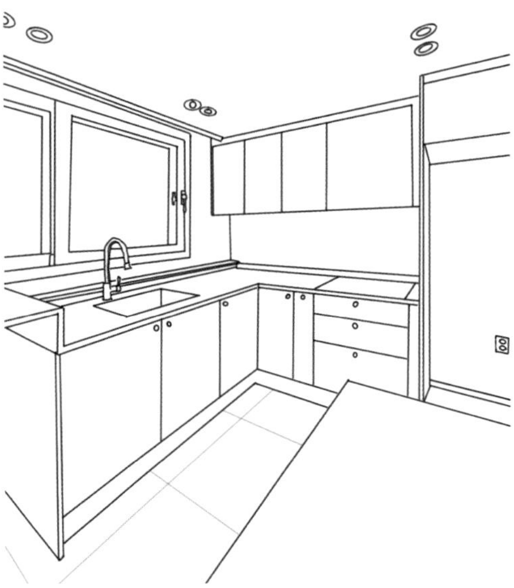

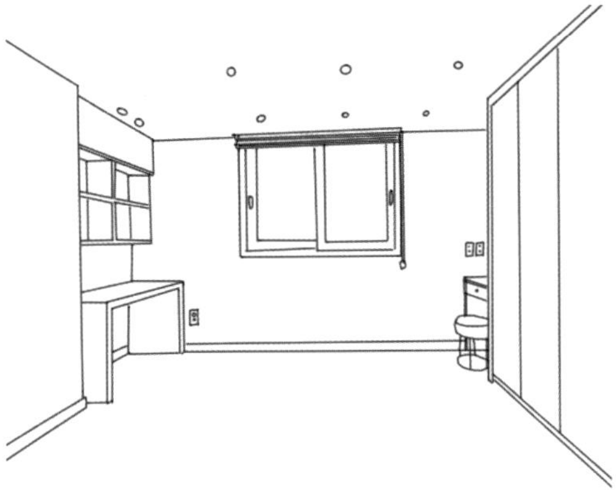

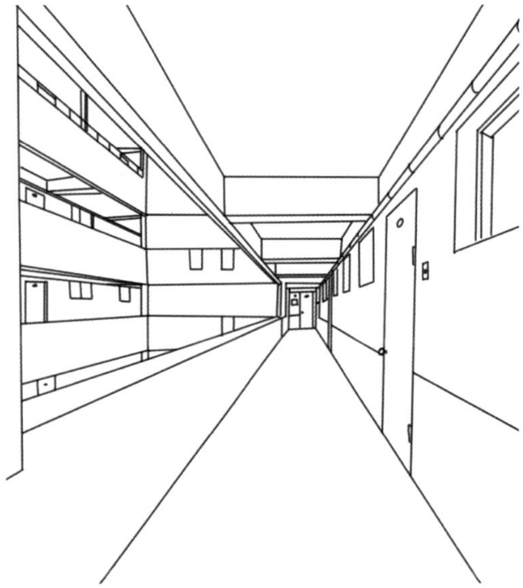

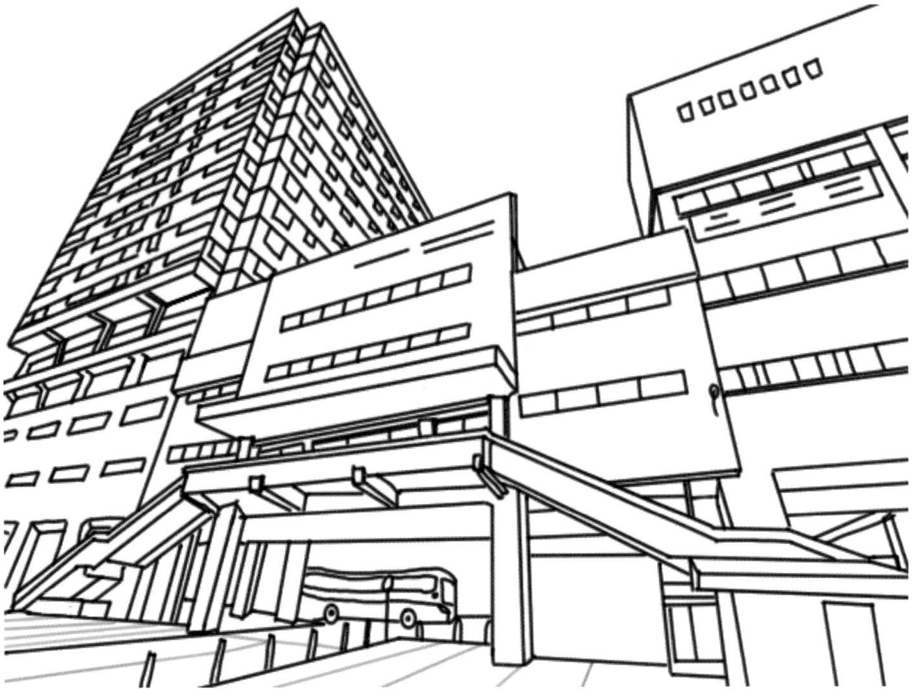

MRMU

ARC307 BU SECTION FALL 2022 | Lim Hyeonjin

CAMPUS INFOBIP

Zagreb, Croatia
3LHD Architects

Site area: 2,048sf

Building area: 3,615sf

Completion: 2021

domestic

communal

commercial

public

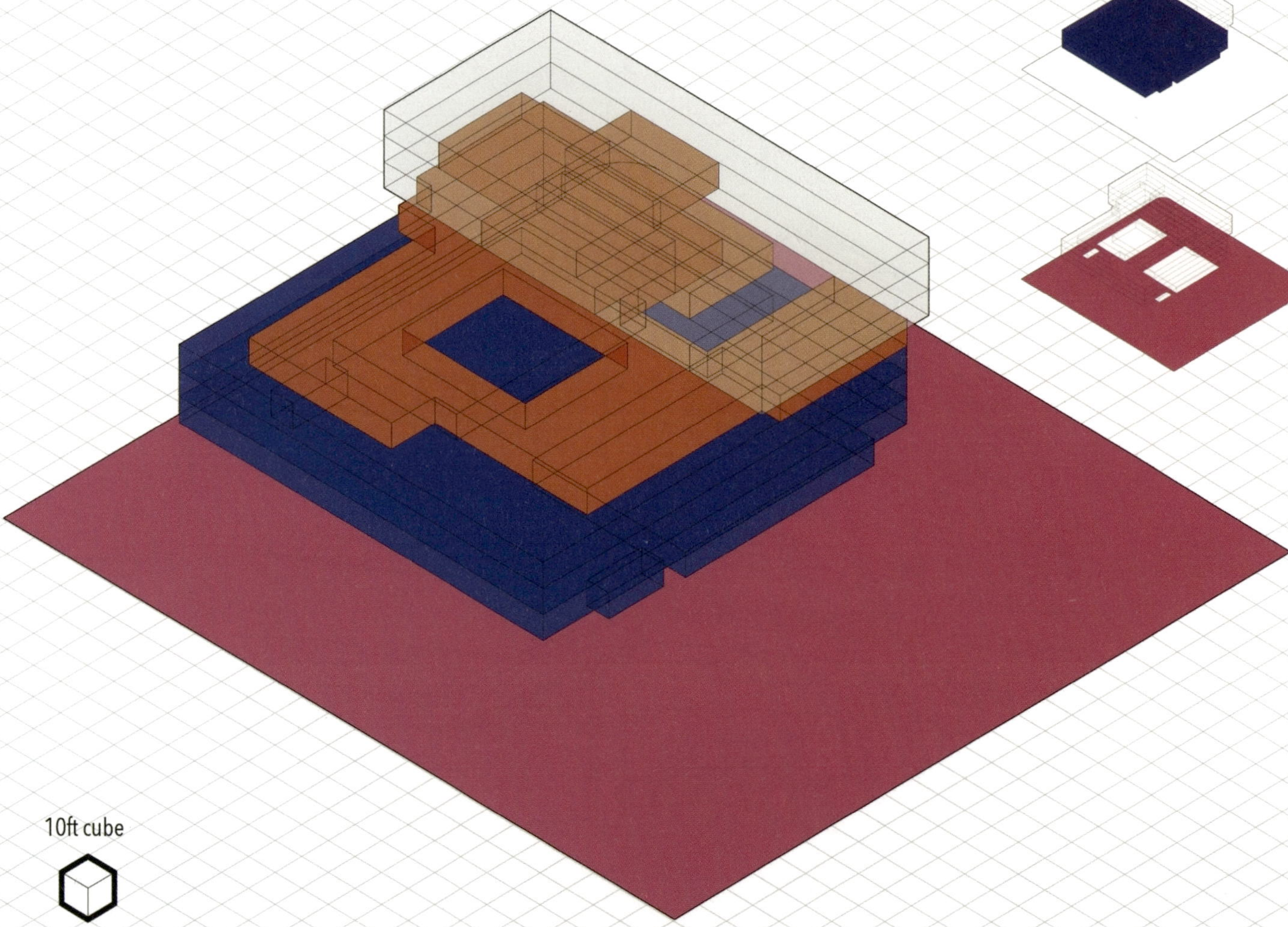

10ft cube

MRMU

ARC307 BU SECTION FALL 2022 | Lim Hyeonjin

FENIX

Rotterdam, Netherlands
Mei architects and planners

Site area: 6,958sf

Building area: 13,613sf

Completion: 2019

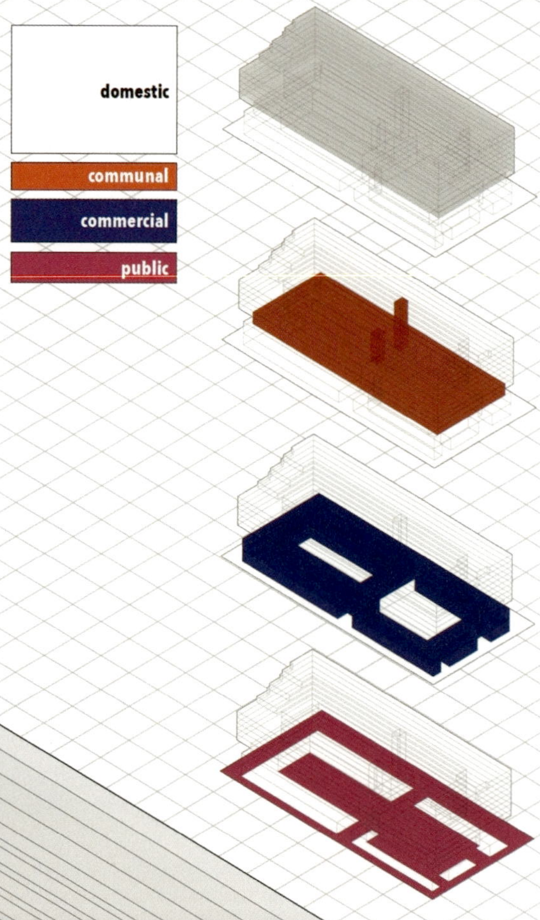

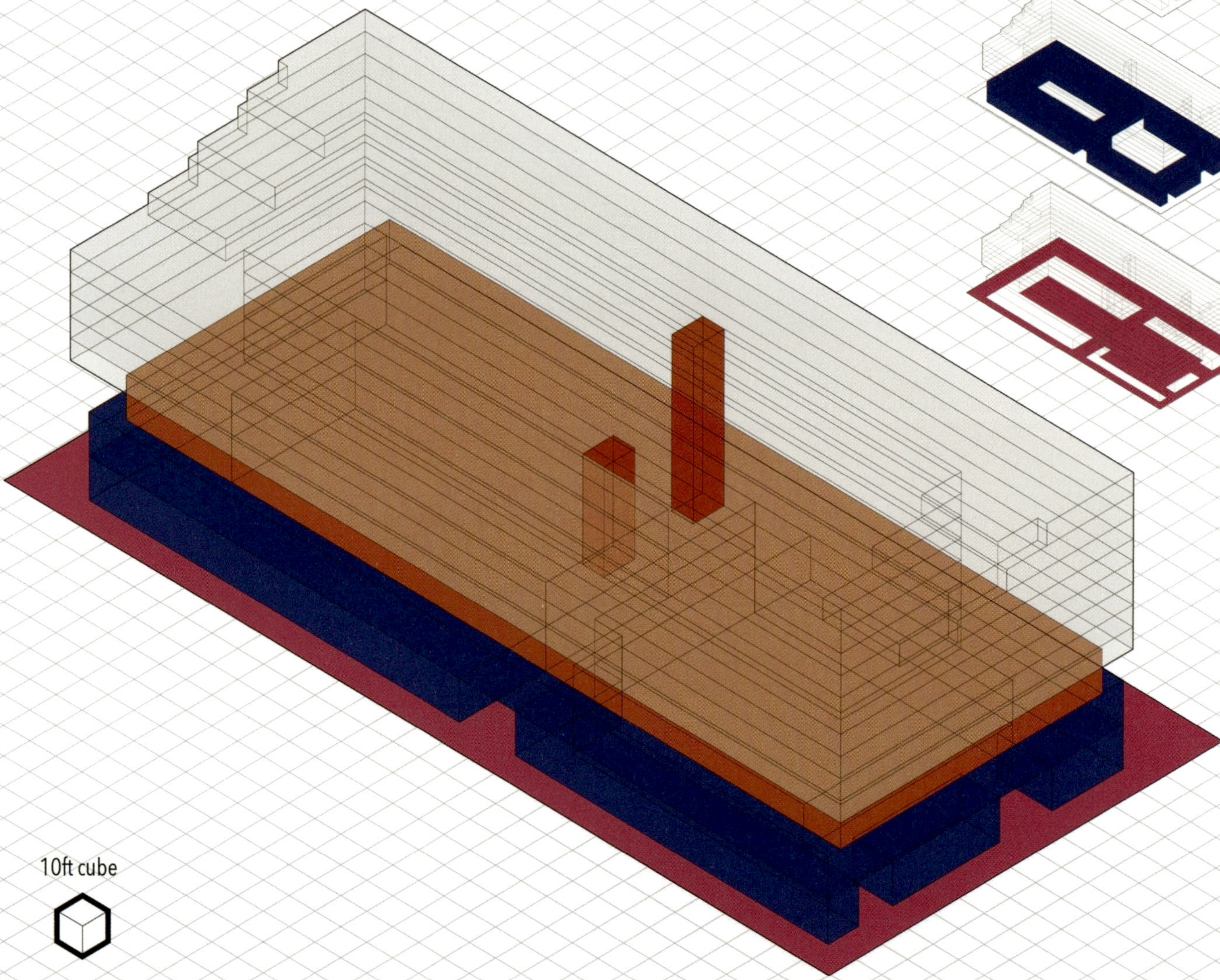

10ft cube

MRMU

ARC045 PNU SECTION FALL 2022 | Seungho Jang

ESMENT INCA BUILDING

Inca

AR3 Gualguasplliteras + MDBA

Site area: 19,141sf

building area: 36,738sf

Completion: 2020

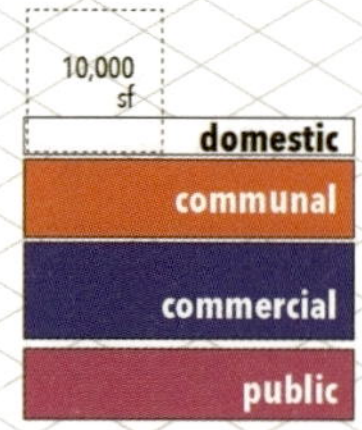

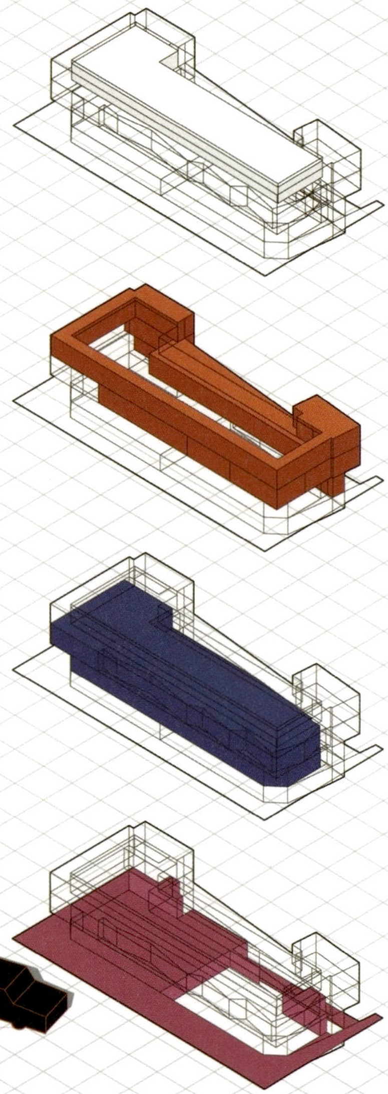

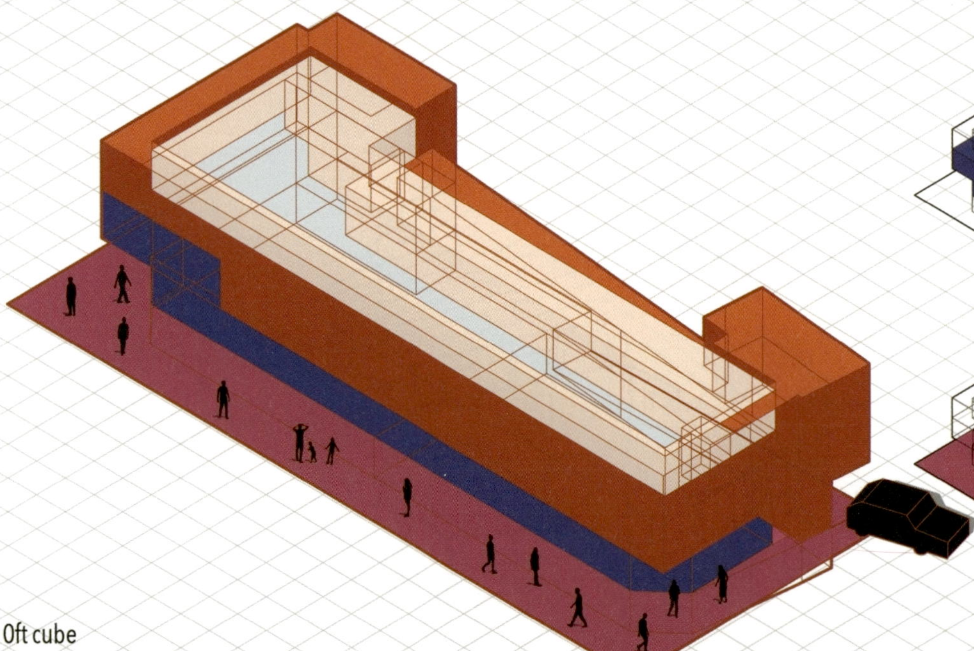

10ft cube

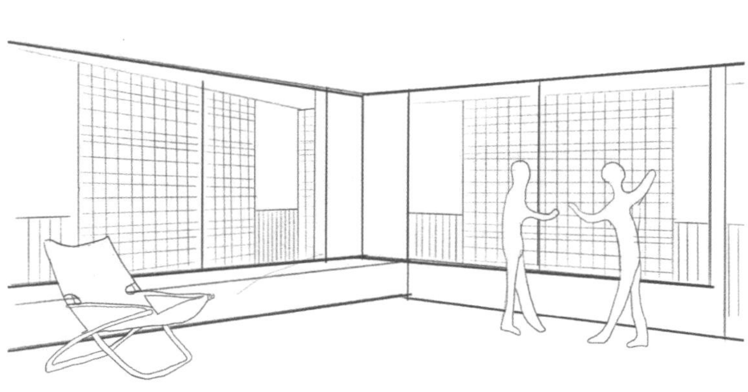

MRMU

ARC045 PNU SECTION FALL 2022 | Seungho Jang

NEW CARVER APARTMENTS

Los Angles
Michael Maltzan Architecture

Site area: 29,828sf

building area: 59,674sf

Completion: 2009

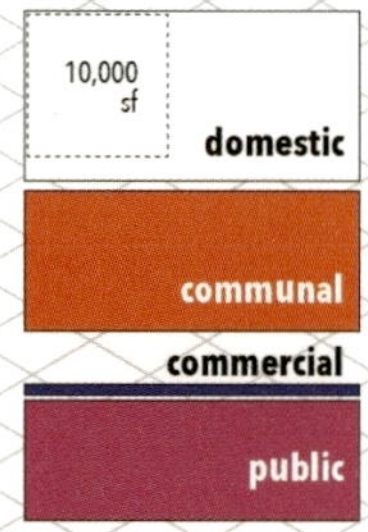

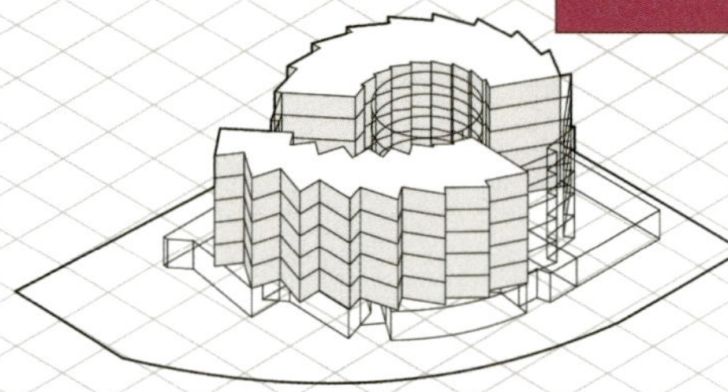

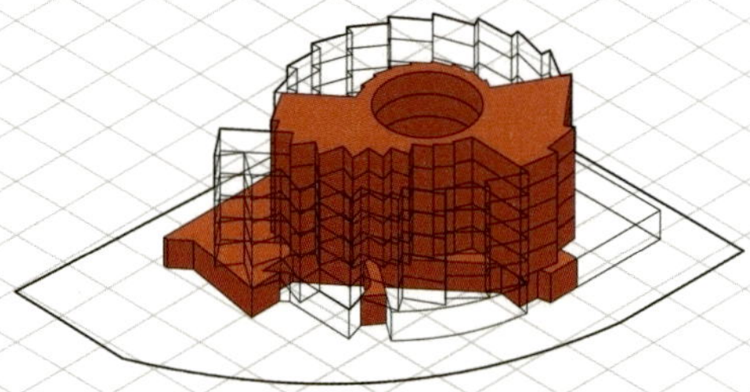

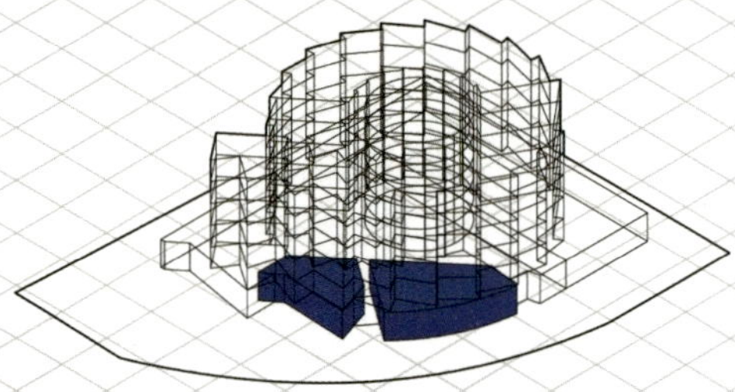

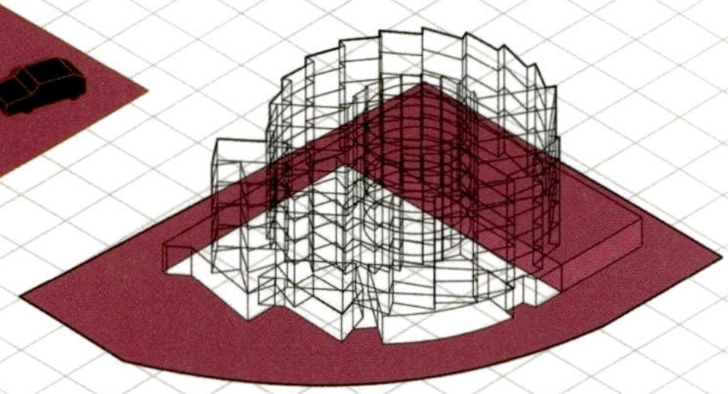

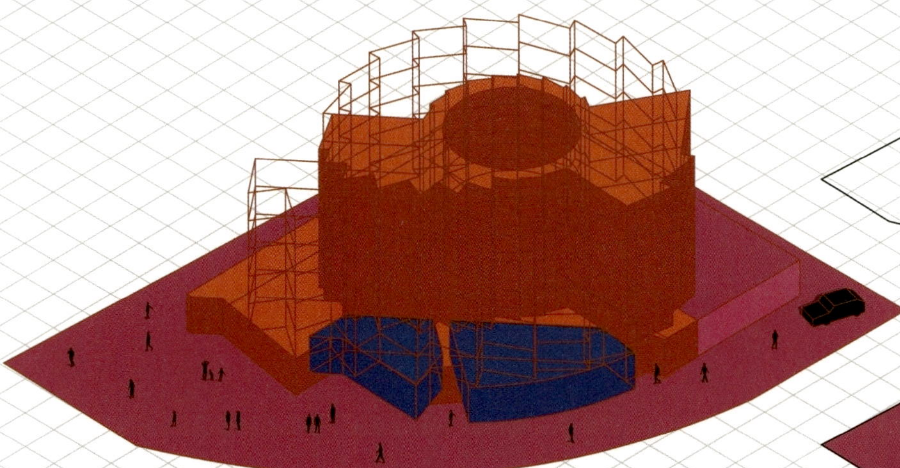

10ft cube

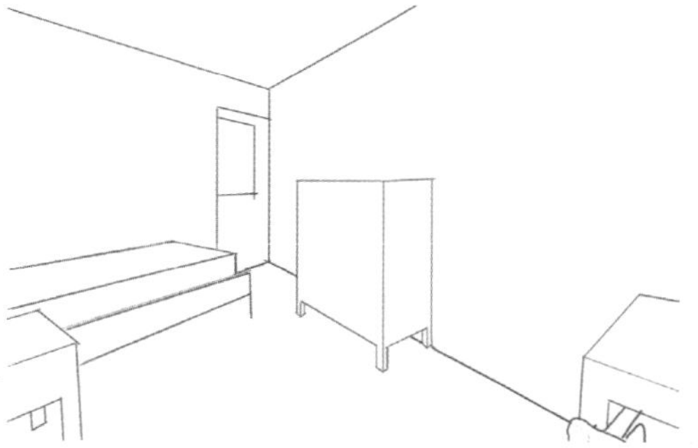

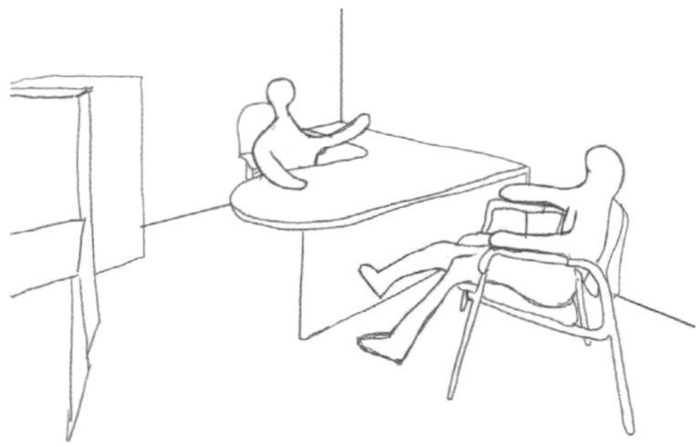

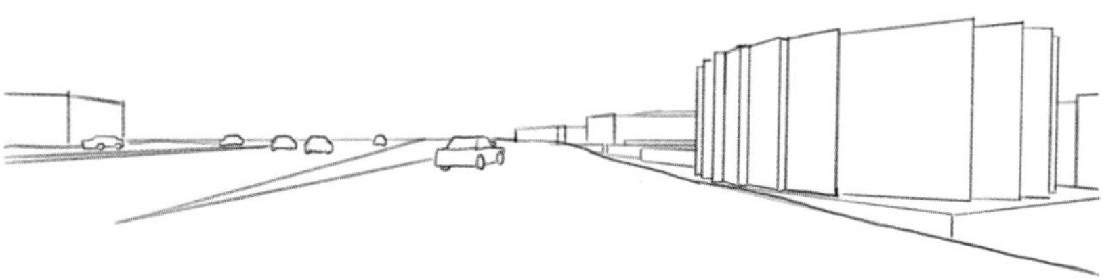

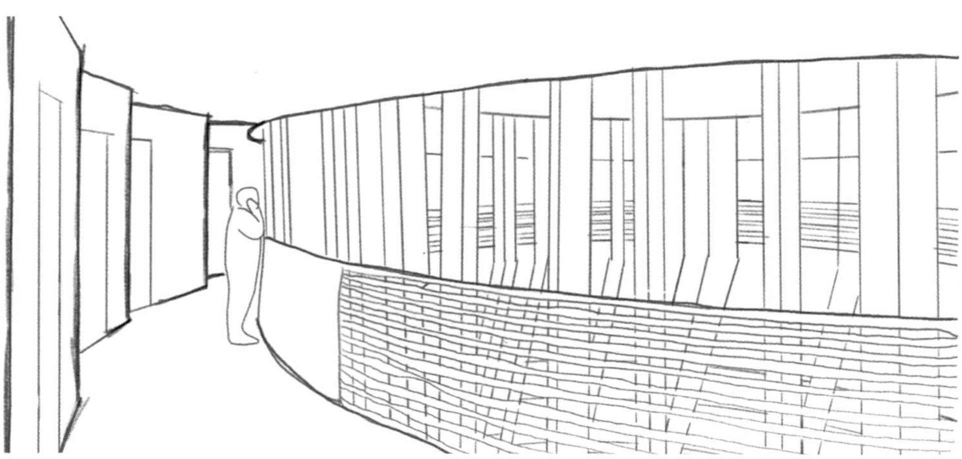

MRMU

ARC307 BU SECTION FALL 2022 | JUNG UNGHUI

MARKTHAL

ROTTERDAM
MVRDV

Site area: 8,400sqm

building area: 95,000sqm

Completion: 2014

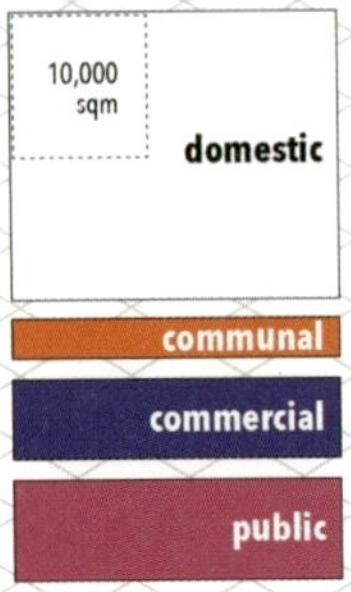

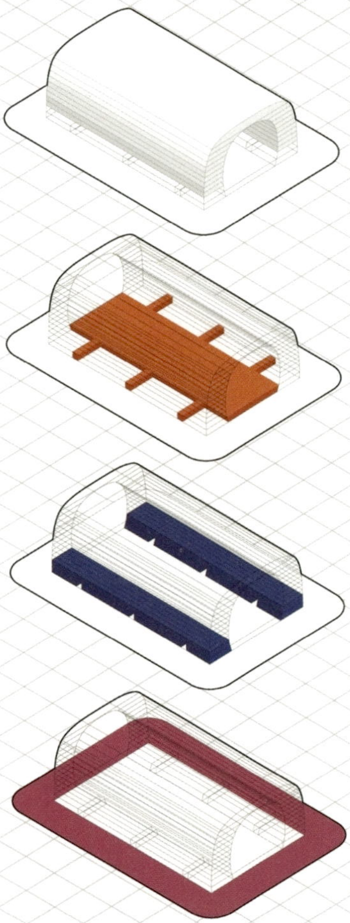

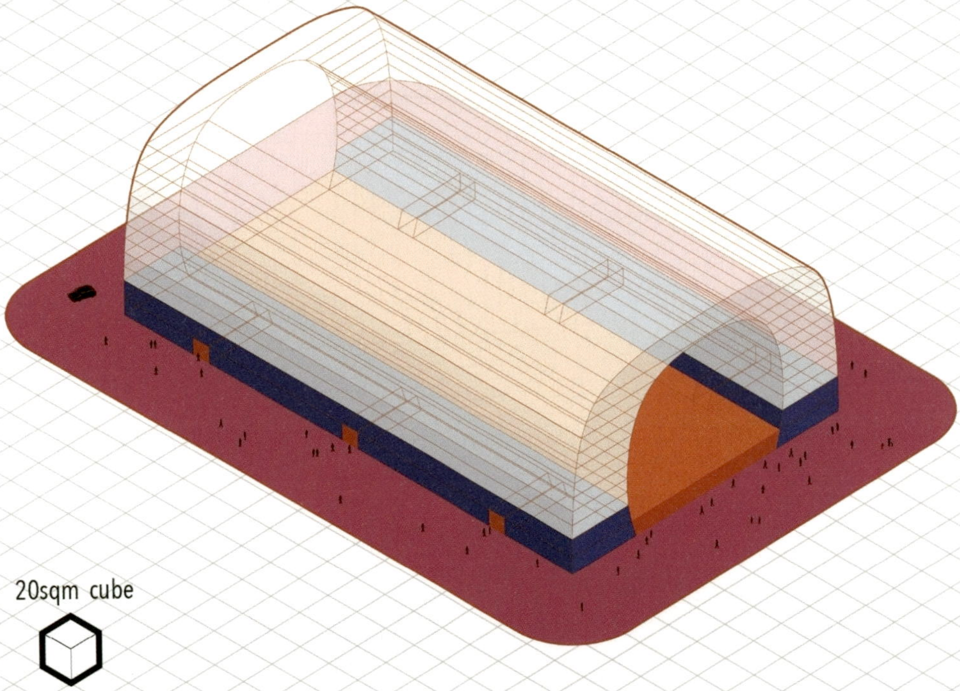

20sqm cube

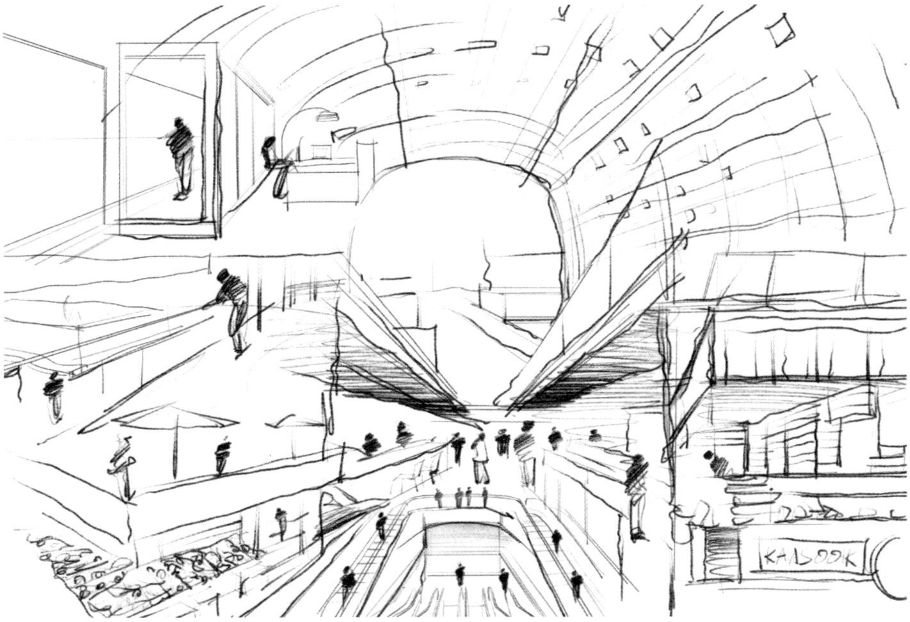

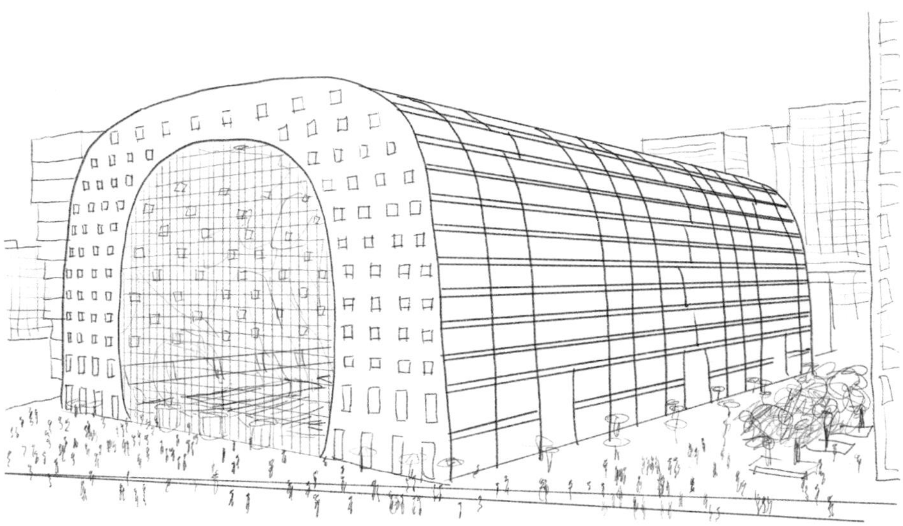

MRMU

ARC307 BU SECTION FALL 2022 | JUNG UNGHUI

150 NORTH THIRD STREET RESIDENTIOAL COMPLEX

COLUMBUS
JBAD

Site area: 3,480sqm
building area: 8.000sqm
Completion: 2021

1,000 sqm **domestic**

communal

commercial

public

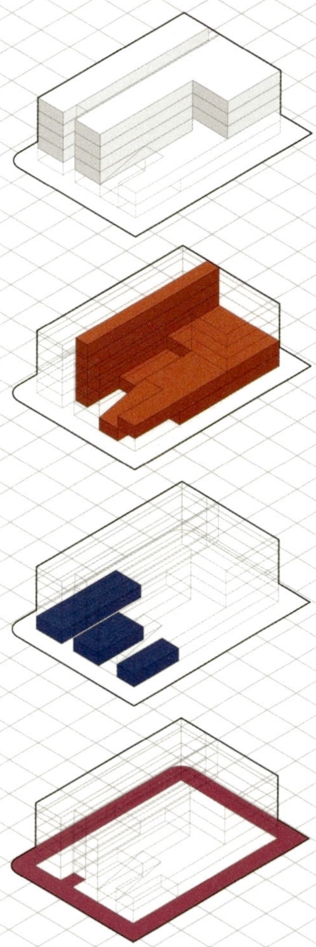

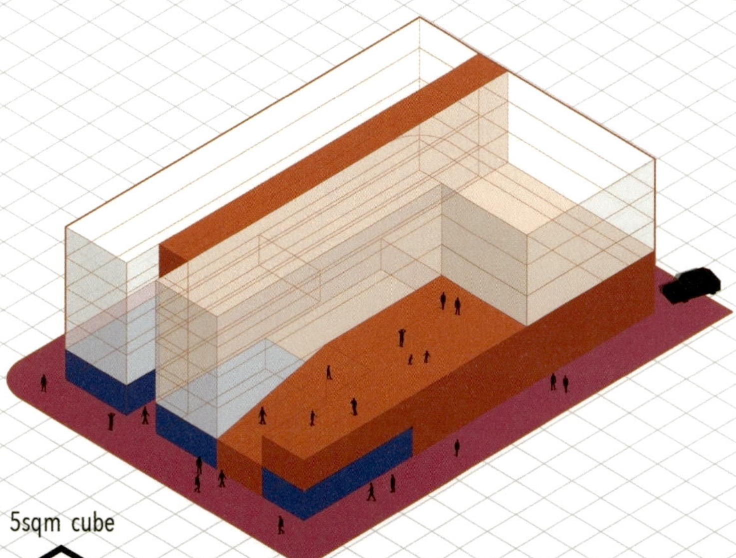

5sqm cube

@150

MRMU

ARC307 BU SECTION FALL 2022 | Matthew Williams

DO YOU HEAR ME WHEN YOU SLEEP

London
DOGMA

Site area: TBD

building area: TBD

Completion: TBD

domestic

communal

commercial

public

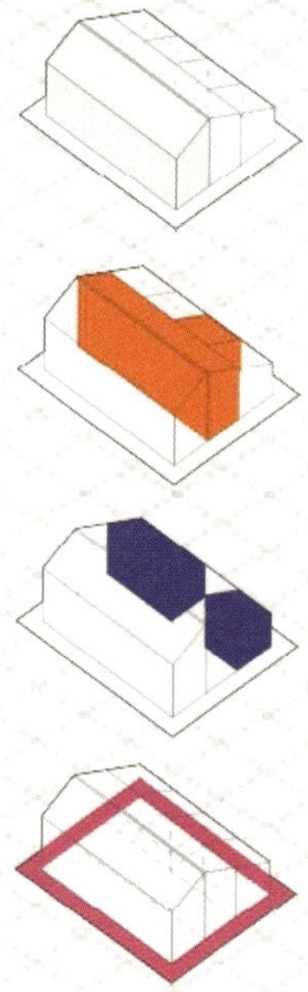

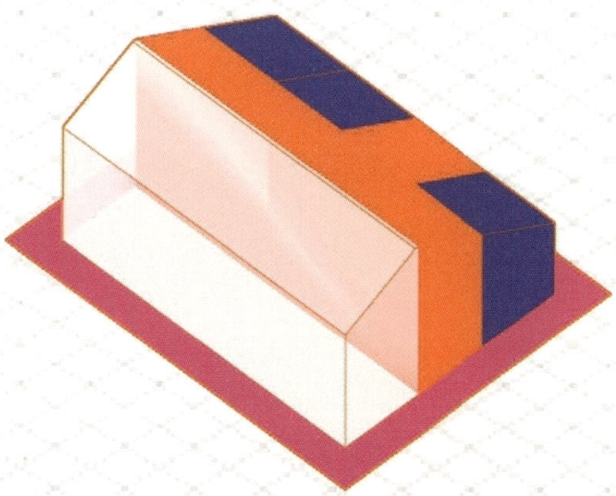

10ft cube

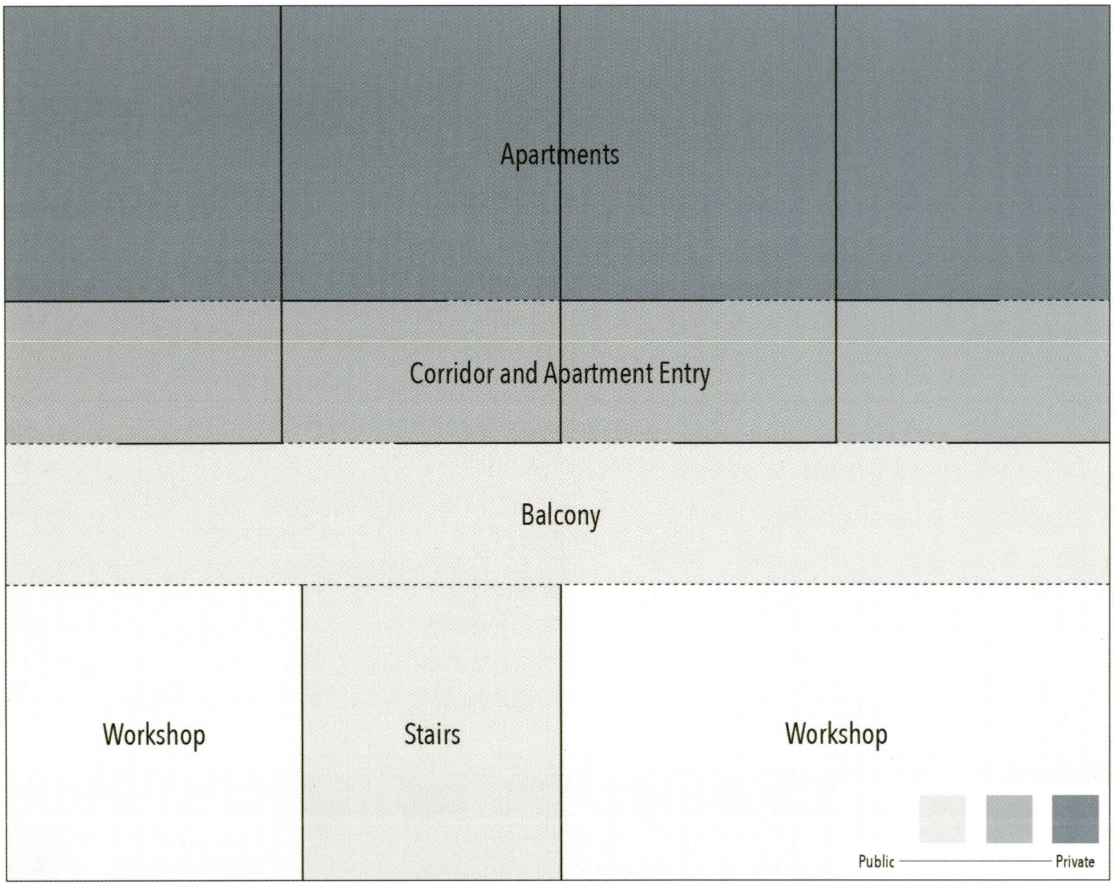

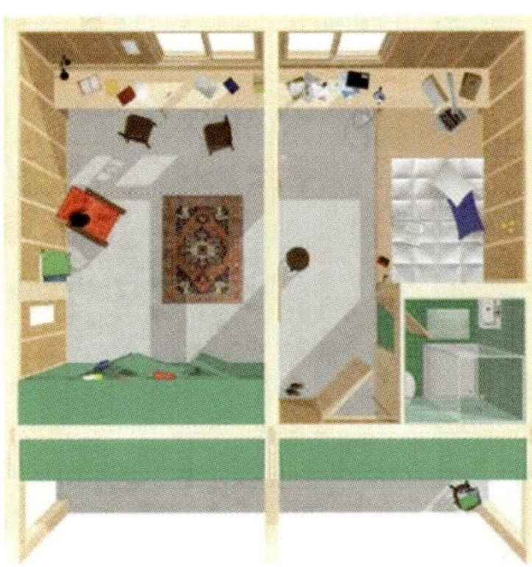

The day begins in one of several apartment configurations.

You then exit from the apartment into the corridors. The corridors are sectioned off to give residents more privacy.

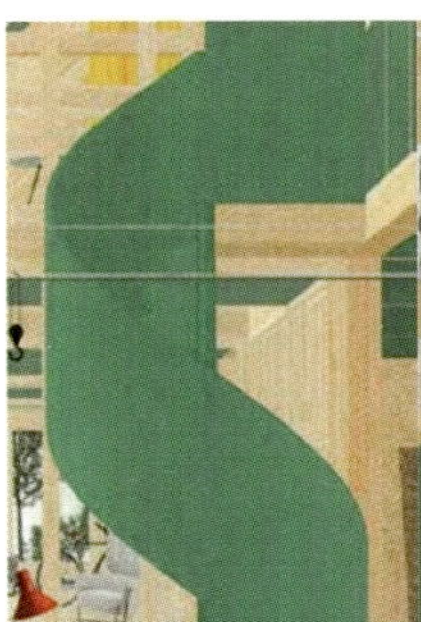

Subsequently, you descend down the stairs.

The stairs lead you into the commercial areas where people can set up workshops, stores, or anything they may want.

Finally, you exit the building through the main corridor.

MRMU

ARC307 BU SECTION FALL 2022 | Matthew Williams

FIREHOUSE RENOVATION

Denver
Meridian 105 Architecture

Site area: 8,660sf
building area: 12,000sf
Completion: 2021

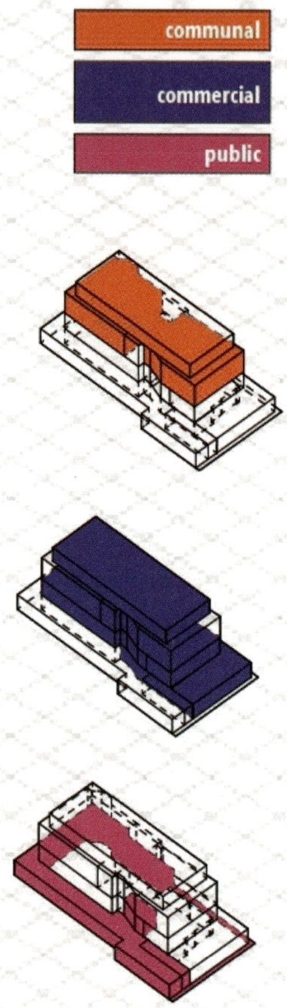

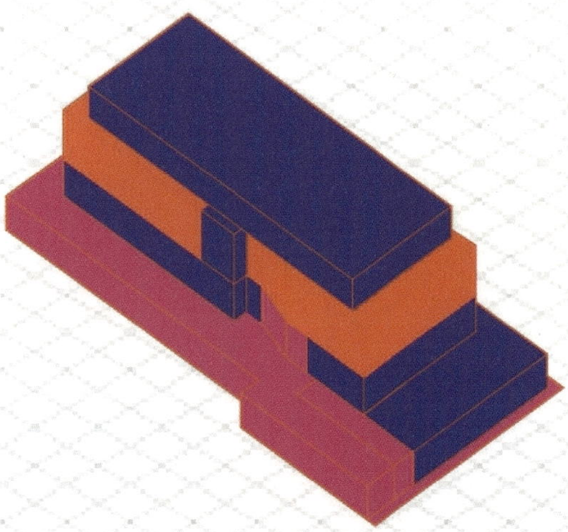

10ft cube

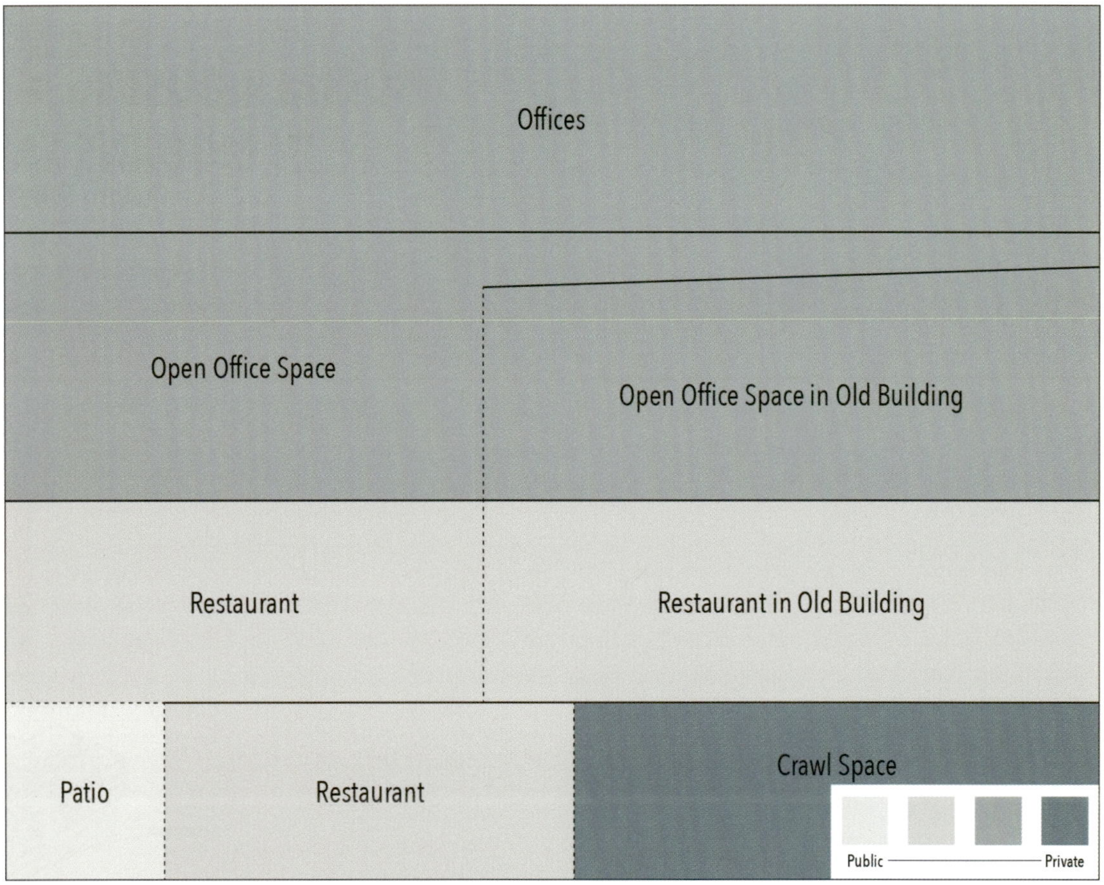

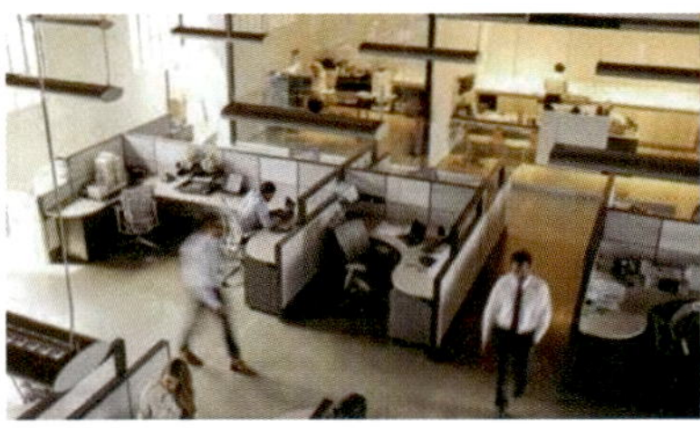

The most private area of the firehouse are the offices on the top floor.

The floor below is home to more open-space offices as well as coworking areas.

On the bottom two floors there's a restaurant.

The open patio area in the southwest corner of the building can be opened to the public or it can also be used as extra seating for the restaurant.

The front door of the building leads out to a public space.

MRMU

ARC307 BU SECTION FALL 2022 | Sarah Bannan

SOCIAL HOUSING + SHOPS IN MOUANS SARTOUX

Comte & Vollenweider

Site area: 14,348 sf

Number of Residential Units: 17 units

Completion: 2015

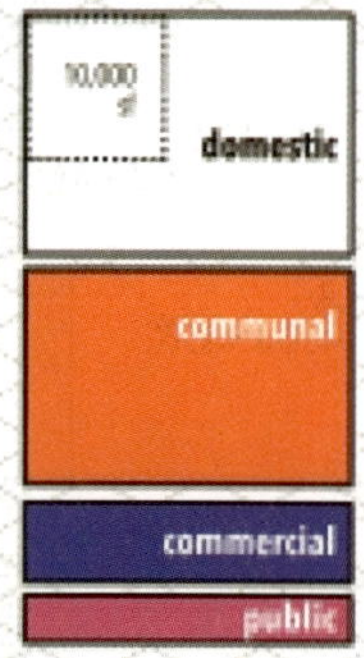

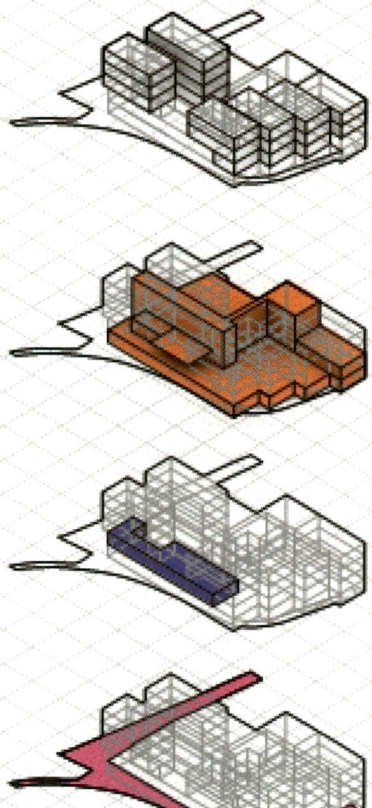

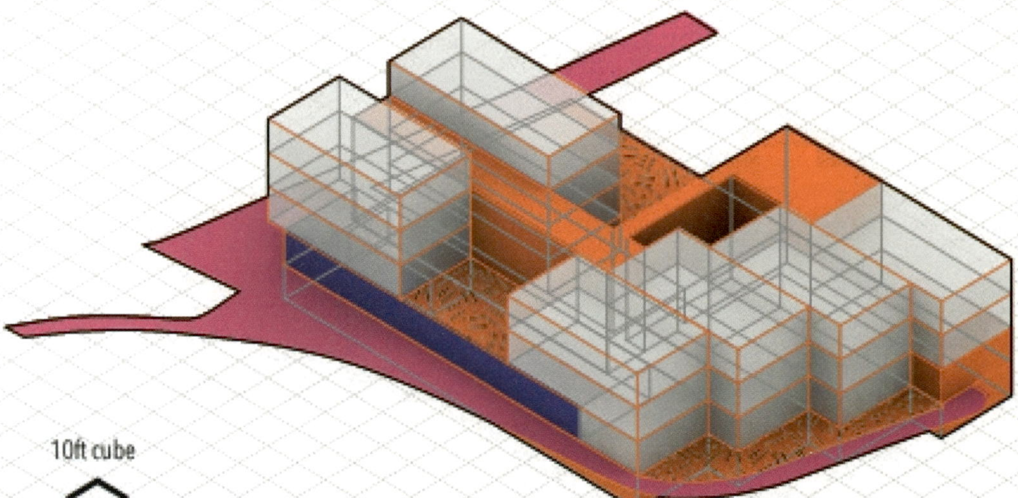

10ft cube

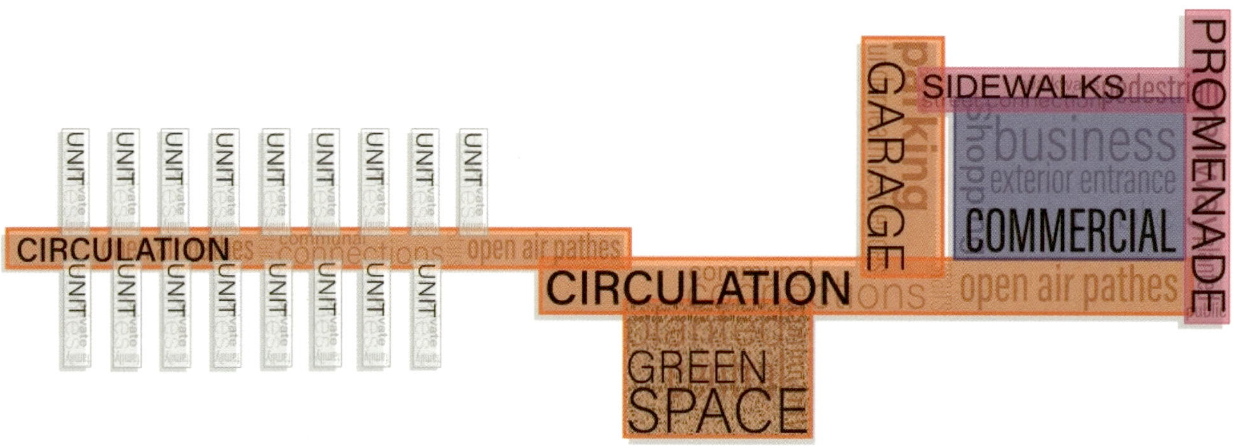
UNIT
CIRCULATION
connections
open air pathes
CIRCULATION
GREEN
SPACE
parking
GARAGE
SIDEWALKS
business
exterior entrance
COMMERCIAL
open air pathes
PROMENADE

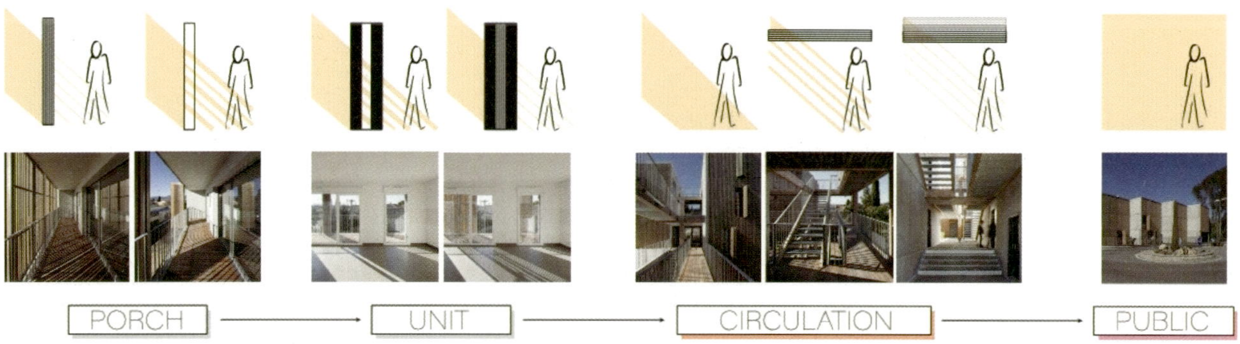
PORCH
UNIT
CIRCULATION
PUBLIC

MRMU

ARC307 BU SECTION FALL 2022 | Sarah Bannan

ERNIE DAVIS HALL

Syracuse, New York
Mack Scogin Merrill Elam Architects

Building area: 145,000sf
Number of residents: 250 students
Completion: 2010

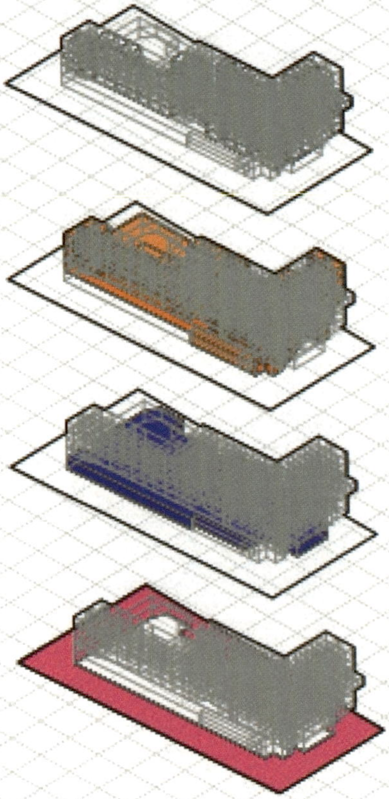

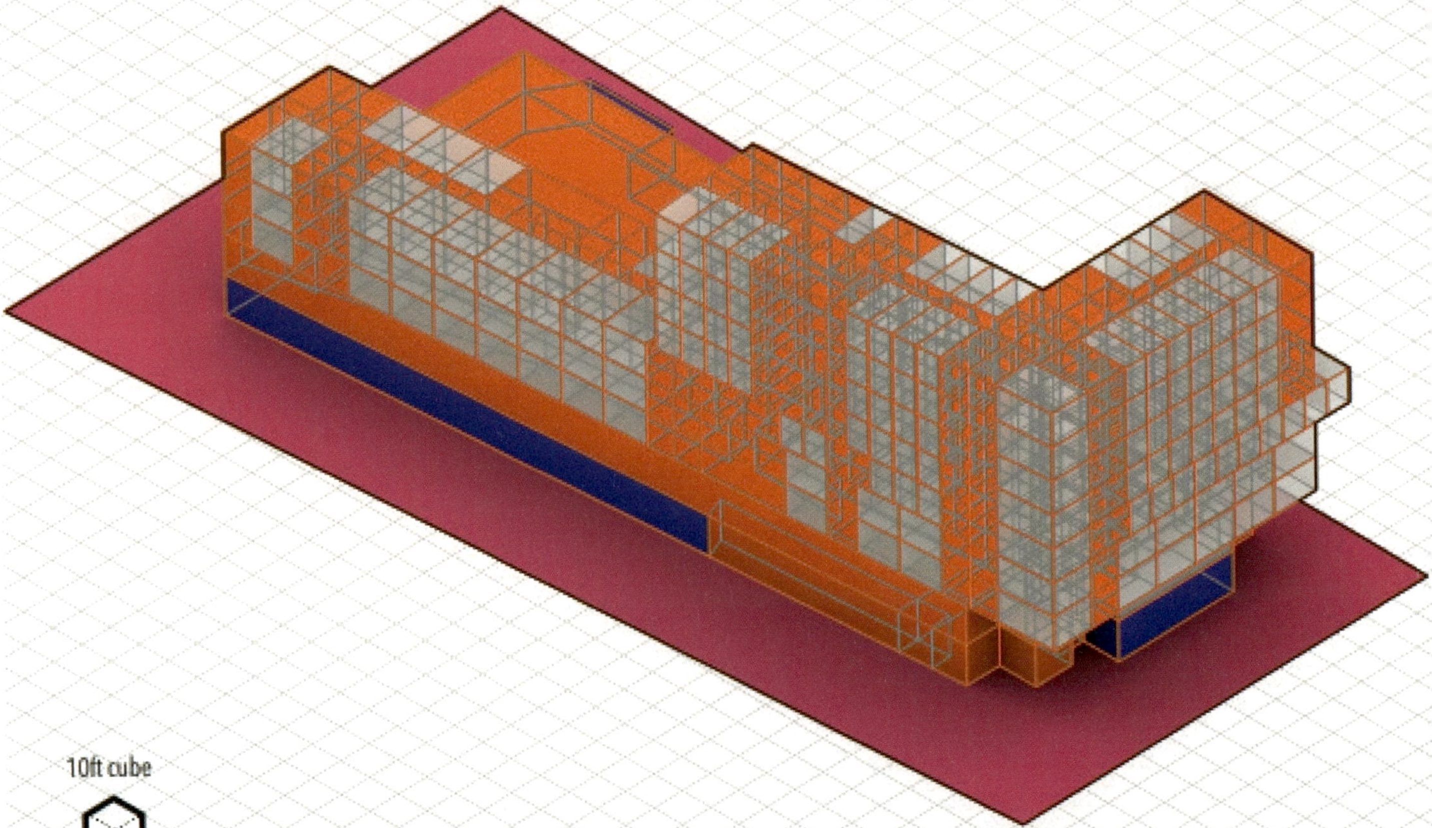

10ft cube

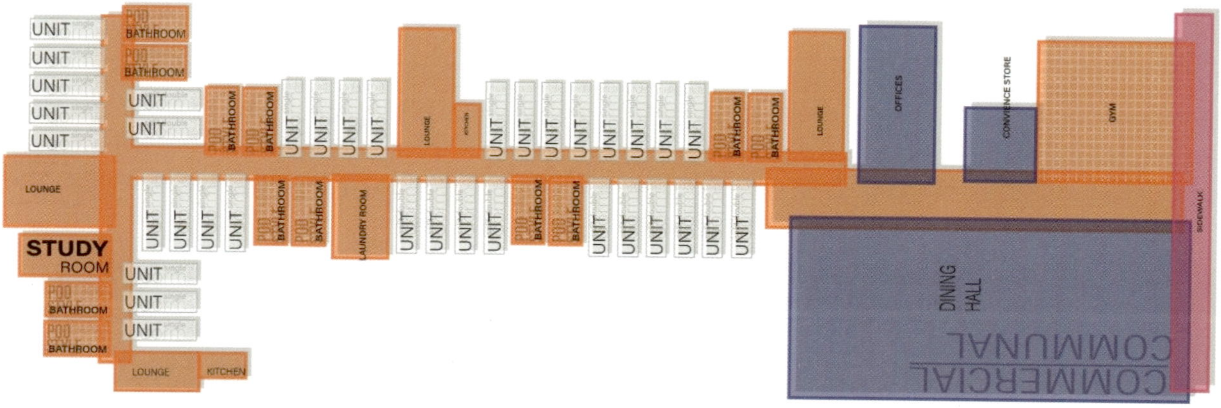
UNIT
POD
BATHROOM
LOUNGE
STUDY
ROOM
LAUNDRY ROOM
KITCHEN
OFFICES
CONVIENCE STORE
GYM
SIDEWALK
DINING
HALL
COMMUNAL
COMMERCIAL

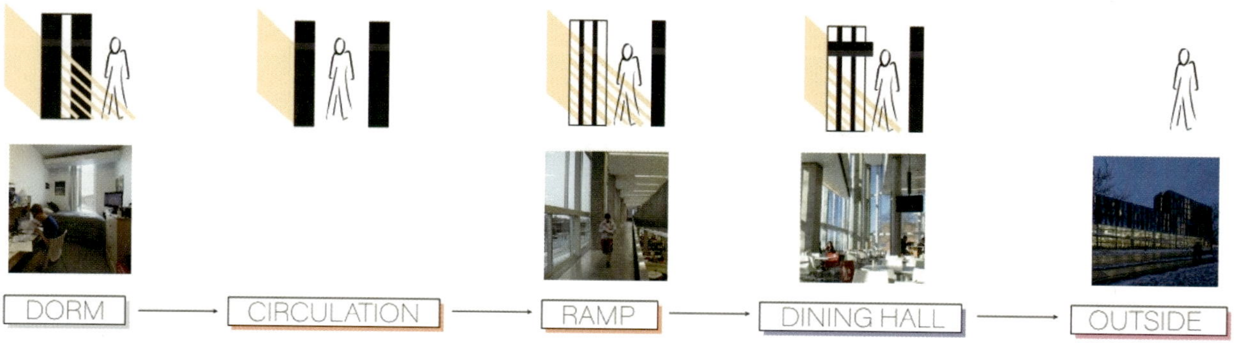
DORM
CIRCULATION
RAMP
DINING HALL
OUTSIDE

MRMU

ARC307 BU SECTION FALL 2022 | Sarah Bannan

WILLIAMS TERRACE

Charleston, South Carolina
David Baker Architects

Site area: 18,274sf
Building area: 47,851sf
Number of Residential Units: 42 units
Completion: 2017

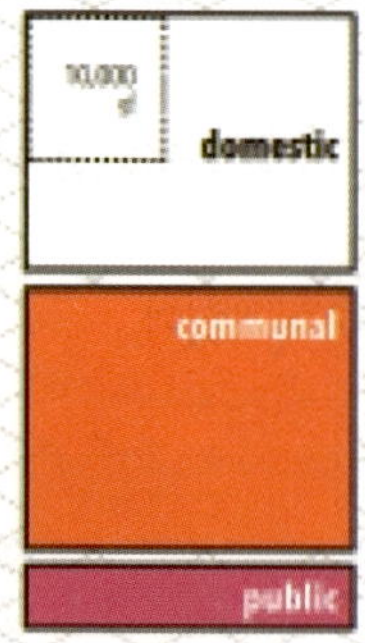

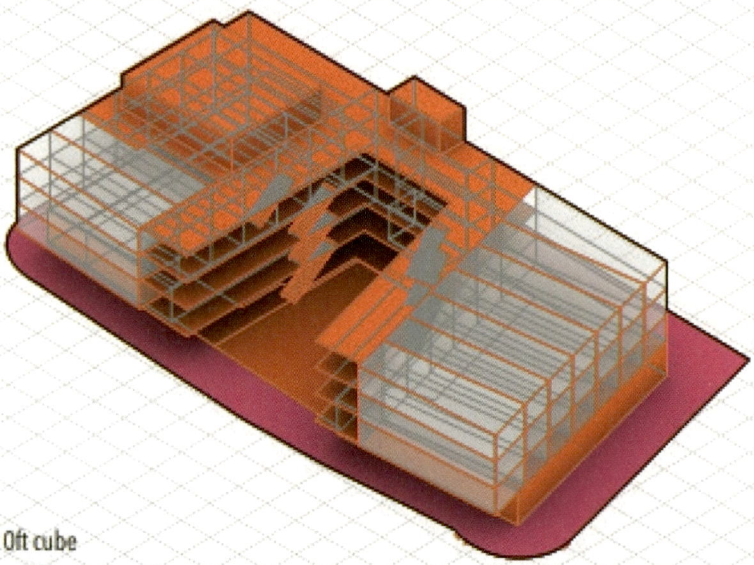

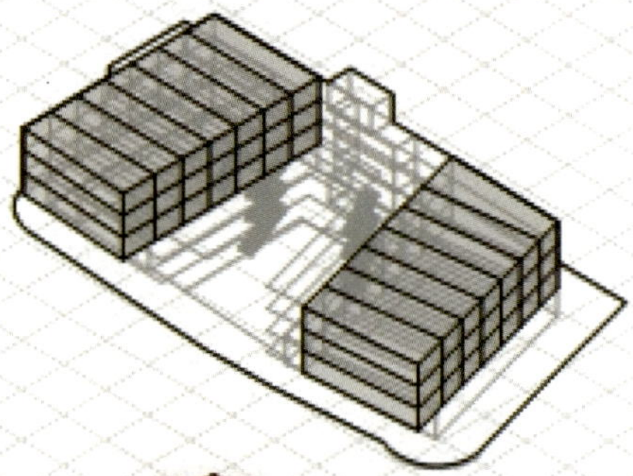

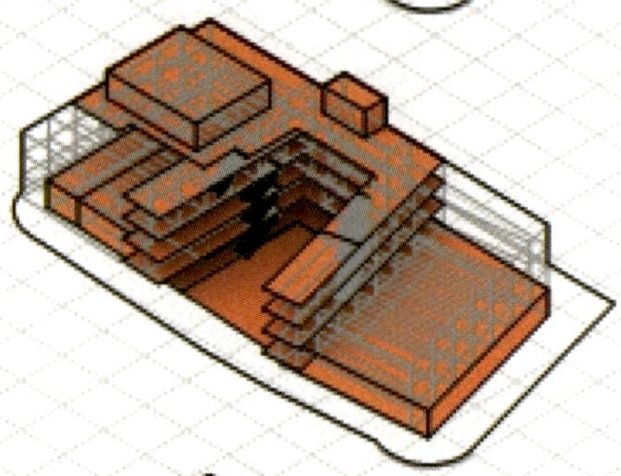

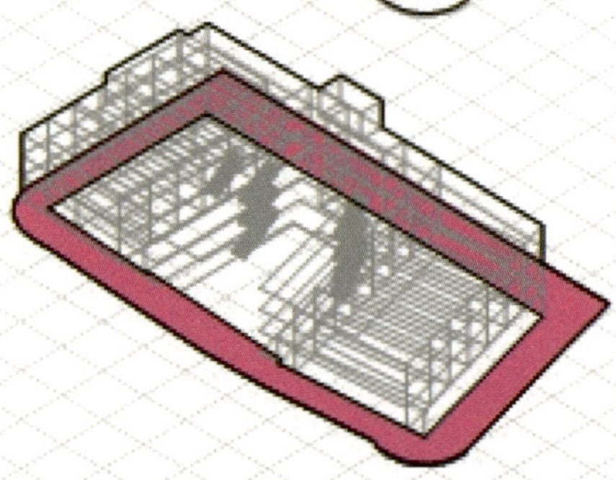

10ft cube

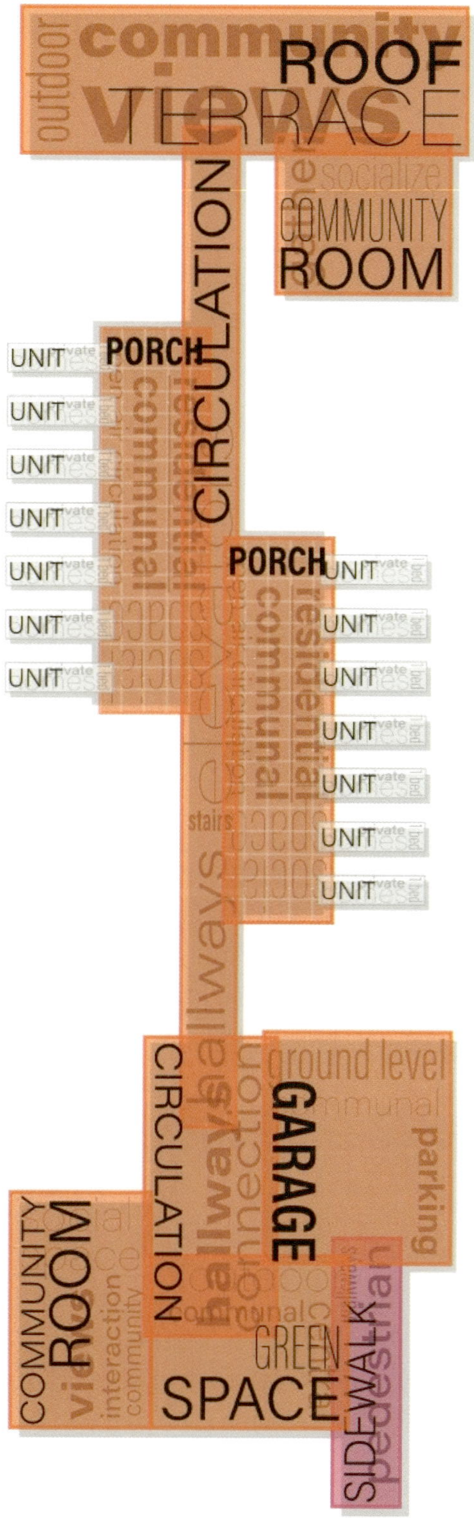
ROOF
TERRACE
outdoor
community
views
socialize
COMMUNITY
ROOM
CIRCULATION
PORCH
UNIT
UNIT
UNIT
UNIT
UNIT
UNIT
UNIT
communal
PORCH
UNIT
UNIT
UNIT
UNIT
UNIT
UNIT
UNIT
residential
communal
stairs
hallways
ground level
communal
GARAGE
parking
CIRCULATION
hallways
connection
COMMUNITY
ROOM
interaction
community
GREEN
SPACE
SIDEWALK
pedestrian

MRMU

ARC307 BU SECTION FALL 2022 | Yuelin "Peter" Wang

8 TALLET

Copenhagen
Bjarke Ingels Group

Site area: 18,300sf
building area: 71,000sf
Completion: 2010

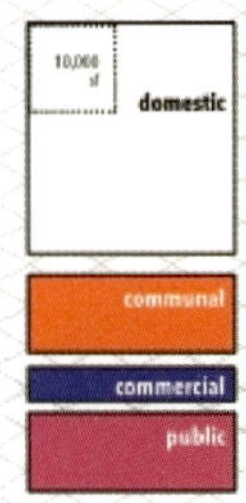

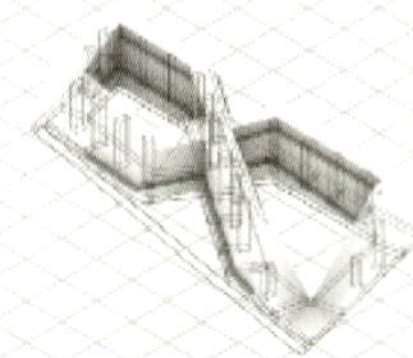

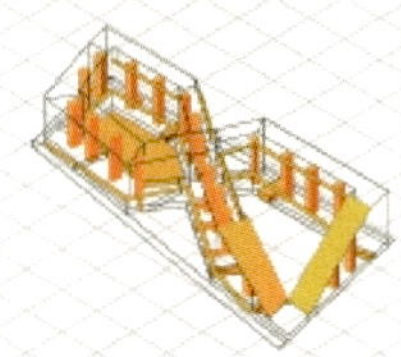

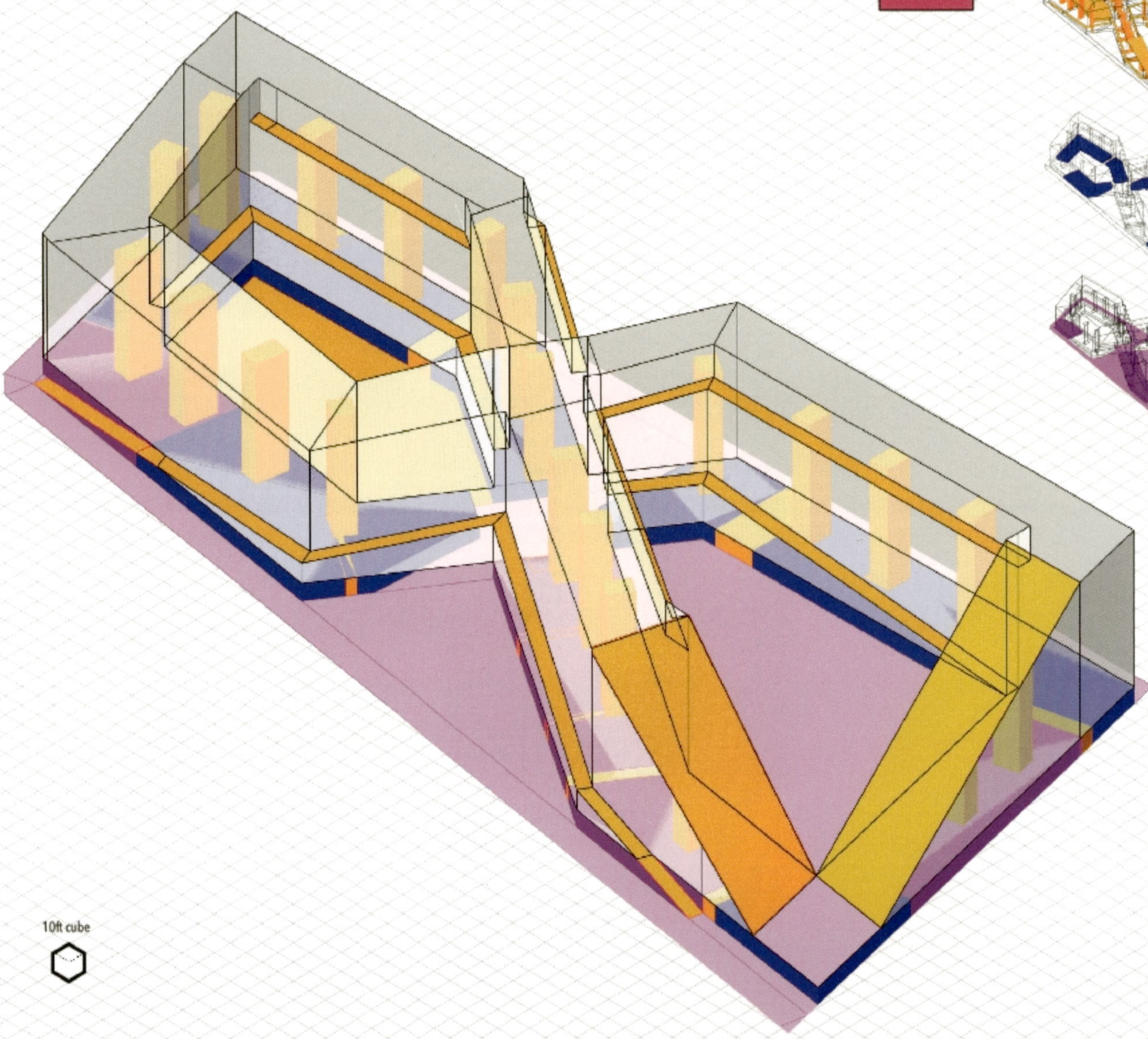

10ft cube

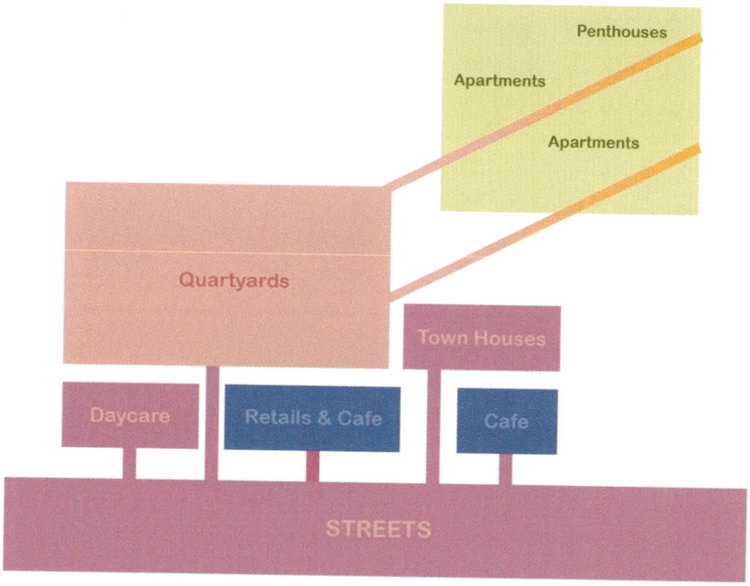
Penthouses
Apartments
Apartments
Quartyards
Town Houses
Daycare
Retails & Cafe
Cafe
STREETS

Front Garden
Front Garden & Ramp
Ramp & Building
Lower Ramp & Courtyard
Courtyard
Courtyard Entrance
Entrance & Street
Waterfront Cafe

MRMU

ARC307 BU SECTION FALL 2022 | Yuelin "Peter" Wang

MALAYSIAN SHOPHOUSES

Southeastern Asia folk architecture

Site area: 17,920sf

building area: 53,700sf

Completion: n/a

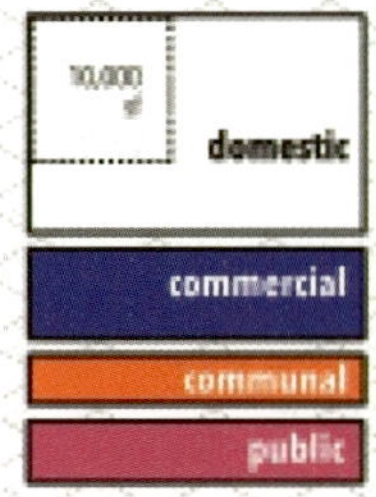

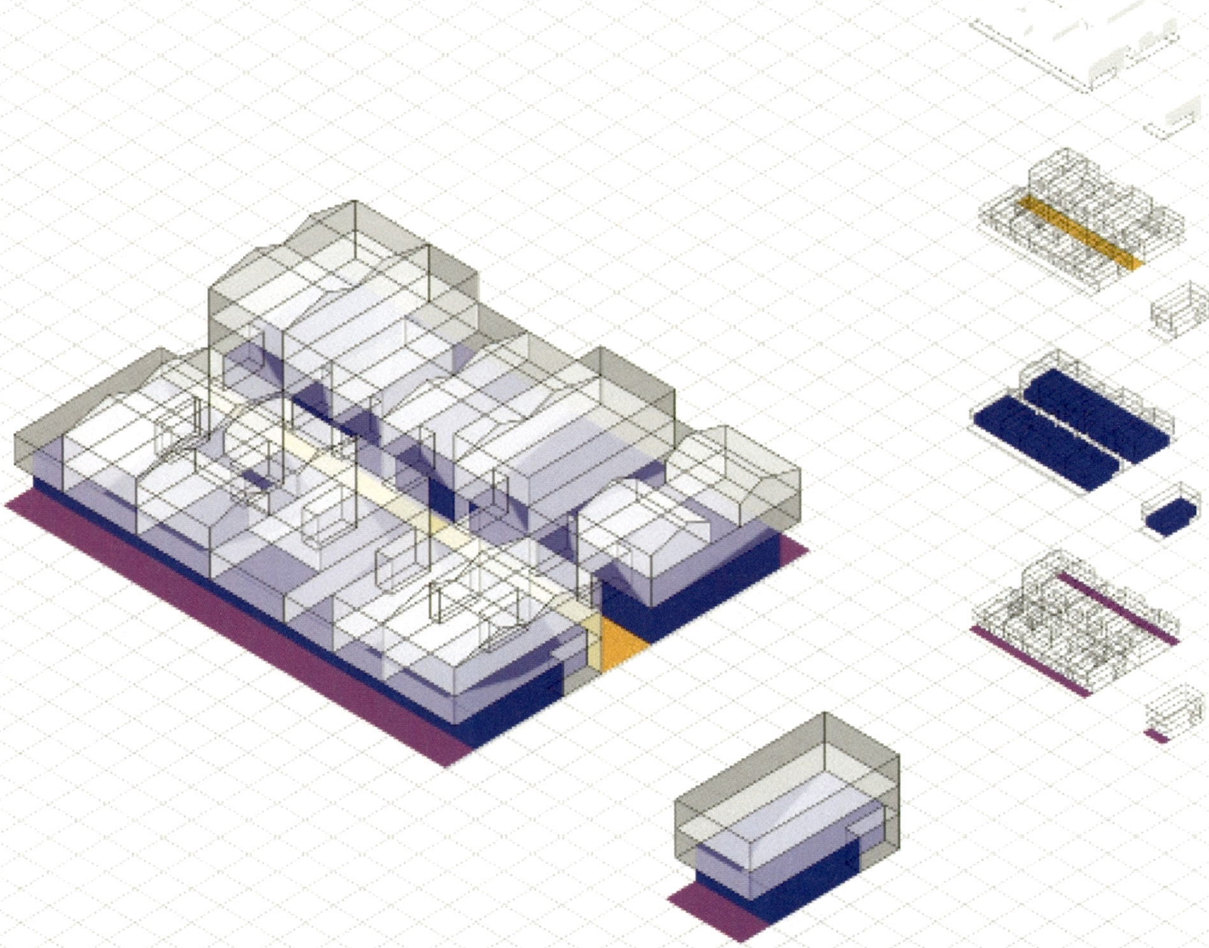

10ft cube

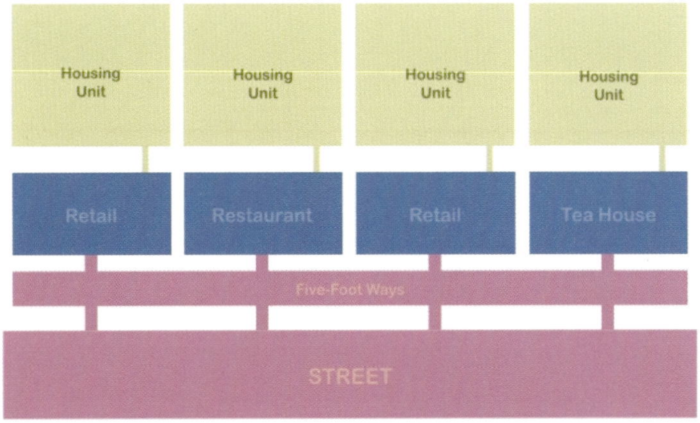
Housing Unit
Housing Unit
Housing Unit
Housing Unit
Retail
Restaurant
Retail
Tea House
Five-Foot Ways
STREET

MRMU

ARC307 BU SECTION FALL 2022 | Yiting Zhong

RCD HOUSE & OFFICE
Bangkok,Thailand
MasterMind Studio

Area: 8265sm

Resident Units:7

Completion: 2020

10,000 sf

domestic

communal

commercial

public

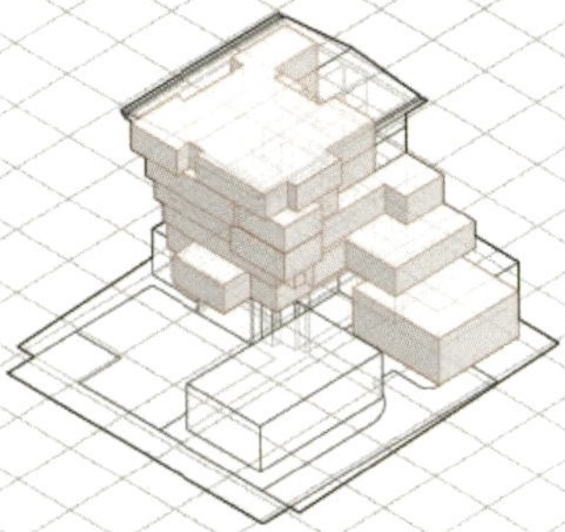

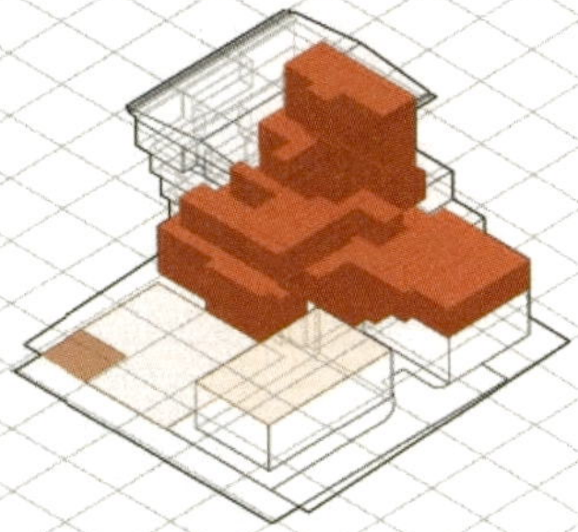

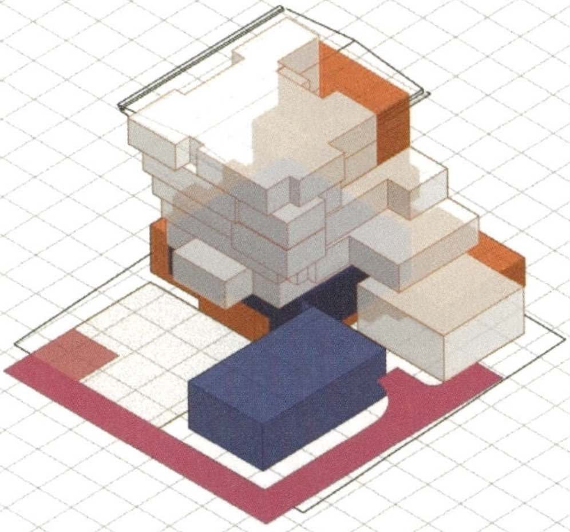

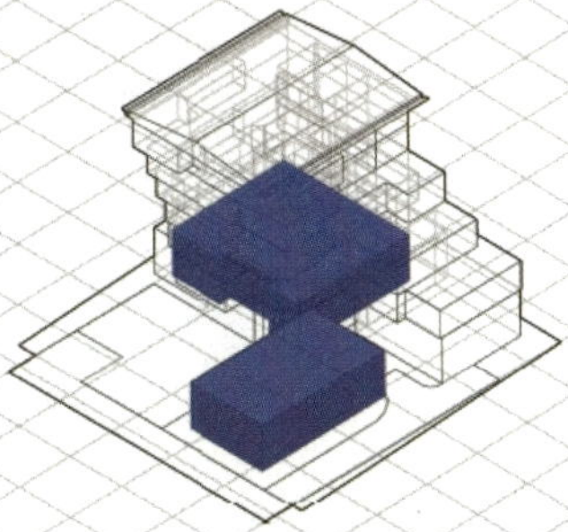

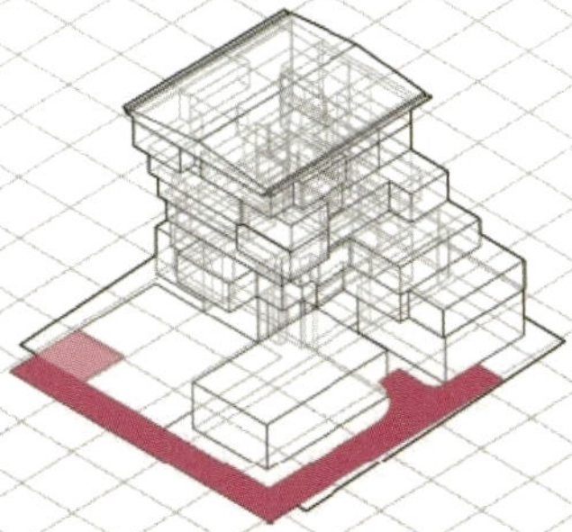

10ft cube

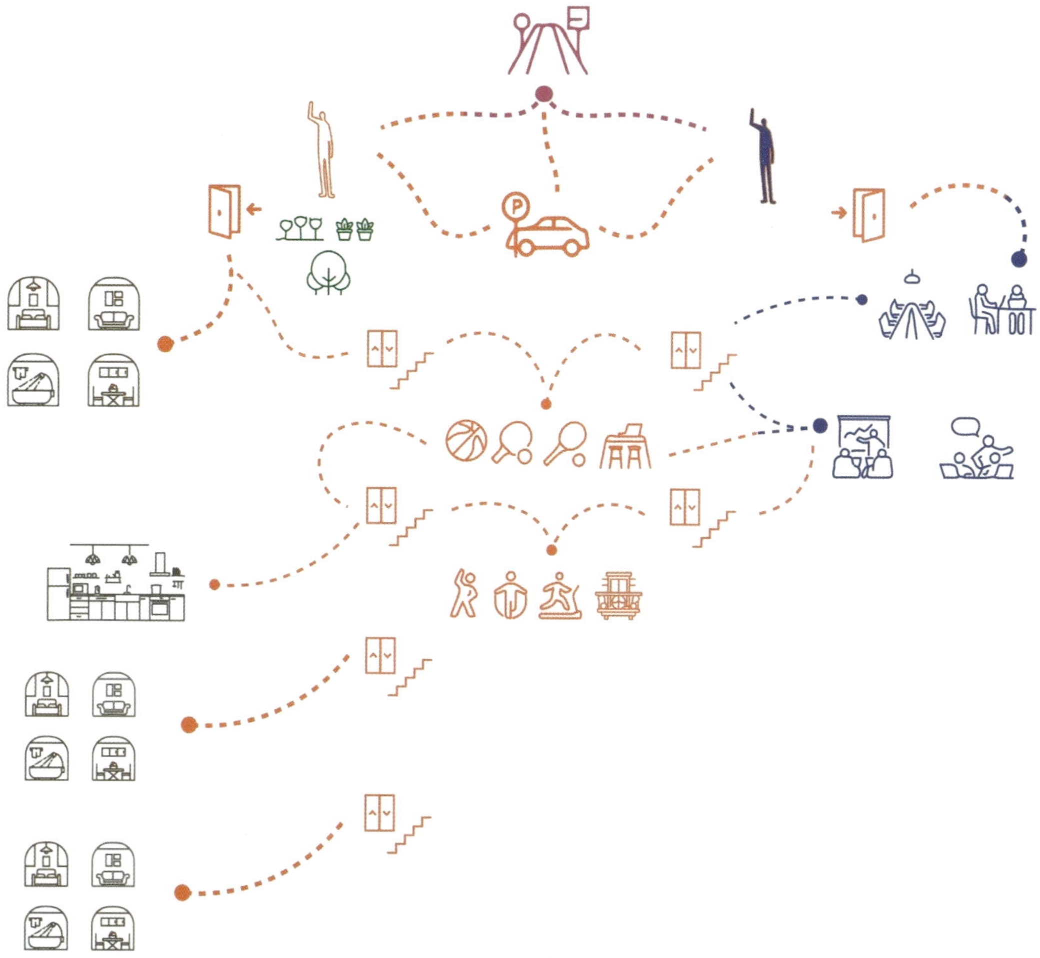

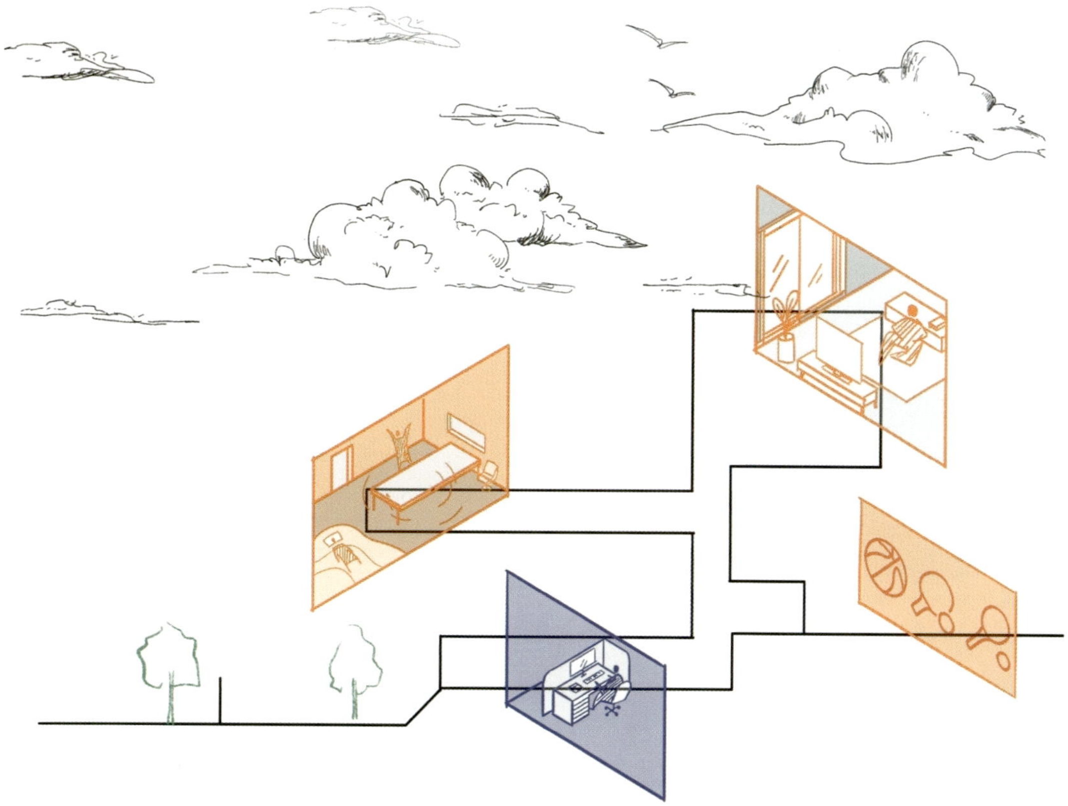

MRMU

ARC307 BU SECTION FALL 2022 | Yiting Zhong

VAJNORSKA 21

Bratislava,Slovaka
Cakov+Partners

Area: 2284sm

Resident Units:55

Completion: 2021

10,000 sf

domestic

communal

commercial

public

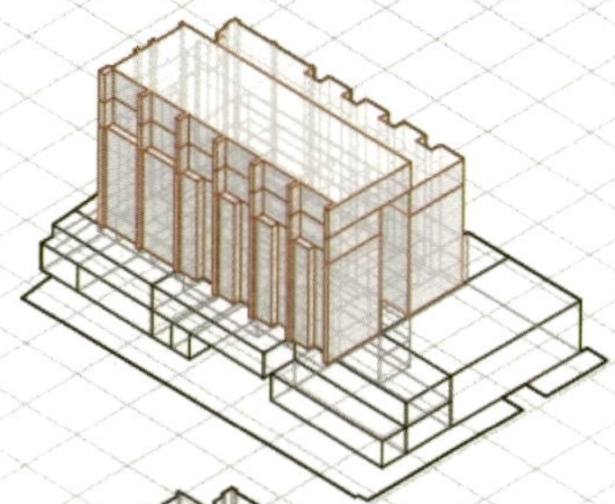

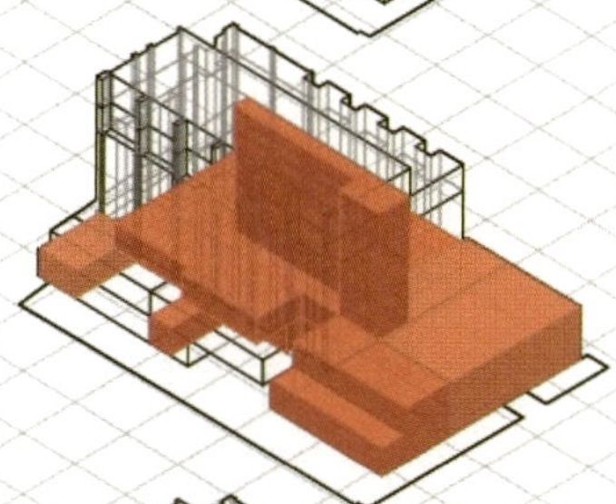

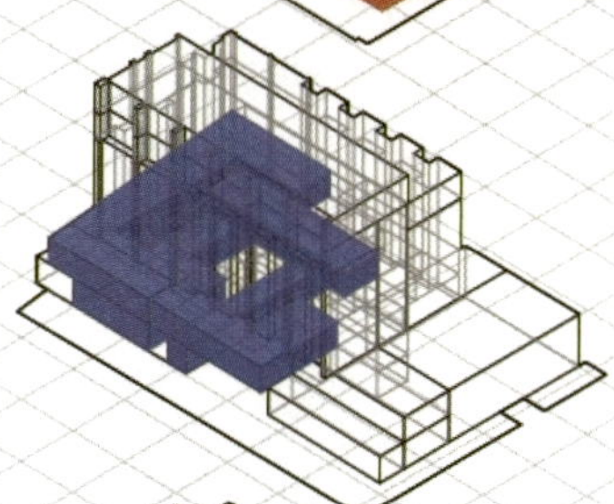

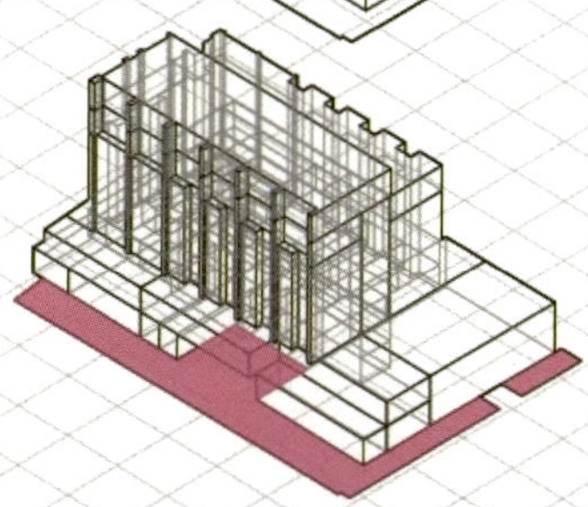

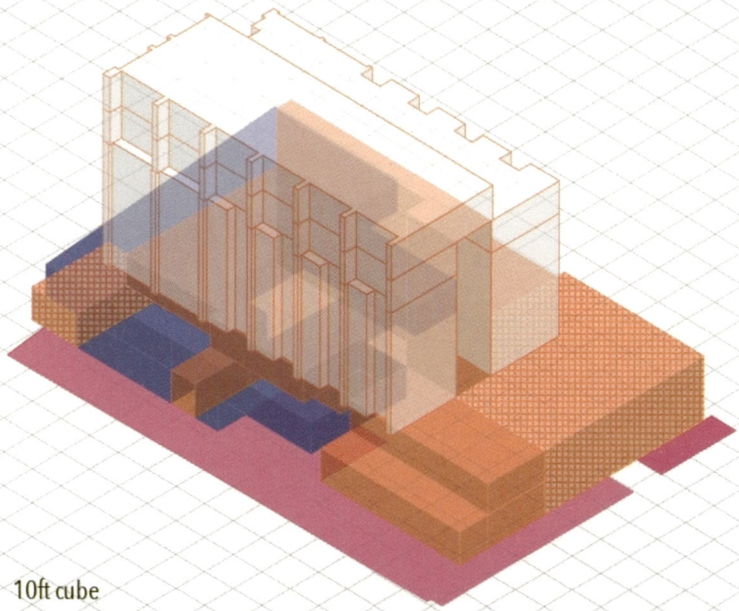

10ft cube

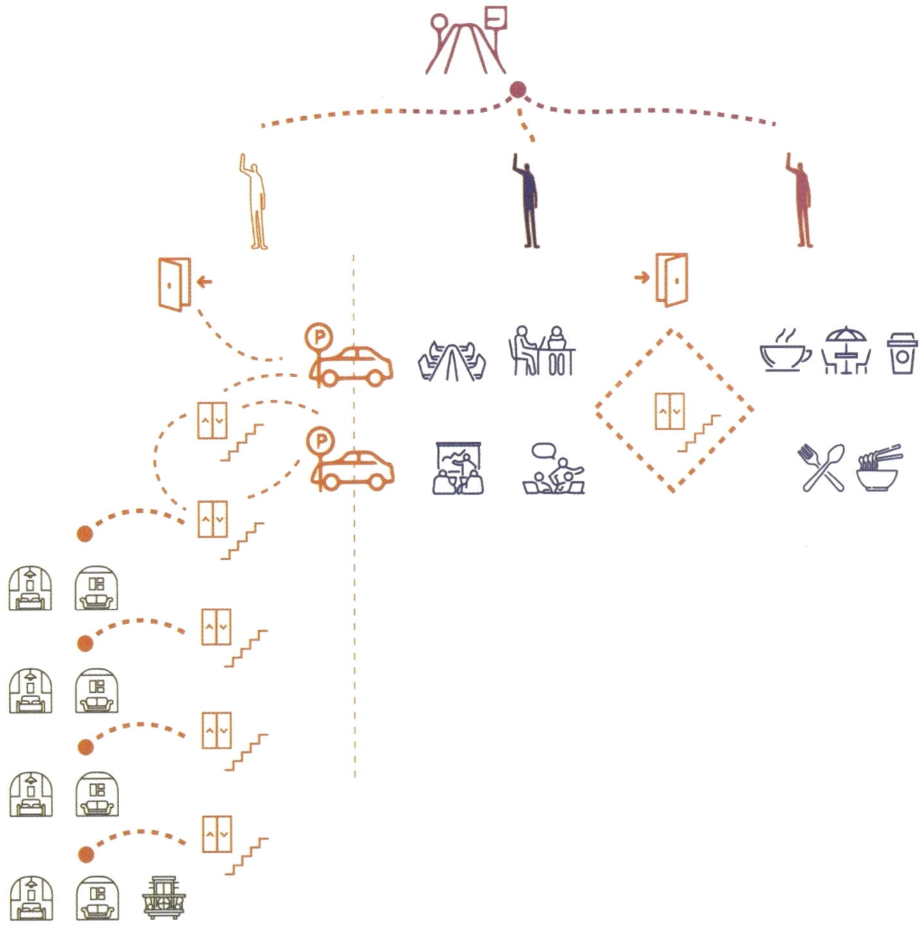

MRMU

ARC307 BU SECTIONFALL 2022 | Michela Btittis-Tannenbaum

Cities & Gardens

Ho Chi Minh City, Vietnam

ROOM+ Design & Build

Site area: 1,798sf
Building area: 8,719sf
Completion: 2018
Residential Units: 8

10,000 sf

domestic

communal

commercial

public

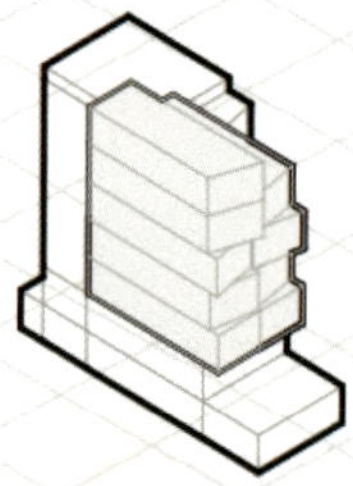

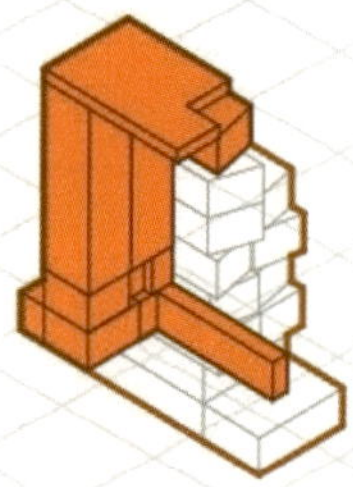

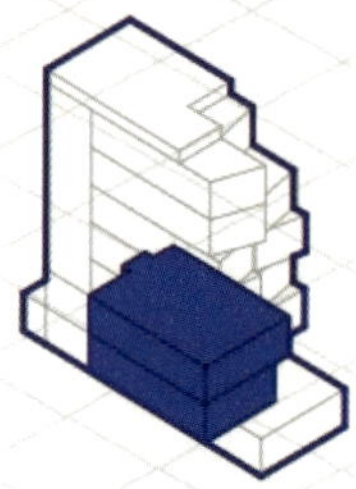

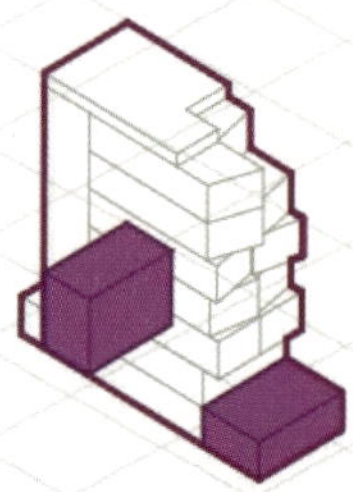

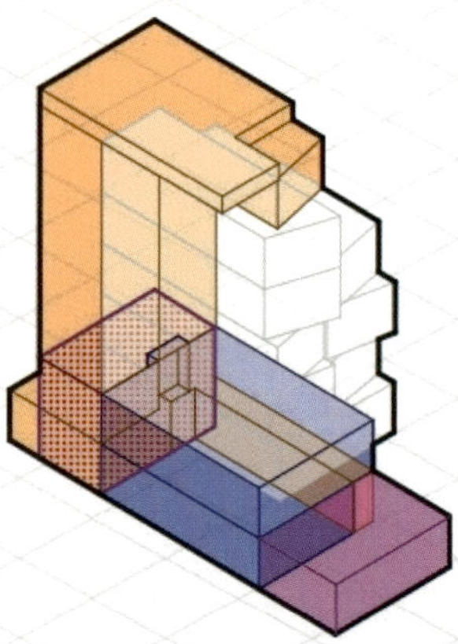

10ft cube

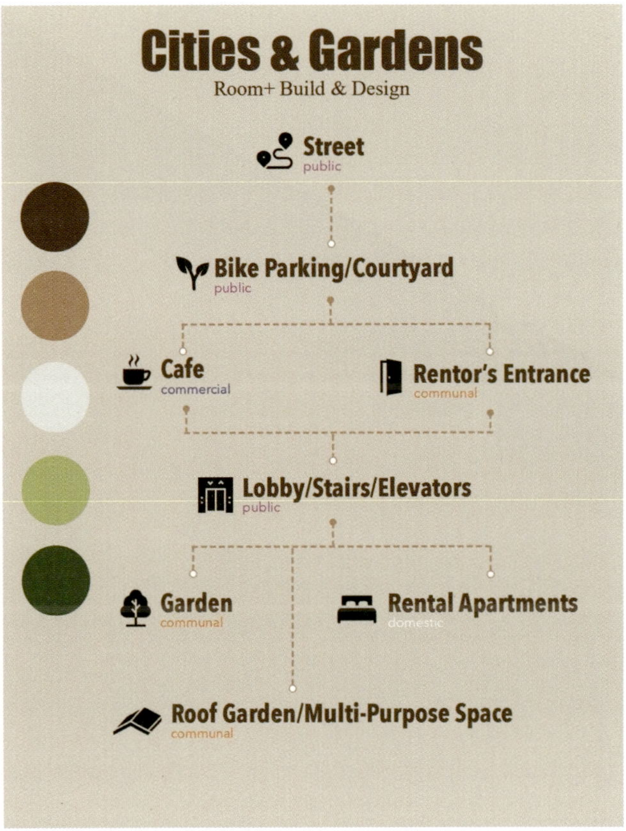
Cities & Gardens
Room+ Build & Design
Street
public
Bike Parking/Courtyard
public
Cafe
commercial
Rentor's Entrance
communal
Lobby/Stairs/Elevators
public
Garden
communal
Rental Apartments
domestic
Roof Garden/Multi-Purpose Space
communal

cities & gardens

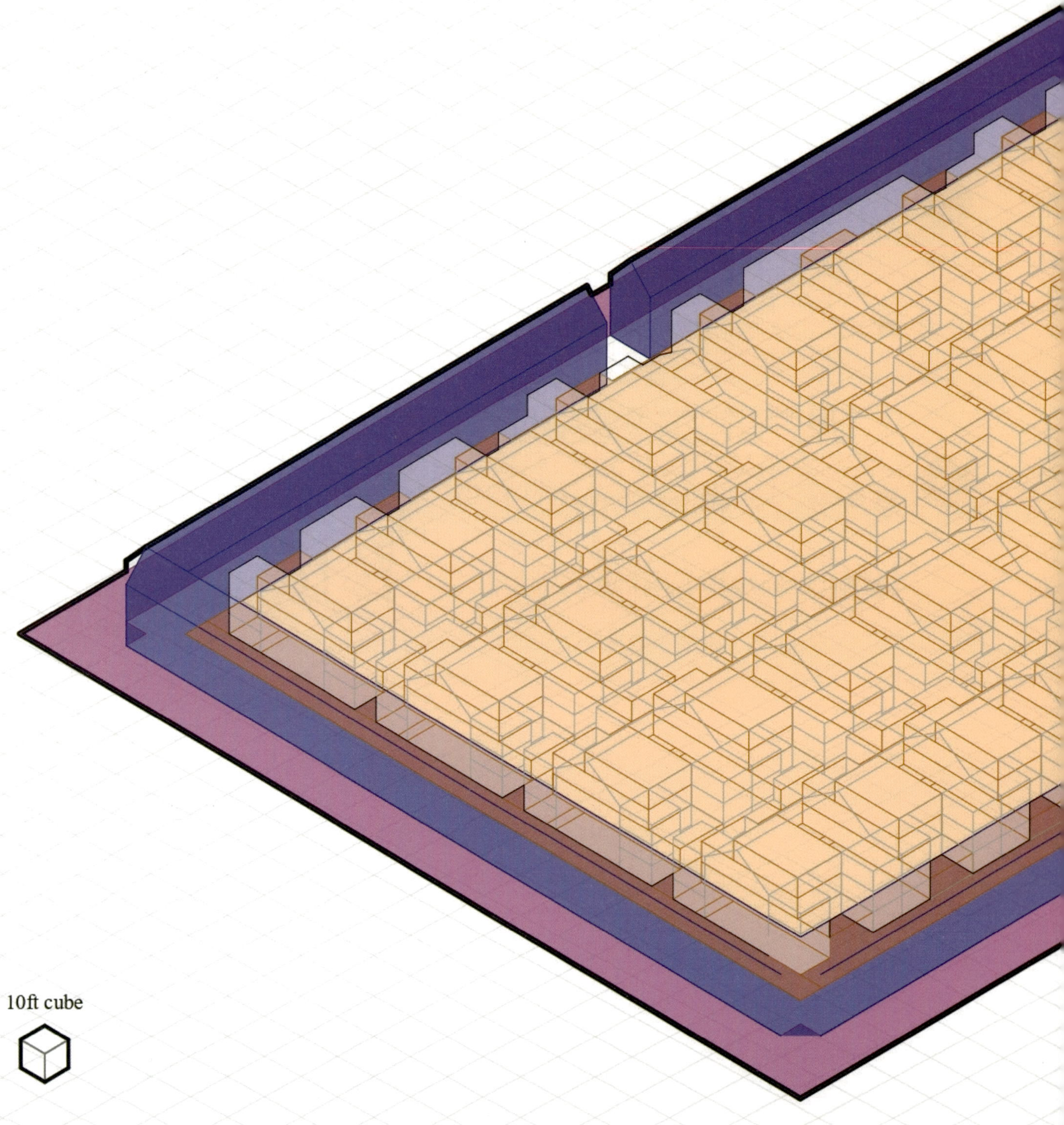
10ft cube

MRMU

ARC307 BU SECTIONFALL 2022 | Michela Btittis-Tannenbaum

Typical Shikumen Block

Shanghai, China

Traditional shanghai architectural style

Site area: varies
Building area: varies
Completion: first appeared 1860s
Residential Units: varies

domestic 10,000 sf

communal

commercial

public

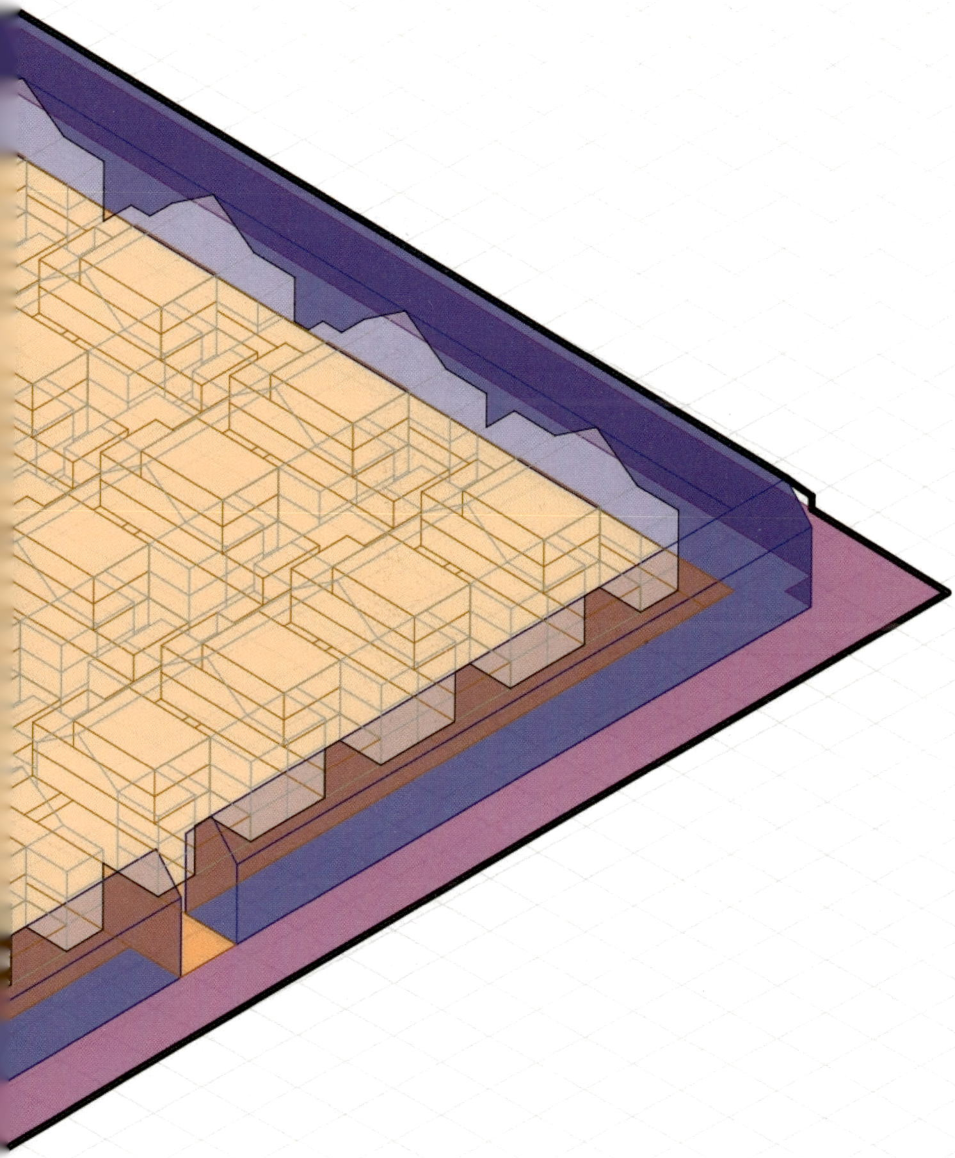

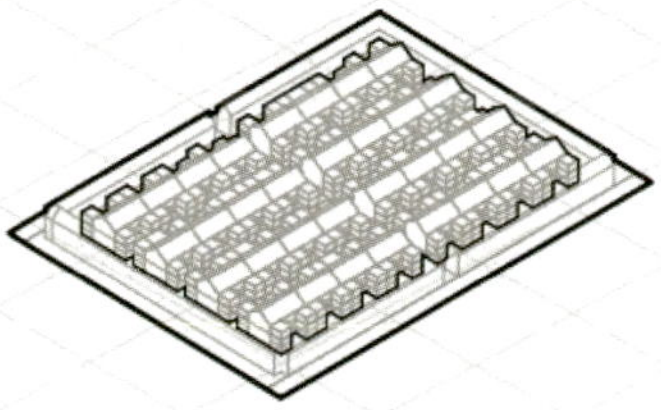

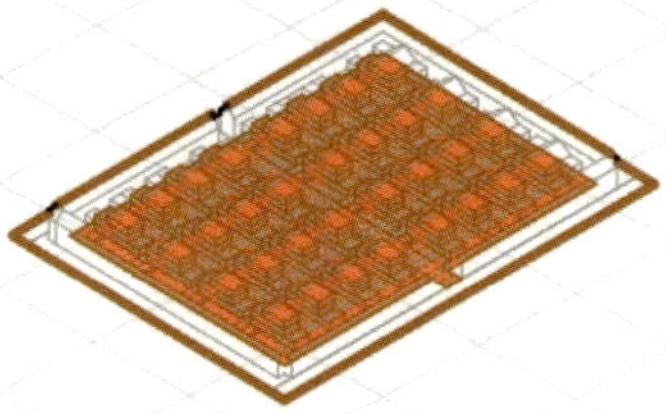

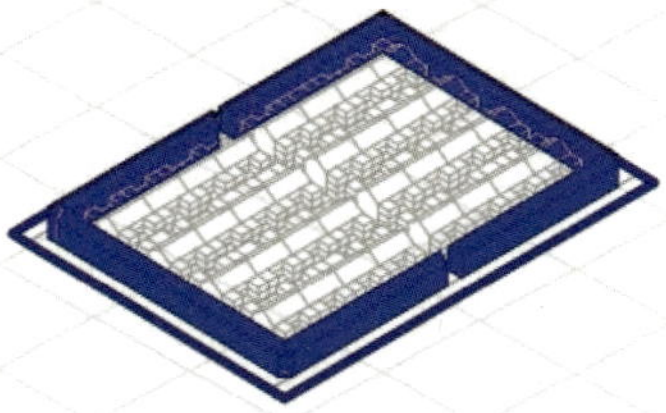

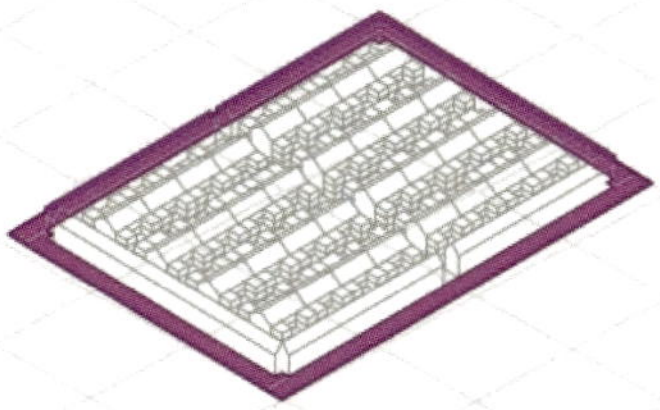

MRMU

ARC307 BU SECTION FALL 2022 | Tianchonghui Fang

TIAM COFFEE SHOP & HOME

Hue

Nguyen Khai Architects & Associates

Site area: 376sf

building area: 2700sf

Completion: 2020

10,000 sf

domestic

communal

commercial

public

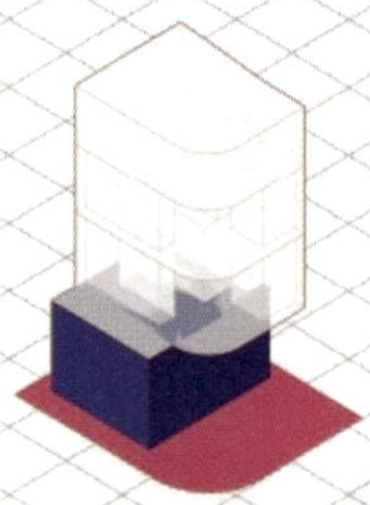

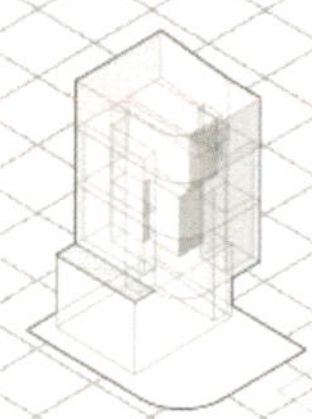

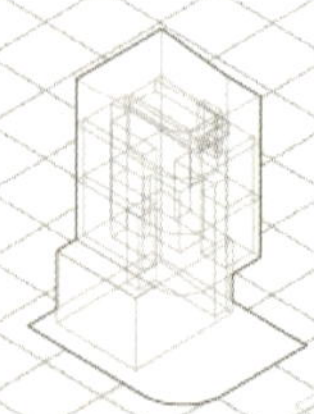

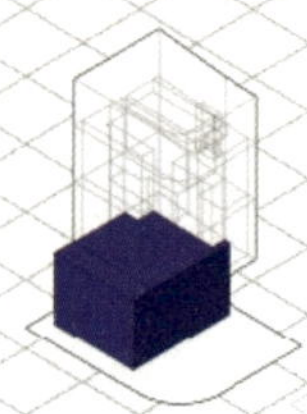

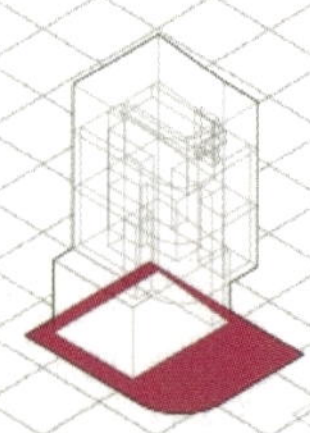

10ft cube

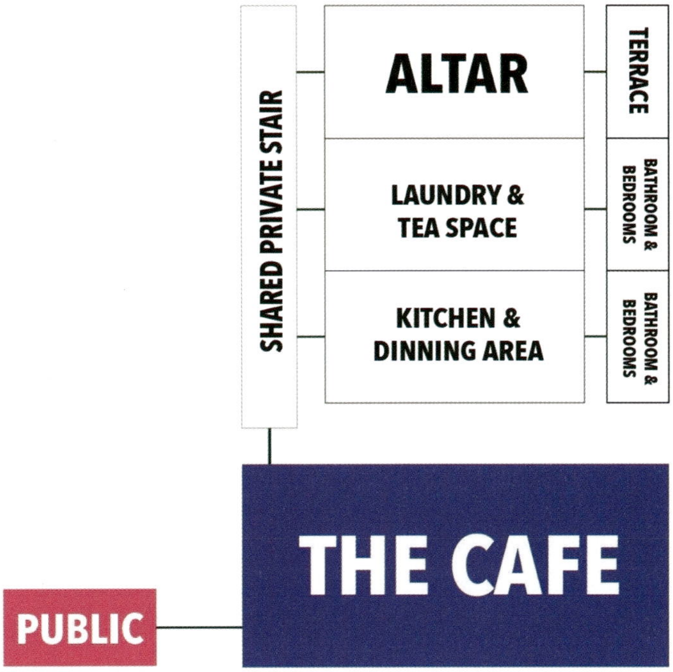
SHARED PRIVATE STAIR
ALTAR
TERRACE
LAUNDRY &
TEA SPACE
BATHROOM &
BEDROOMS
KITCHEN &
DINNING AREA
BATHROOM &
BEDROOMS
THE CAFE
PUBLIC

MRMU

ARC307 BU SECTION FALL 2022 | Tianchonghui Fang

FH OFFICE

Bangkok Noi
TA-CHA Design

Site area: 9,000sf
building area: 43,610sf
Completion: 2020

10,000 sf

domestic

communal

commercial

public

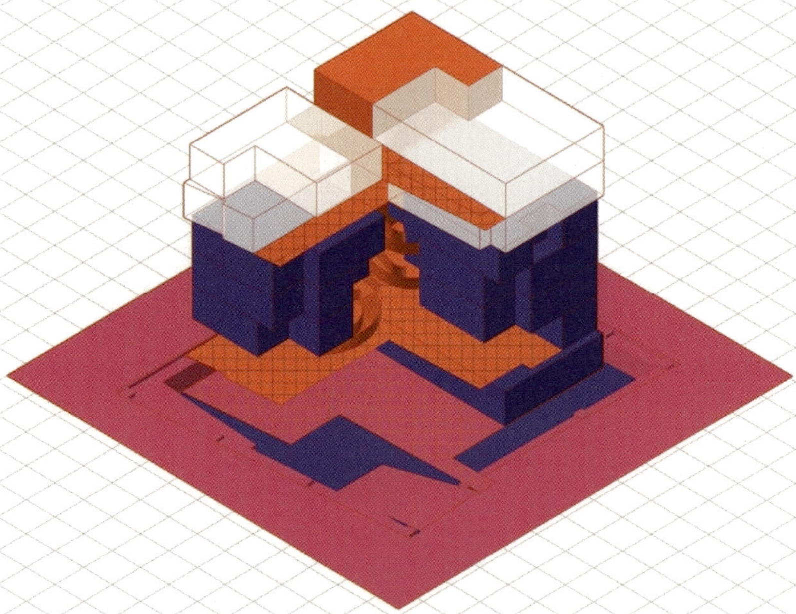

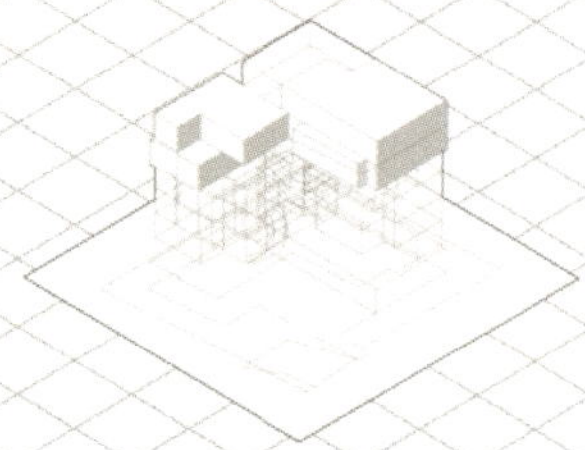

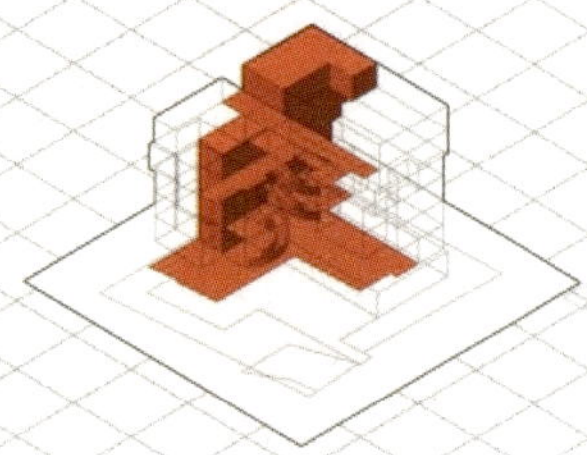

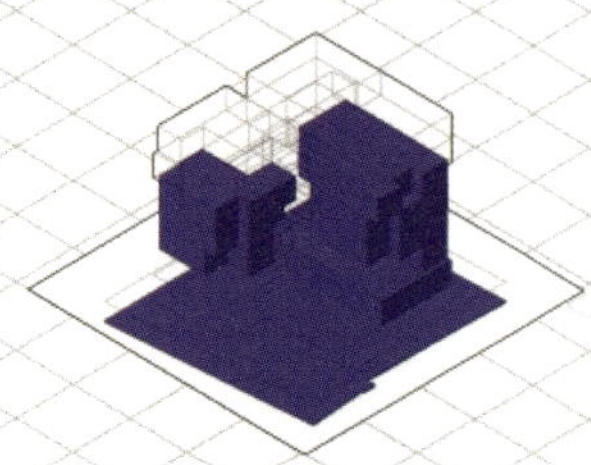

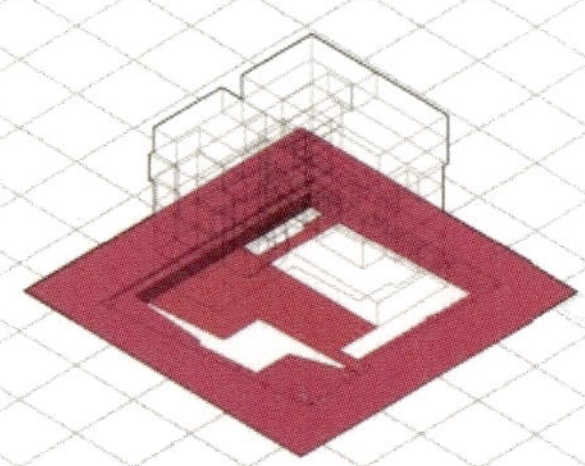

10ft cube

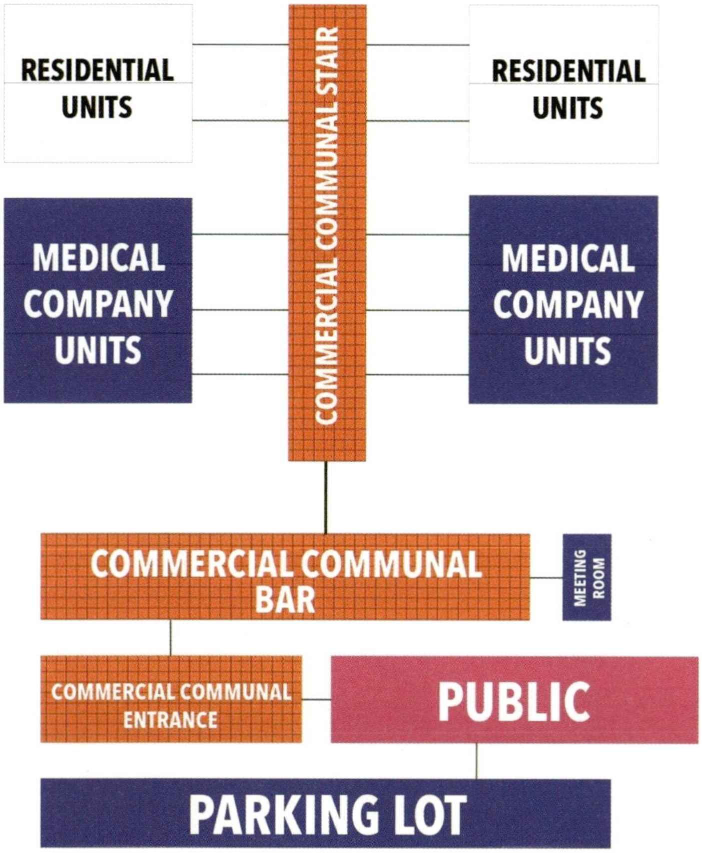
RESIDENTIAL UNITS
COMMERCIAL COMMUNAL STAIR
RESIDENTIAL UNITS
MEDICAL COMPANY UNITS
MEDICAL COMPANY UNITS
COMMERCIAL COMMUNAL BAR
MEETING ROOM
COMMERCIAL COMMUNAL ENTRANCE
PUBLIC
PARKING LOT

BUSAN

THE SECOND CAPITAL OF SOUTH KOREA

Busan Metropolitan City is located in southeastern South Korea. It is the second and largest maritime (port) city in Korea, and maritime trade and logistics industries have developed around Busan Port. It faces Japan with the Korean Strait and Daemado Island in between.

부산광역시는 대한민국 동남부에 위치한다. 대한민국의 제2의 도시이자 최대의 해양(항구) 도시이며, 부산항을 중심으로 해상 무역과 물류 산업이 발달하였다. 일본과는 대한해협과 대마도를 사이에 두고 마주하고 있다.

Photo by Christopher Lee on Unsplash

Based on the specificity and physical proximity of Korea's second largest metropolitan area with a population of 3.5 million, we chose the Busan metropolitan area as our site for two reasons: the surge in interconnectivity between Busan, Ulsan, Gyeongju and many other small cities. There are many urban problems such as lack of parking spaces, aging transportation and public infrastructure, and lack of urban green areas. A typical type of construction is a single-use concrete apartment complex, which does not provide many opportunities and inspiration-generating urban benefits.

우리는 인구 350만 명의 대한민국 제2의 대도시권의 특수성과, 물리적 근접성을 바탕으로 부산, 울산, 경주 그리고 많은 다른 소도시들 사이의 상호 연결성의 급증이라는 두 가지 이유에서 우리의 부지로 부산 수도권을 선택했다. 주차 공간 부족과 노후화된 교통 공공 인프라, 도시녹지 부족 등 도시문제가 많다. 건설되고 있는 대표적인 유형으로는 1회용 콘크리트 아파트 단지가 있는데, 이는 많은 기회와 영감을 창출하는 도시적 이점을 제공하지 못한다.

황두진 Doojin Hwang 지정우 Jungwoo Ji 이병욱 Byoungwook Lee 이기철 Kichul Lee 임저스틴 Justin Lim 송민경 Minkyung Song

YONGHO-DONG, NAM-GU, BUSAN

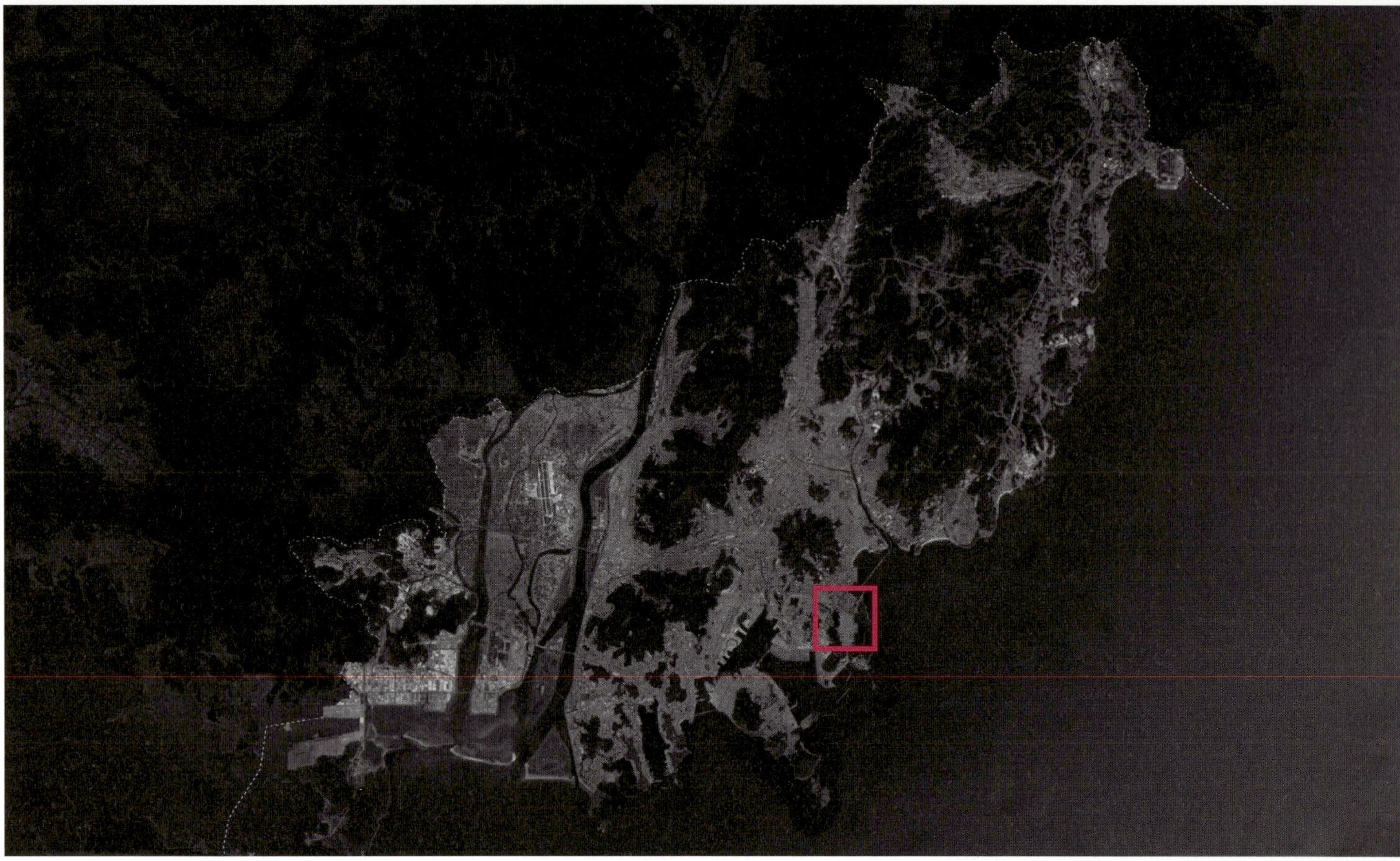

RESEARCH BY
CHANGMIN PARK
HYEONJIN LIM
EUNHYE KIM

Yongho-dong is a legal dong in Nam-gu, Busan. The administrative building is divided into Yongho 1-4 dongs. Located at the southeastern end of the Korean Peninsula, it is an international port city that connects the continent and the ocean. At the foot of Hwangnyeongsan Mountain, middle and low-income families are mixed in the location space leading to the coastline, and recently, pleasant residential areas have been formed by the development of large-scale apartment complexes, and there are recreational spaces for citizens such as Oryukdo Island, a symbol of Busan.

용호동(龍湖洞)은 부산광역시 남구의 법정동이다. 행정동 용호1~4동으로 나뉜다. 한반도의 동남단에 위치하여 대륙과 해양을 잇는 관문인 국제 항만도시이다. 황령산 자락이 해안선으로 이어지는 입지공간 속에 중산층과 저소득층이 혼재하고 최근에는 대단위 아파트단지 개발 등으로 쾌적한 주거지를 형성하고 있으며, 부산의 상징인 오륙도 등 시민 여가선용휴식공간이 소재한다.

© Google Earth

Photo by Heamosoo Kim on Unsplash

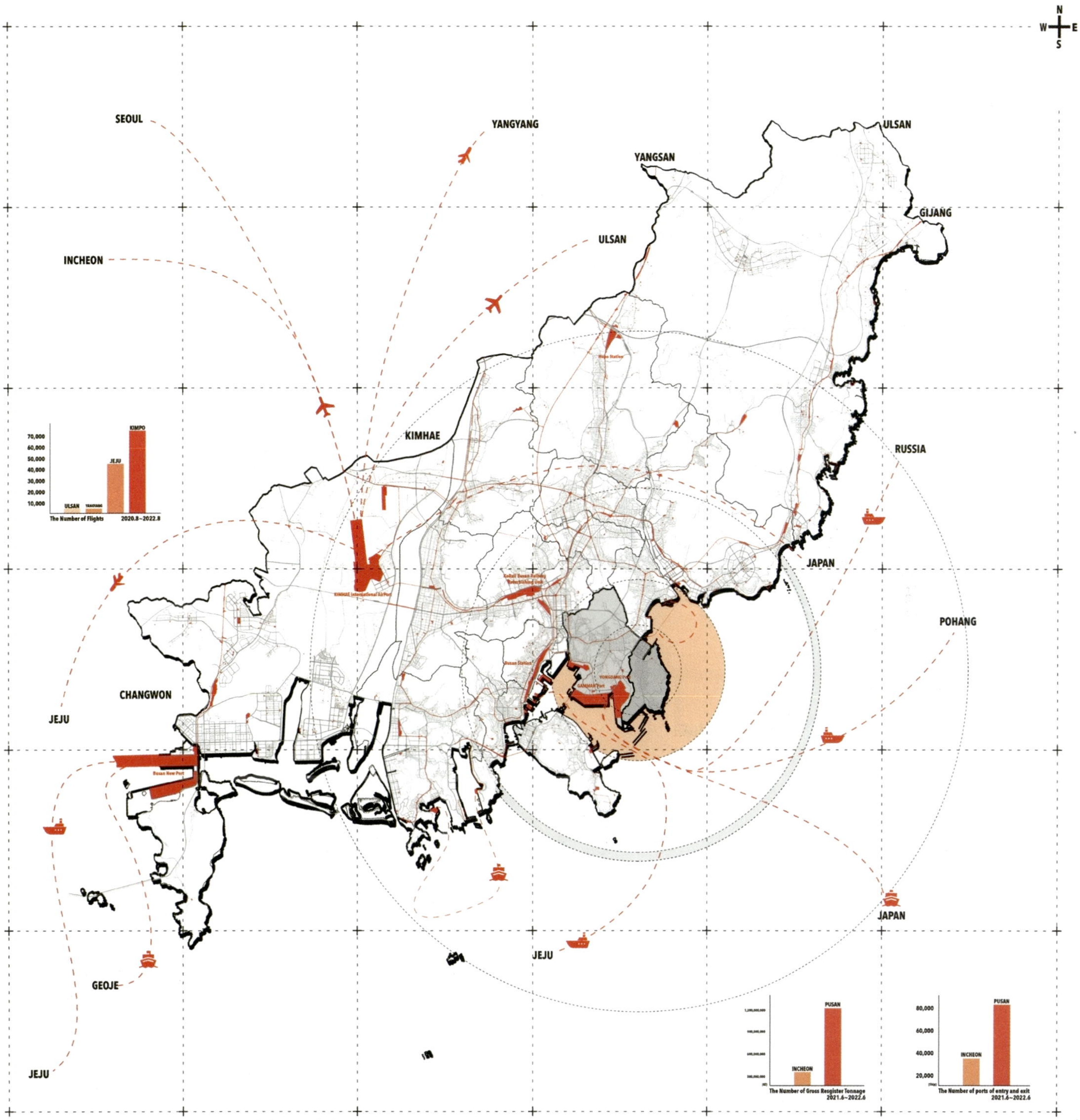

MASS TRANSIT : THE CITY SCALE (L)

MASS TRANSIT

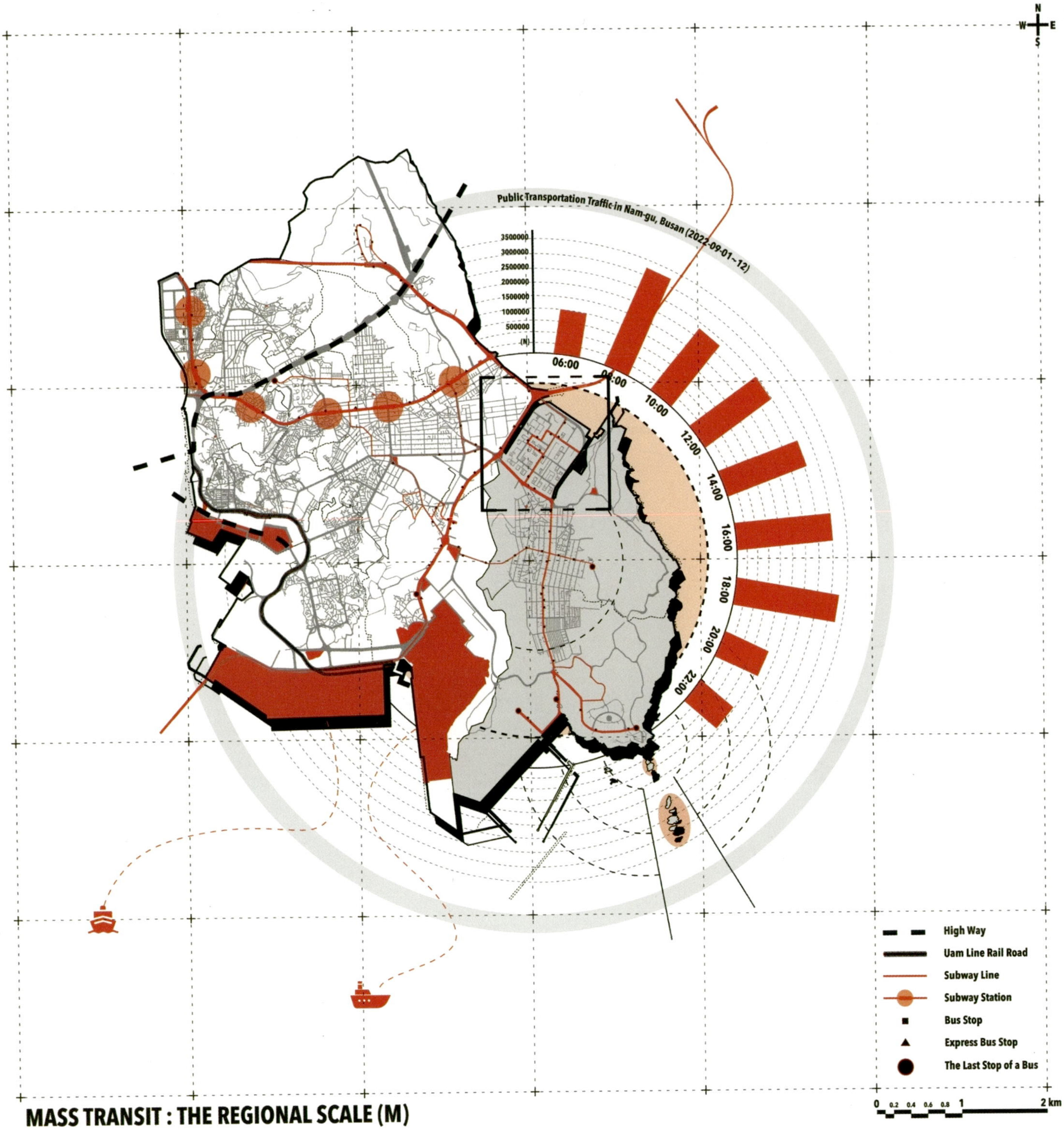

MASS TRANSIT : THE REGIONAL SCALE (M)

MASS TRANSIT

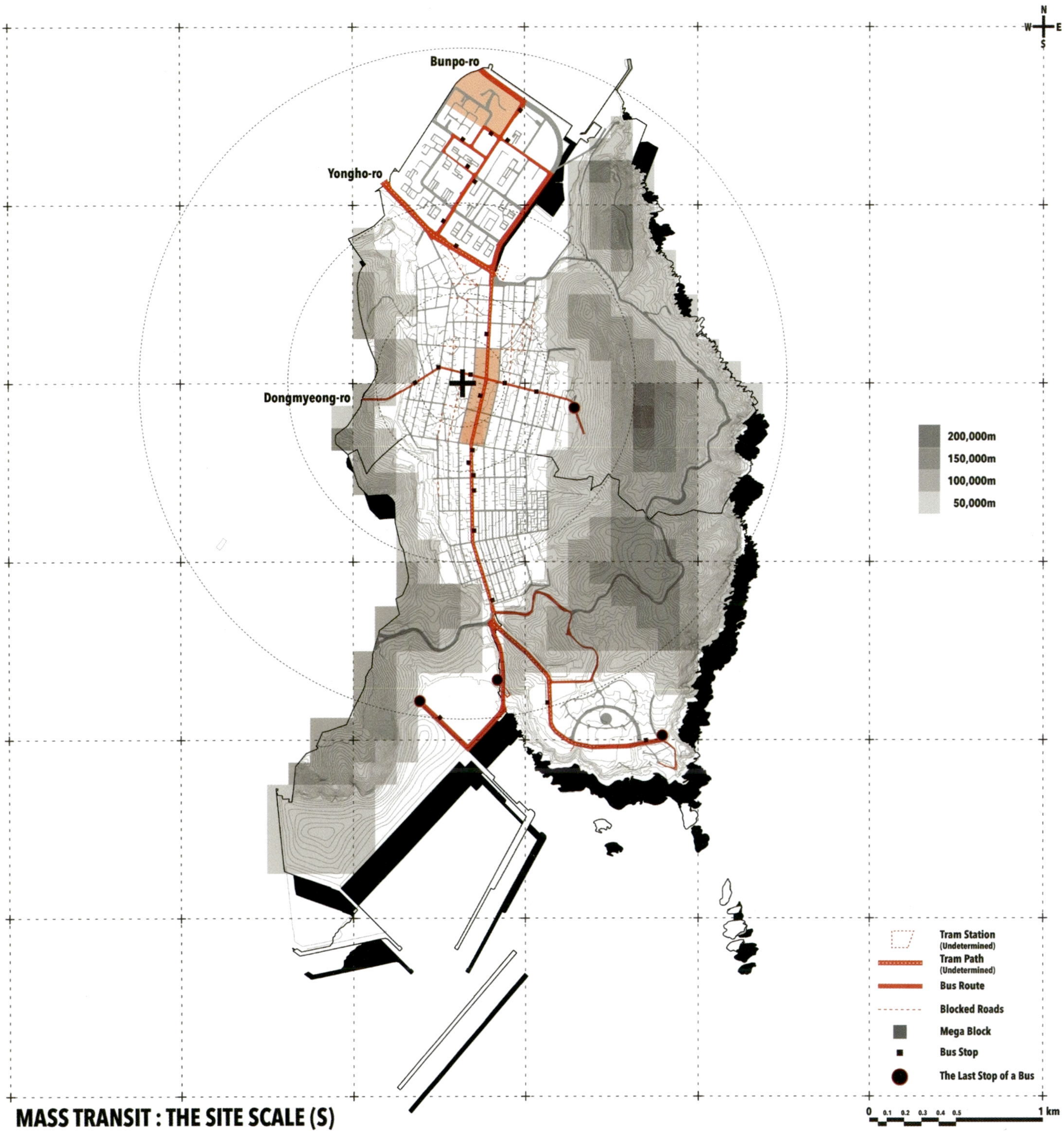

MASS TRANSIT : THE SITE SCALE (S)

MASS TRANSIT

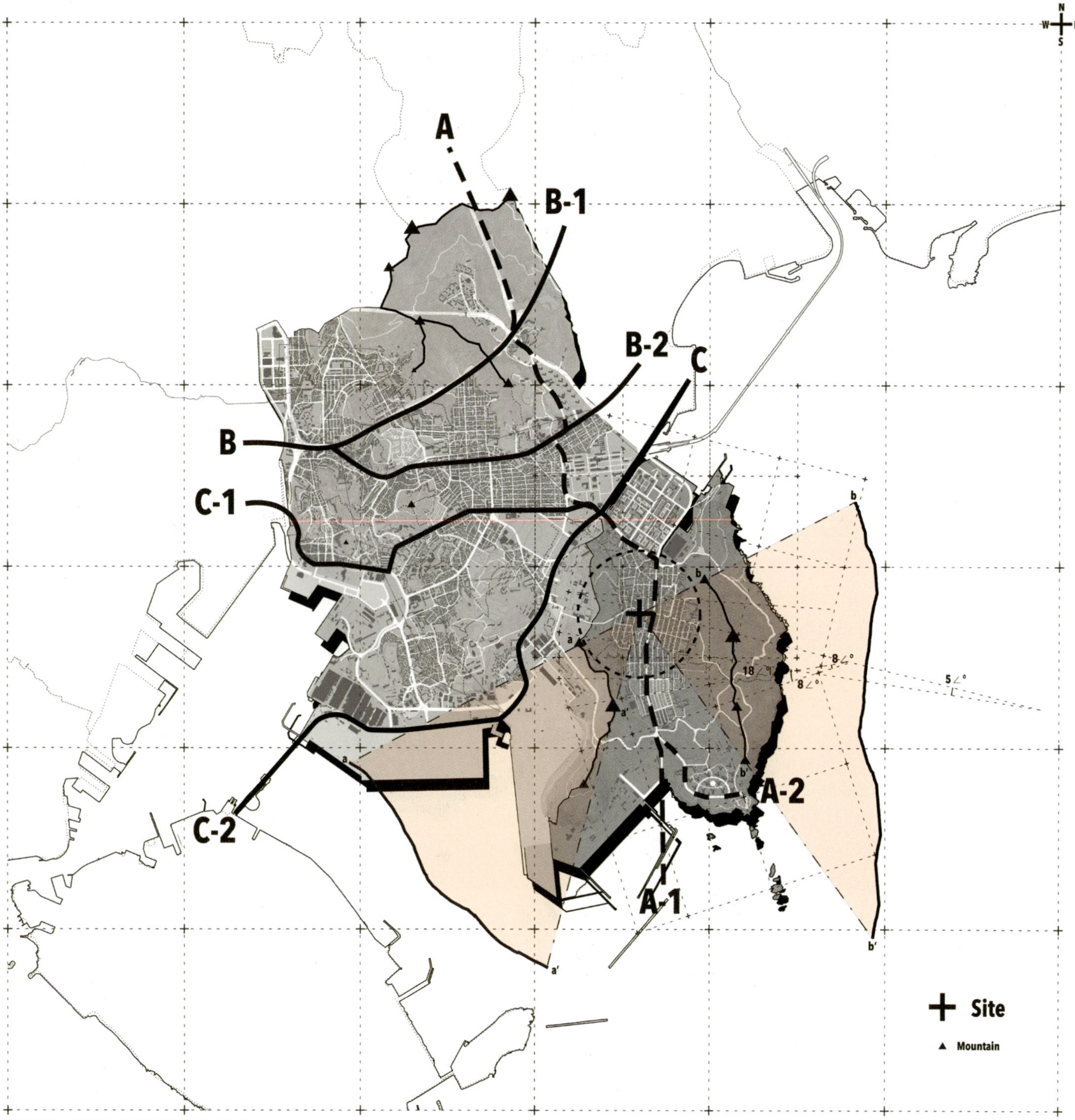

AXIS : THE CITY SCALE (L)

AXIS

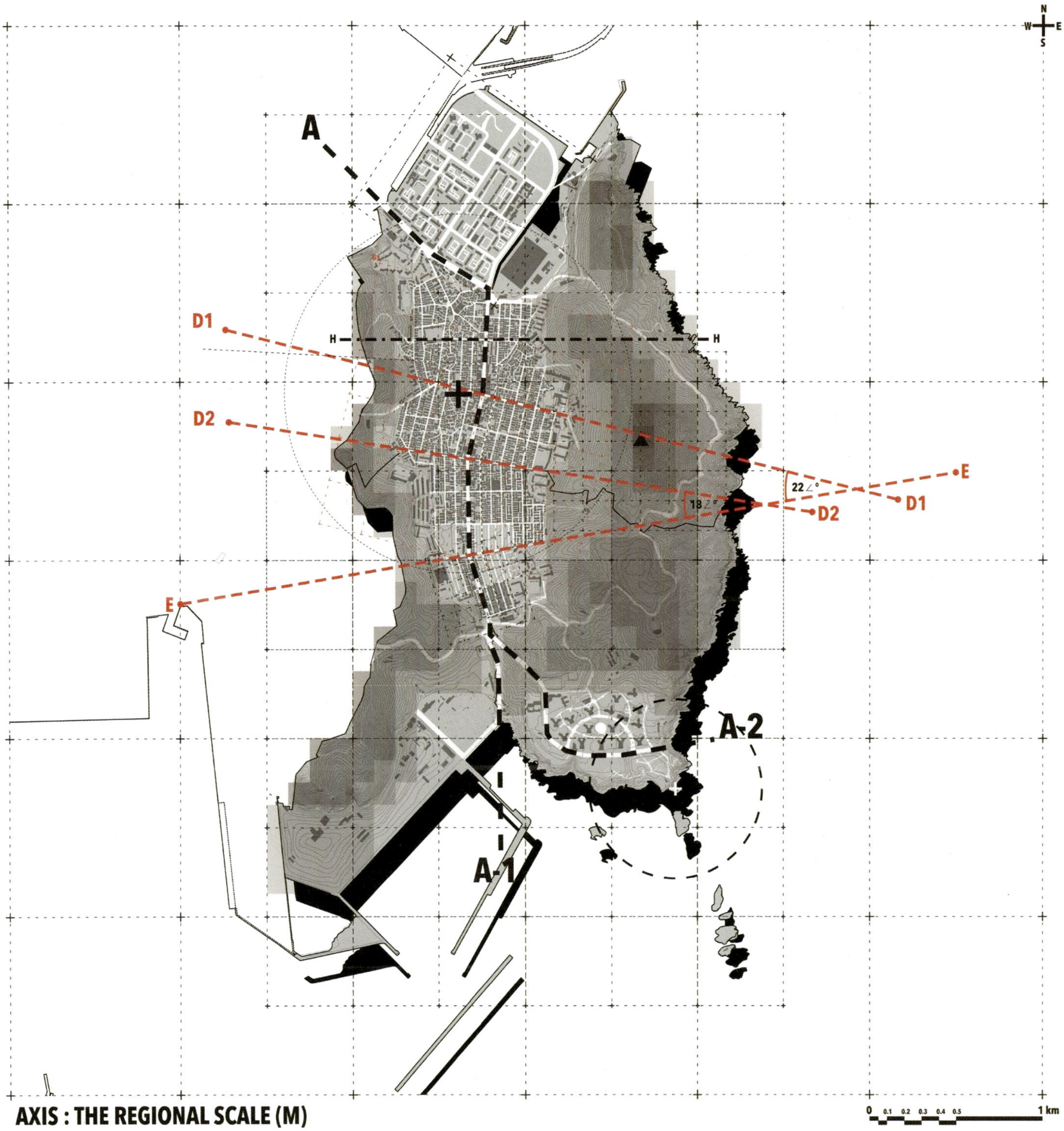

AXIS : THE REGIONAL SCALE (M)

AXIS

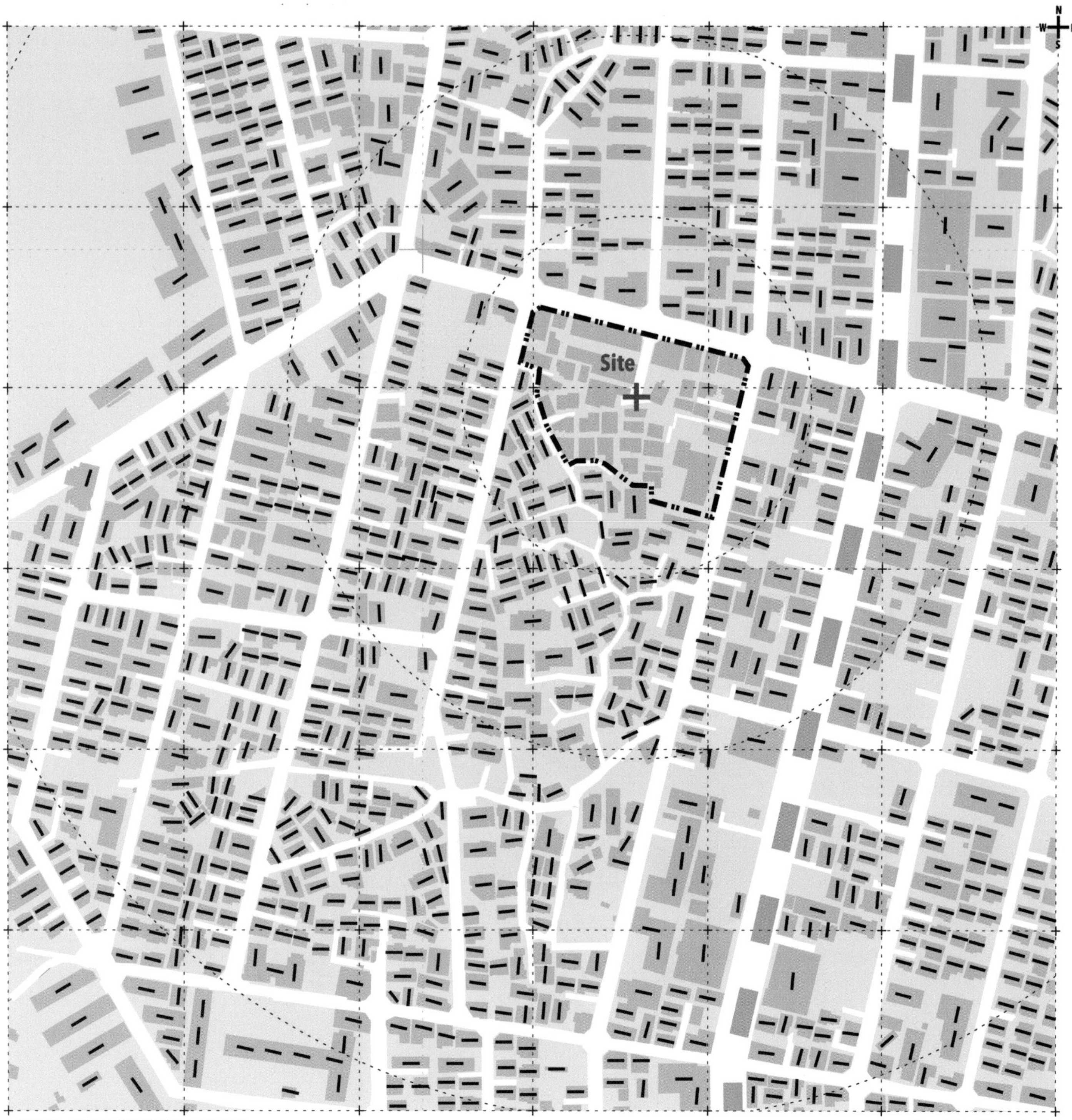

AXIS : THE SITE SCALE (S)

0 50 100 150 200 250 m

AXIS

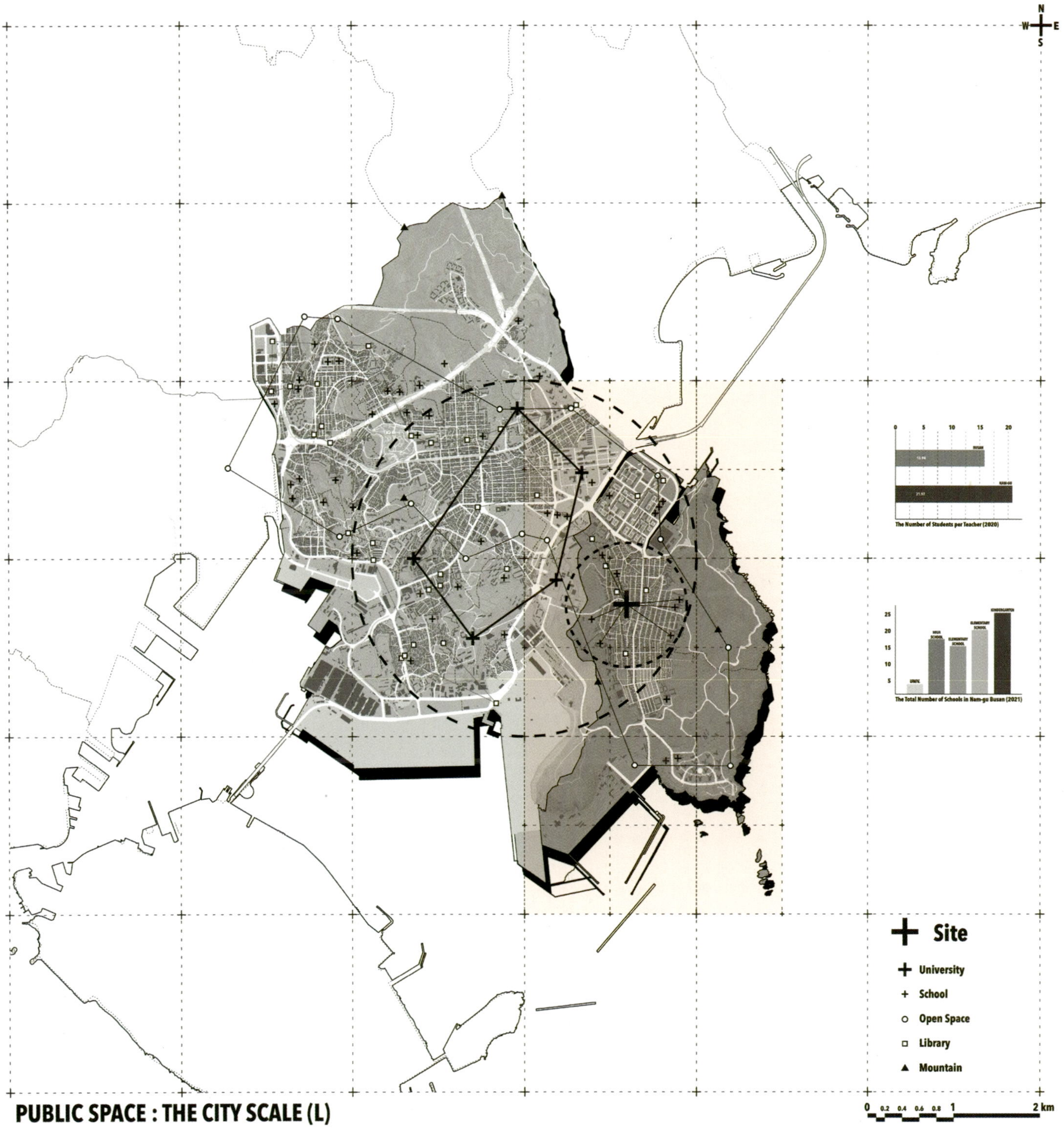

PUBLIC SPACE : THE CITY SCALE (L)

PUBLIC SPACE

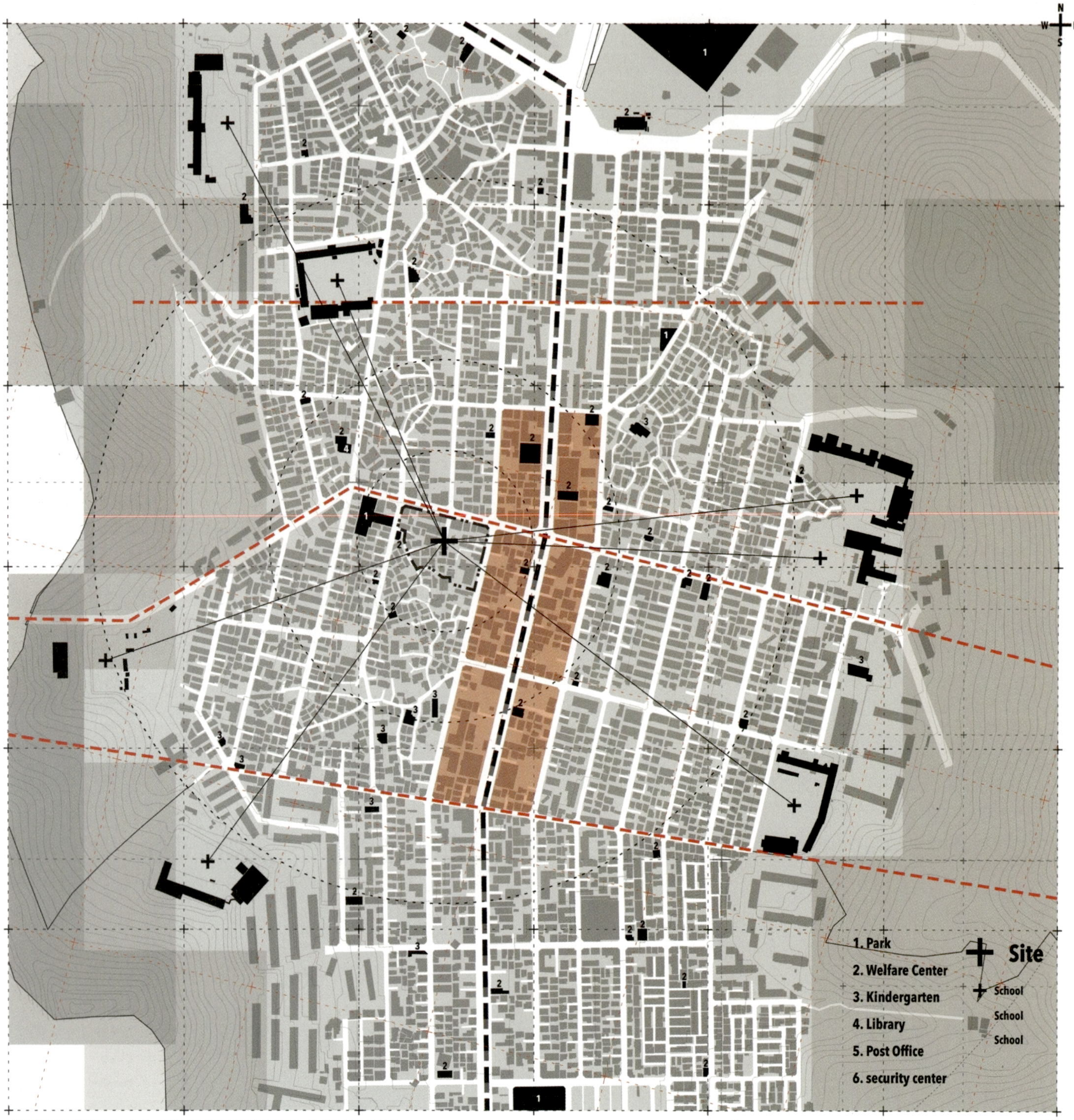

PUBLIC SPACE : THE REGIONAL SCALE (M)

PUBLIC SPACE

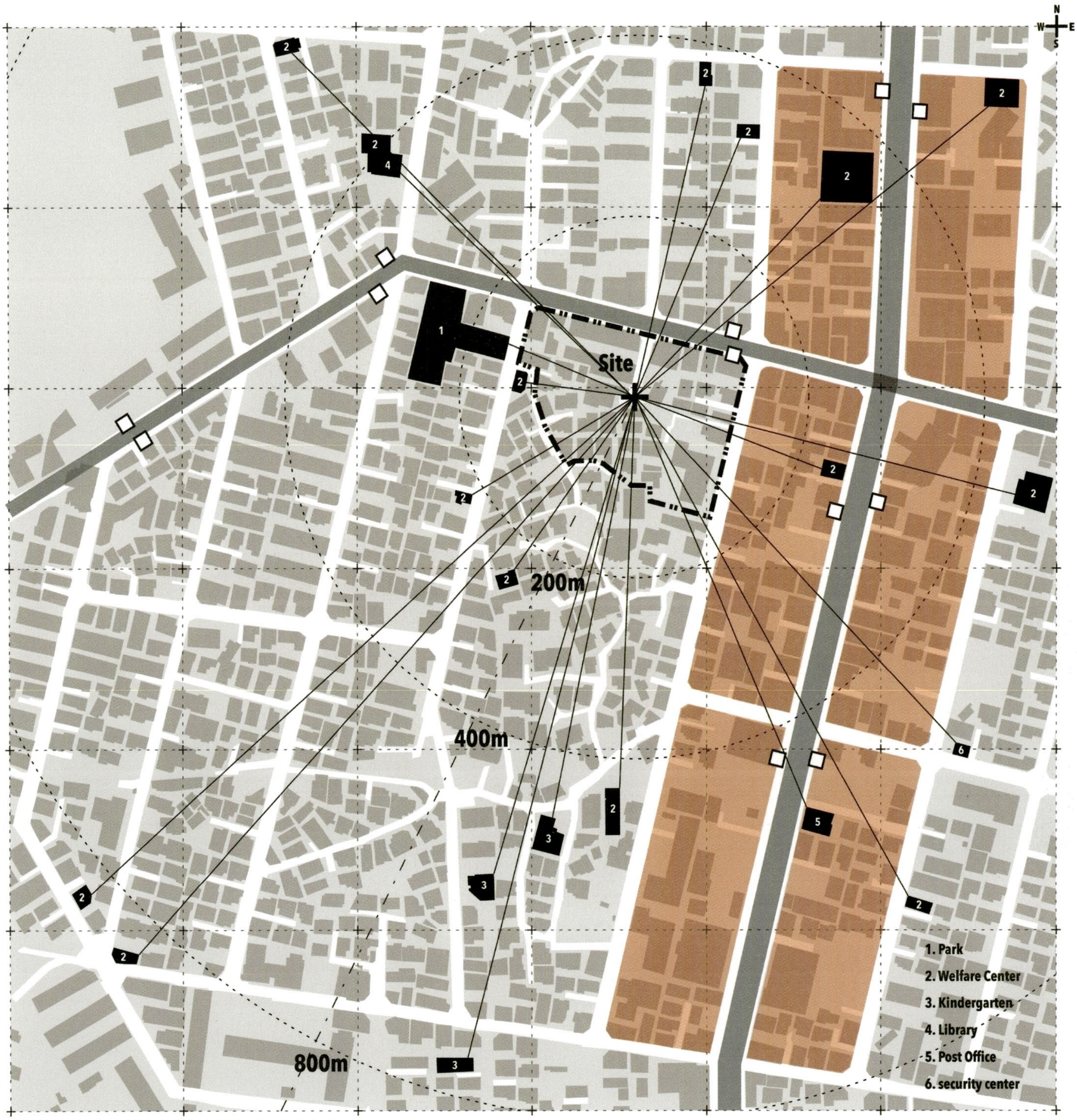

PUBLIC SPACE : THE SITE SCALE (S)

PUBLIC SPACE

Density
SCALE : Regional (Busan)

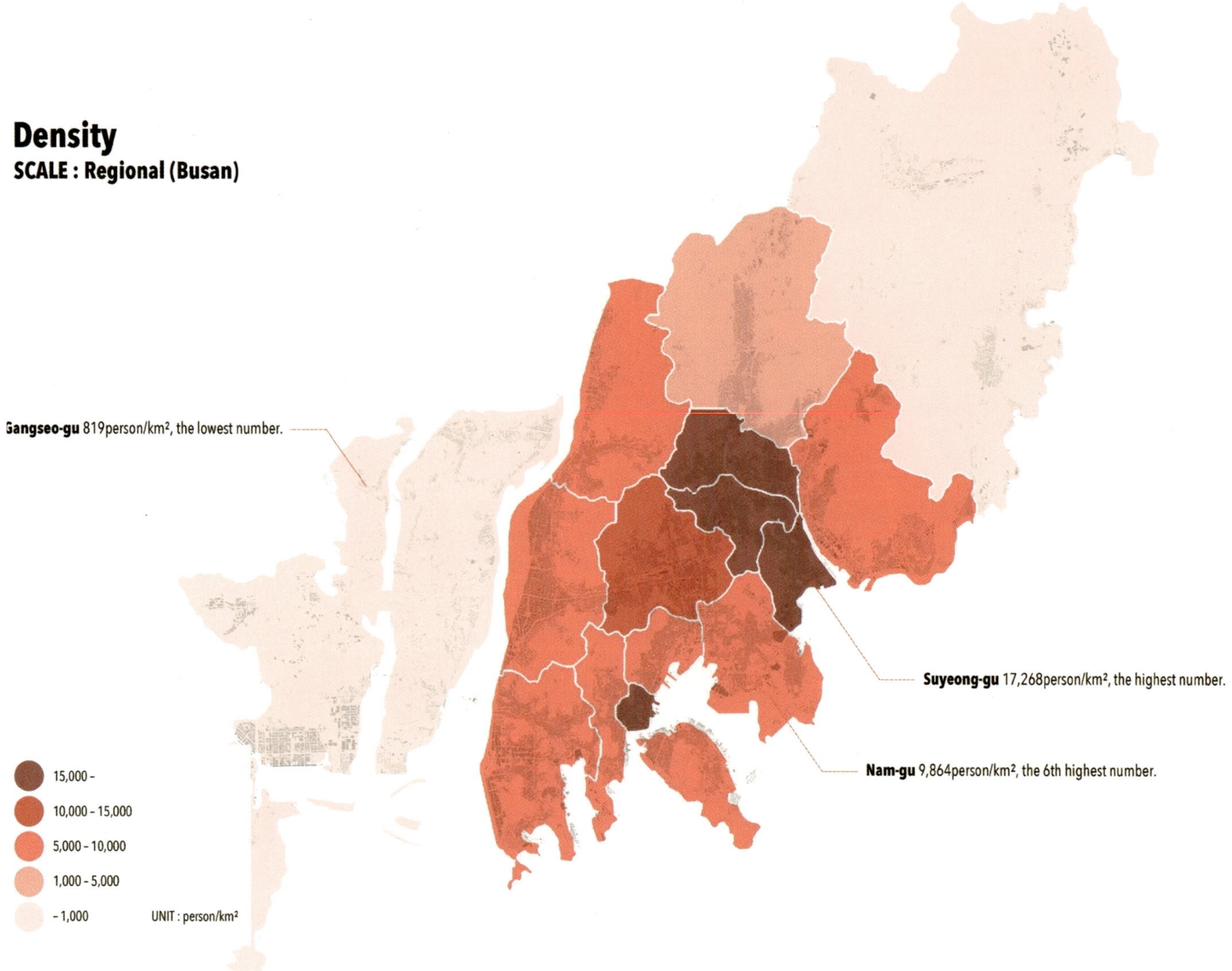

The three most basic and most important things in architectural design are site, program, and user. The reason why density was selected as the first issue of cartography is that density is directly connected to the actual demand of buildings, that is, user. On the regional scale of Busan, the population density of Nam–gu, where the site belongs, was 9,864 people/km², the sixth highest among 16.

DENSITY

Density
SCALE : City (Nam-gu)

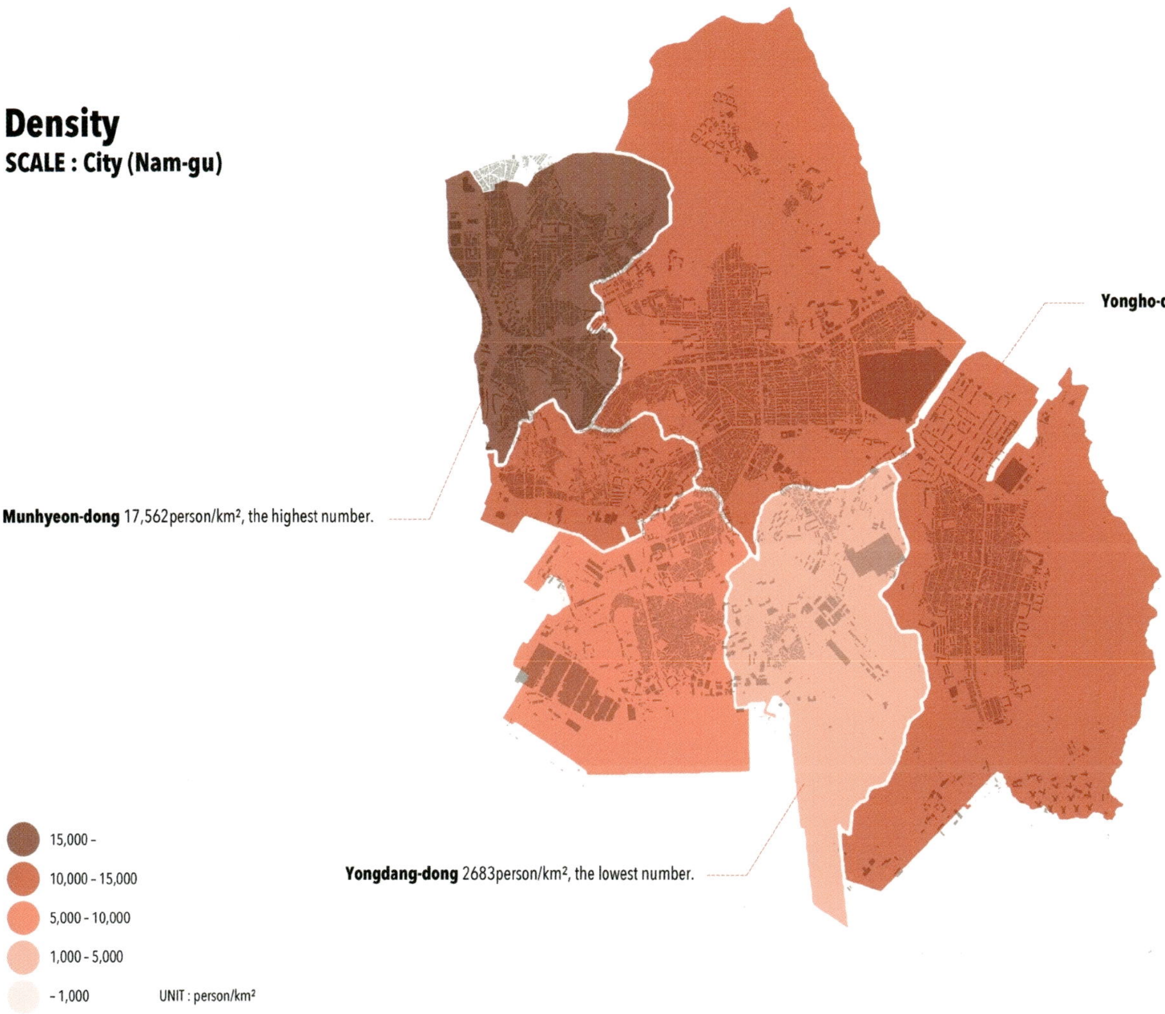

On the urban scale of Nam-gu, the population density of Yongho-dong, where the site belongs, was 12,736 people/km², which is not small in terms of figures, but was the third out of six.

DENSITY

Density
SCALE : Neighborhood (Yongho-dong)

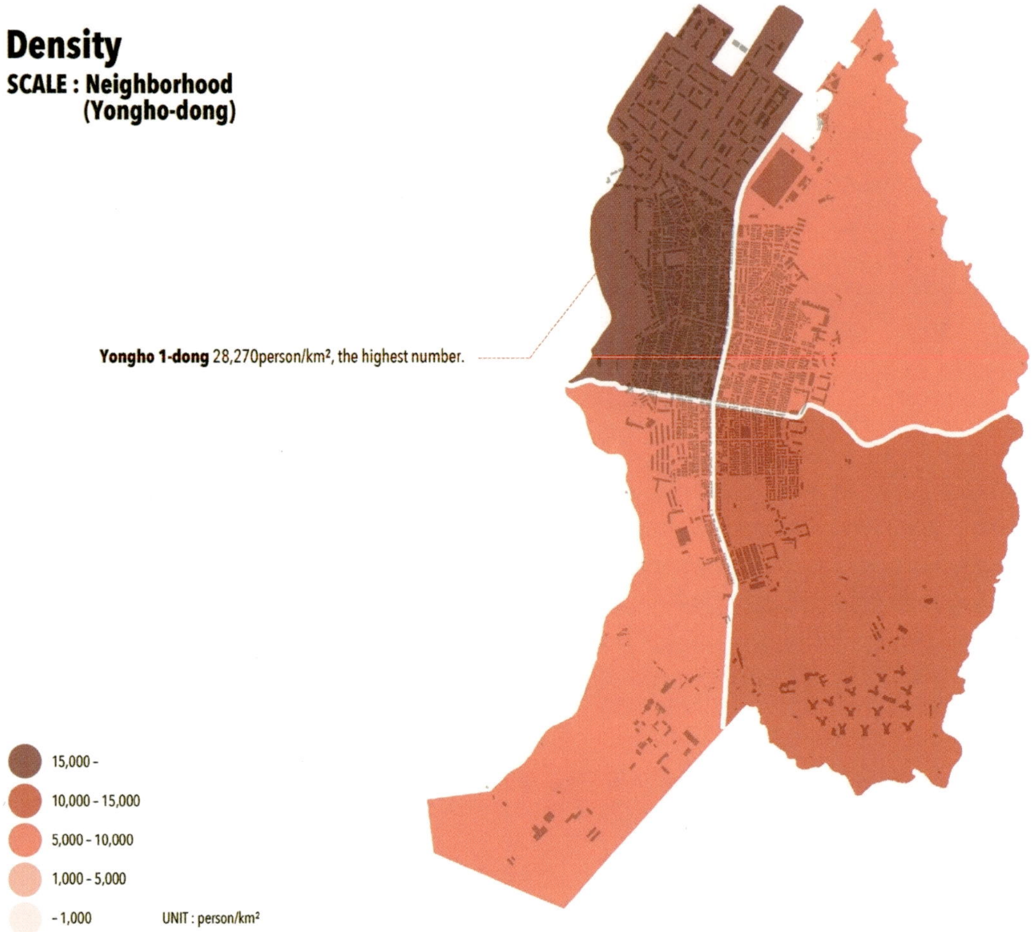

When limited to the neighboring scale of Yongho-dong, Yongho 1-dong, to which the site belongs, was the area with the highest population number and density. Through this, I present the conclusion that a new shared space is needed within the residential block where the recent redevelopment is active and dense around Yongho-ro.

DENSITY

Tourism
SCALE : Regional (Busan)

Busan on a regional scale has two characteristics: a marine city and a movie city in terms of tourism. In addition to Dadaepo Beach and Taejongdae, various tourist attractions using the sea view are lined up along the southern coastline, and there were art-related attractions such as BEXCO, Film Center, and City Museum of Art in certain areas such as Haeundae-gu.

TOURISM

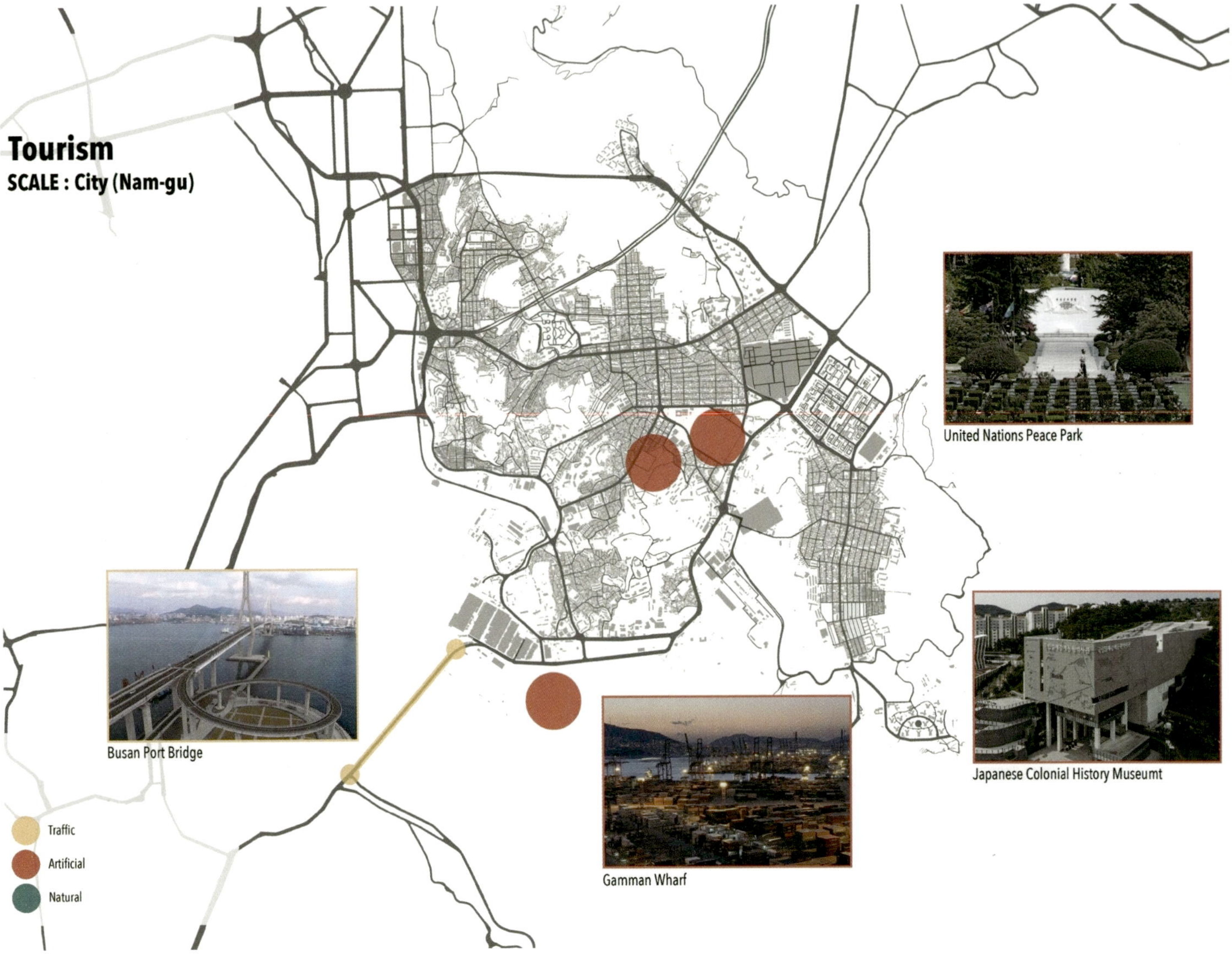

The urban scale of Nam-gu also has two characteristics. Outside, there are a bridge and Gamman Pier connecting Busan Port, and inside, it is a meaningful space designated as a UN special peace and culture zone.

TOURISM

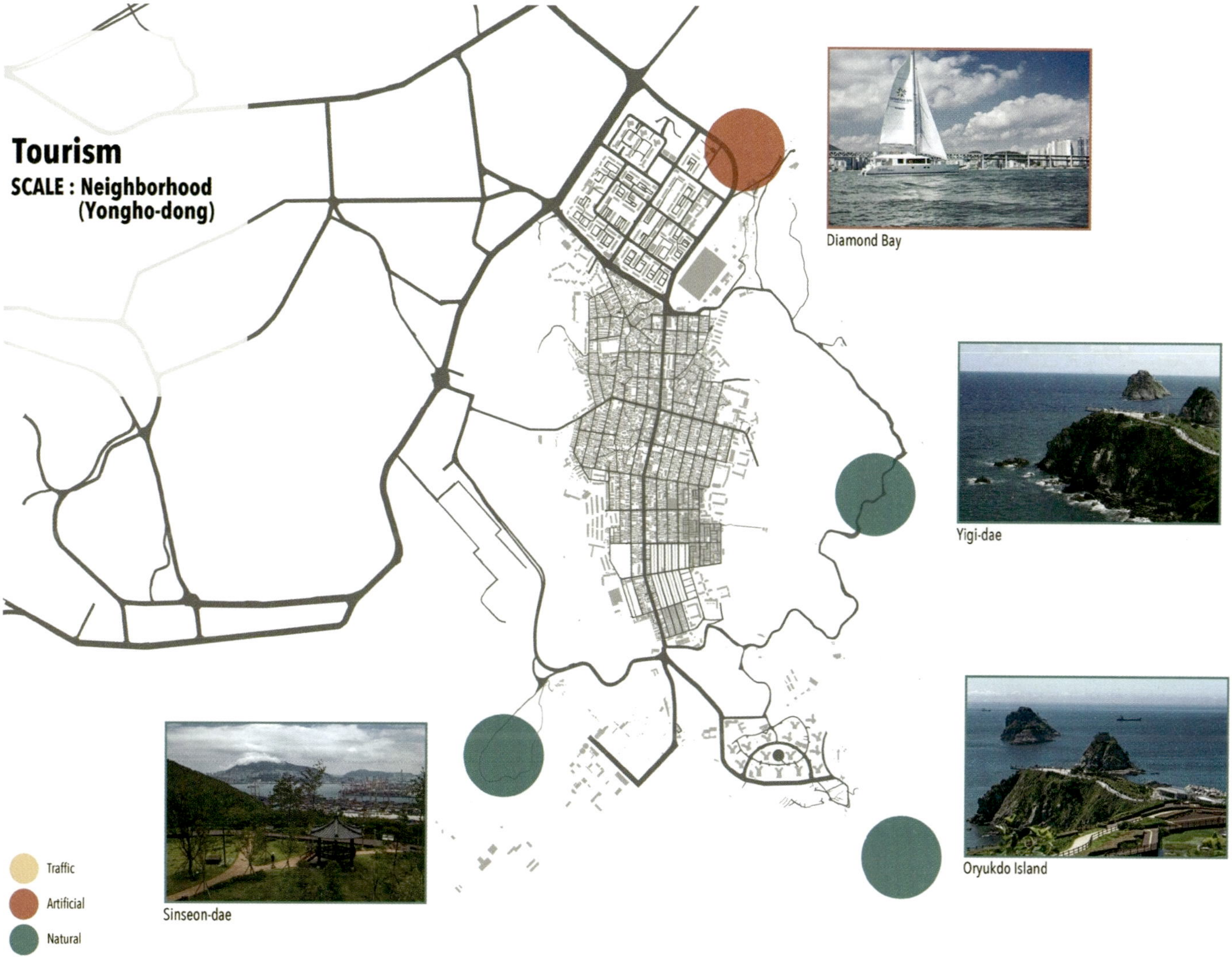

There are many tourist facilities for outsiders in Yongho-dong. This is sufficiently included in the living radius of the residents, but it cannot be seen as a space where a close community of residents can be created and maintained. Therefore, in this MRMU project, the creation of local communities centered on residents should be prioritized.

TOURISM

Local Identity
SCALE : City (Nam-gu)

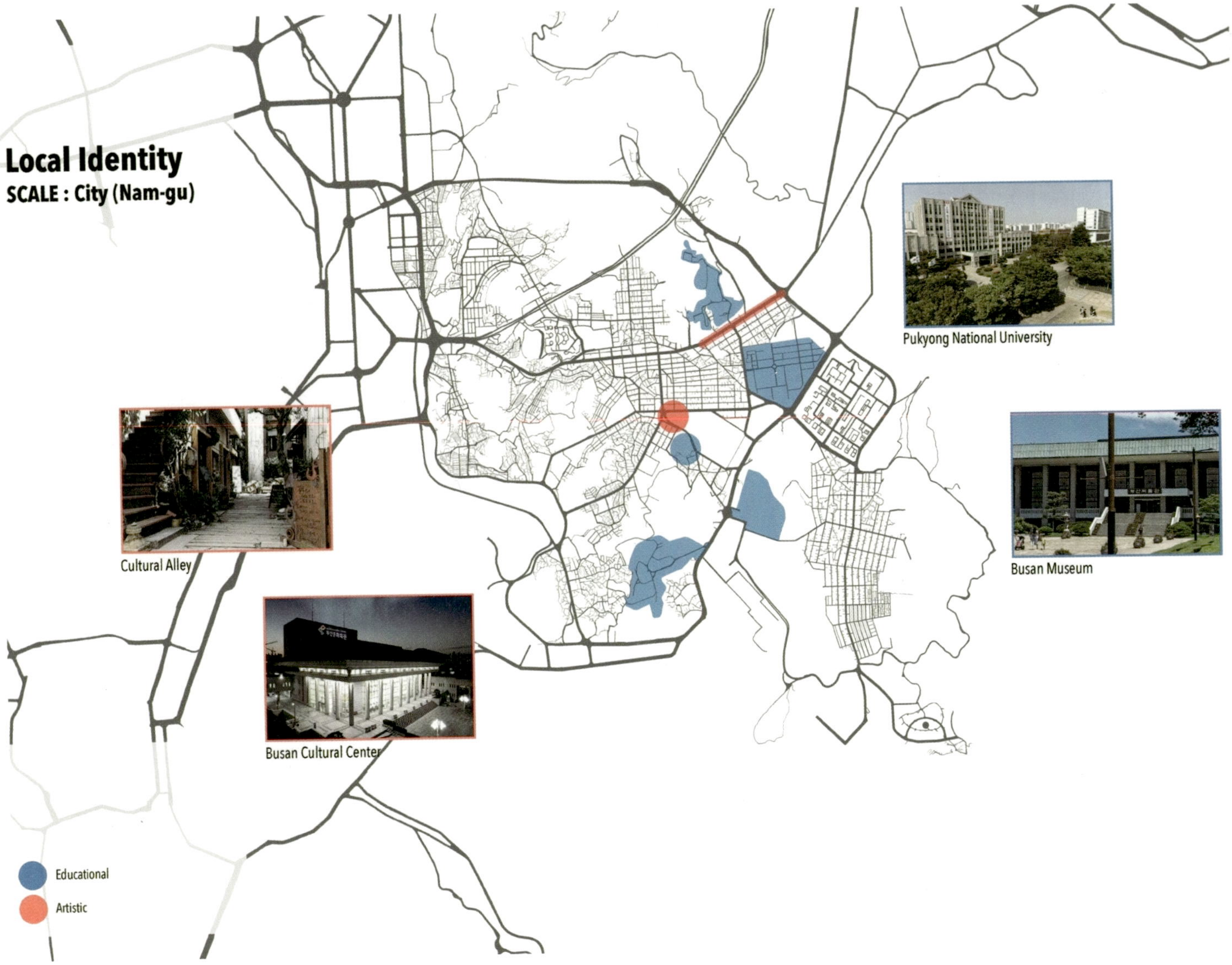

Currently, Nam-gu has four universities and 49 facilities under high school. Among them, Daehak-ro, which was formed between two universities located in Daeyeon-dong, surrounding the subway station, is the busiest space. In addition, there are large cultural facilities such as the Busan Museum and the Busan Cultural Center, so I concluded that Nam-gu is a city of education and culture.

Local Identity
SCALE : Neighborhood (Yongho-dong)

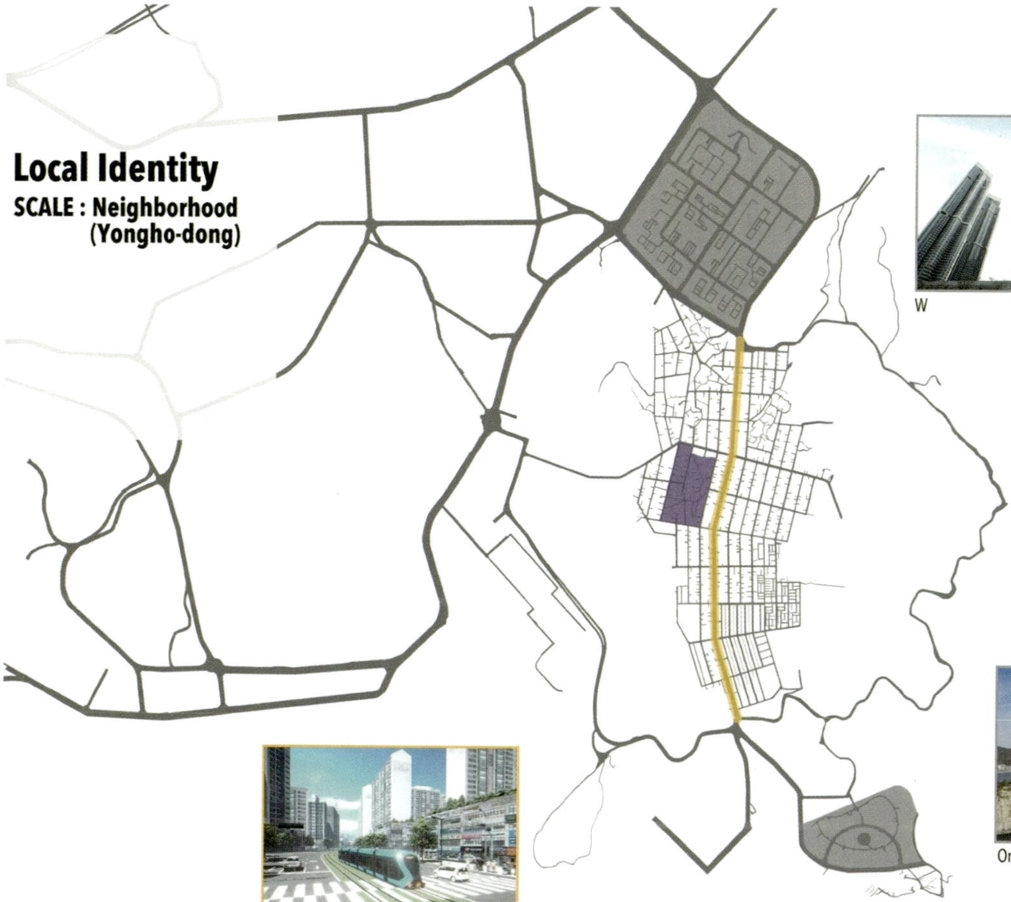

W

Oryukdo SK View

Busan Metro Oryukdo Line

- Traffic
- Gated Community
- Potential

In the late Joseon Dynasty, it was a salt pond, and after liberation, it was a poor neighborhood where patients with infectious diseases lived. Since then, it has entered the ranks of rich villages, but it is still an isolated city. Yongho-dong is also an area that has attracted attention from real estate companies due to redevelopment issues that have continued since early last year (2021).

LOCAL IDENTITY

THE SALT POND COMMUNITY

PROJECT BY
CHANGMIN PARK

PROJECT INFORMATION
Location : 126-4, Dongmyeong-ro, Nam-gu, Busan, Republic of Korea
Use : Mid-Rise Mix-Use
Building area : 4,928sqm
Total Floor area : 21,124sqm
Building coverage ratio : 58%
Floor area ratio : 249%
Structure : SRC
Floor : B2 - 6F
Height : 20.1m
The number of parking spaces : 242
The number of households : 123

The place name of Yongho-dong, Nam-gu, Busan, means that it is the place where a salt lake is located. The village, where the salt farm was located, was called 'Buen,' which means salt-grilling jar, and 'Gaega,' which means water where seawater flows in and out. Salt production began at the end of the Goryeo Dynasty, reaching its heyday after liberation, and became Busan's largest salt production site, but salt production was stopped in 1967 as reclamation projects were carried out and Dongkuk Steel was built in the 60s. Currently, traces of salt fields in the past can be identified only through the road name and place names.
In this project, I intend to form a community (salt crystal) in Yongho-dong by providing an open space (salt pond) to accommodate people (sea water) on the land with the motif of the disappeared salt pond in Yongho-dong. The motif from the spatial composition of salt fields, which is divided into three categories: reservoir, evaporation, and crystallization, is applied to the open space to maximize the inflow and stay of seawater (local residents) and provide a space to crystallize like salt. It also suggests a type of MRMU near work where various people can mix with each other and participate in local community-level activities and participation by reviving the memory of the salt pond in Yongho-dong, a bank that traps seawater and used to bake salt.

부산광역시 남구 용호동의 지명은 염전호가 있는 곳이라는 의미이다. 염전이 있었던 이 마을은 소금을 굽는 동이(盆)를 뜻하는 '분'에 바닷물이 드나드는 곳의 물을 뜻하는 '갯가'에서 분개(盆浦)라 불렀다. 고려 말부터 소금을 생산하기 시작했고 해방 이후 전성기를 맞아 부산 최대의 소금 생산지가 되었지만, 60년대에 매립 사업이 진행되고 동국제강 등이 세워지면서 1967년을 마지막으로 소금생산이 중지되었다. 현재는 도로명과 지명을 통해서만 과거 소금밭의 흔적을 파악할 수 있다.
이 프로젝트에서 용호동의 사라진 염전을 모티브로 대지에 사람들(바닷물)을 수용할 오픈 스페이스(염전)을 제공하여 용호동 지역의 공동체(소금결정)를 형성하고자 한다. 저수지, 증발지, 결정지의 세 단계로 구분되는 염전의 공간구성에서 얻은 모티브를 오픈스페이스에 적용하여 바닷물(지역주민)의 유입과 머무름이 극대화되고 결국에는 소금처럼 결정화(커뮤니티화) 될 수 있는 공간을 제공하고자 한다. 또한 바닷물을 가두는 둑이자 소금을 굽는 동이였던 용호동의 염전에 대한 기억을 되살려, 다양한 사람들이 서로 섞이고 지역 공동체 단위의 행위와 참여가 일어날 수 있는 직주근접의 MRMU 유형을 제안한다.

CONCEPT INSPIRATION

CONCEPT MODEL

CONCEPT METAPHOR

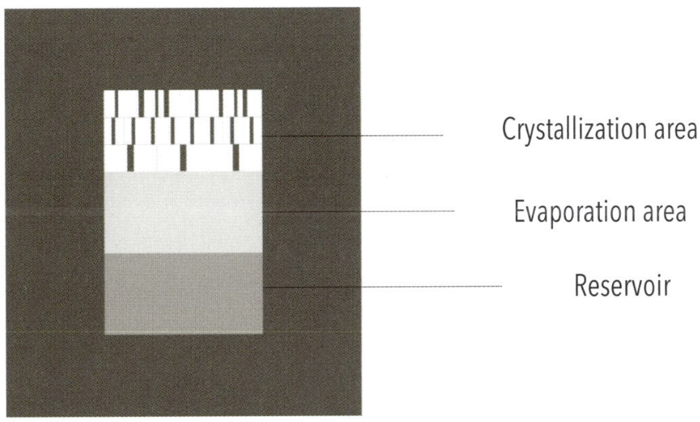

TIMELINE OF SALT POND IN YONGHO-DONG

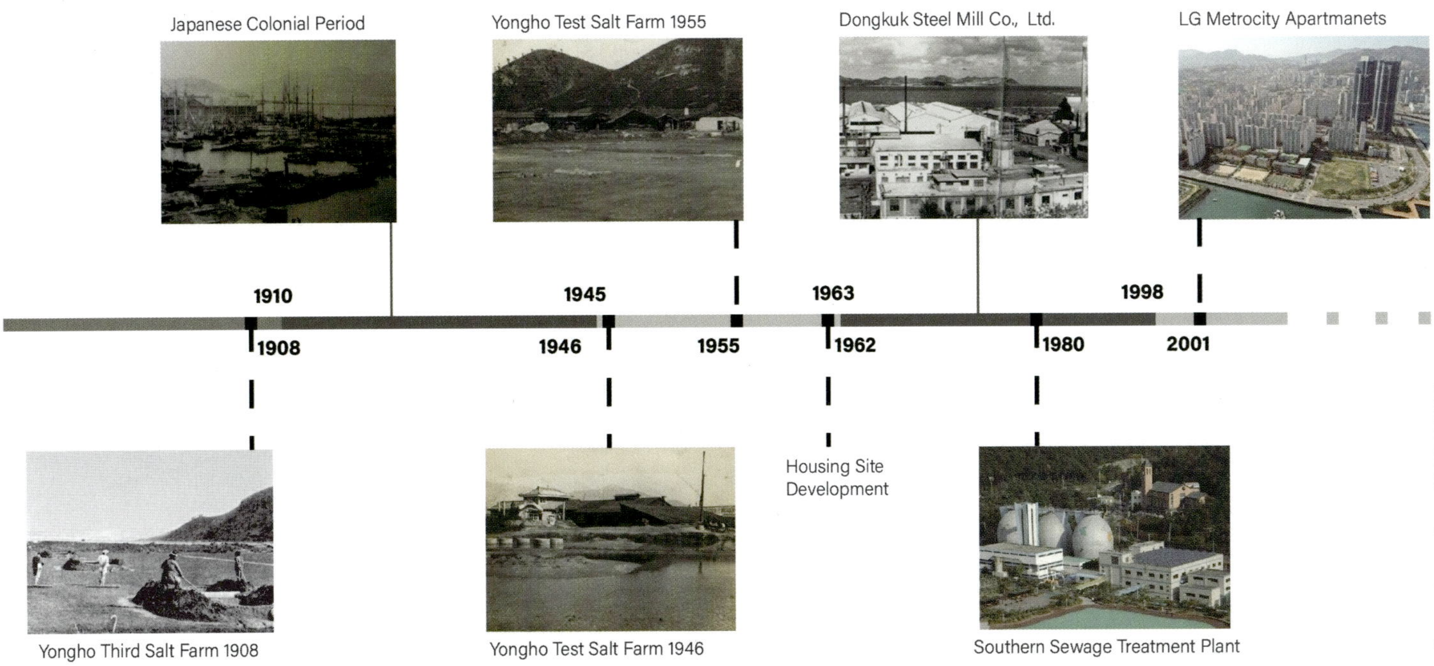

SITE ANALYSIS

Readjustment of Road Sections

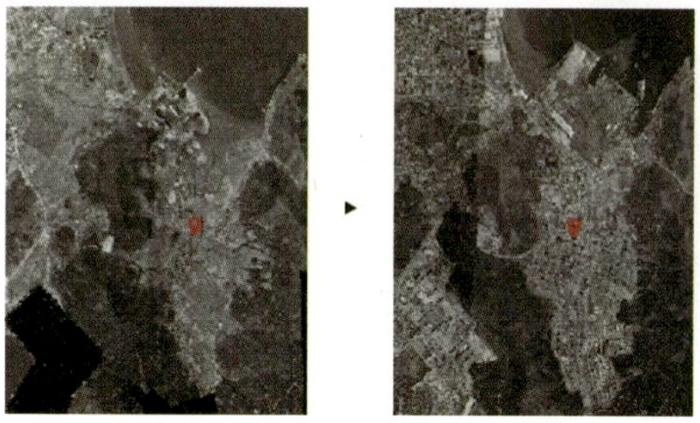

1967 ▸ 1975

Major Element

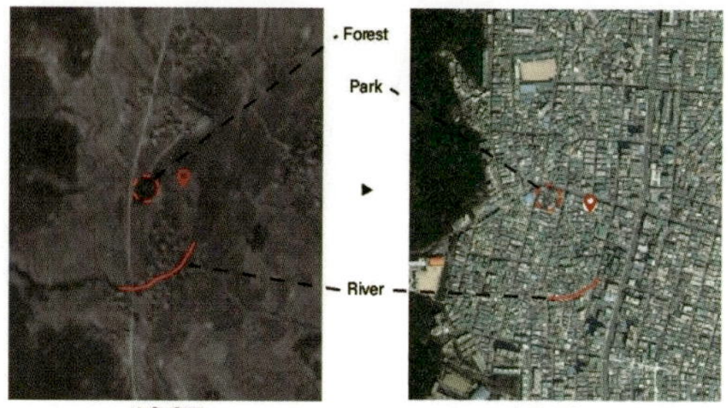

1967 ▸ 2019

Photographs in 1952

1. Forming a village around the river
2. Development in the north-south direction
3. The arrangement of the houses leans southwest

- The accessibility of the workplace (salt pond) and the movement path to the entrance of the village

- Gwangalli Sea Scenery

- Axis facing south

BUILDING IN CHRONOLOGICAL ORDER

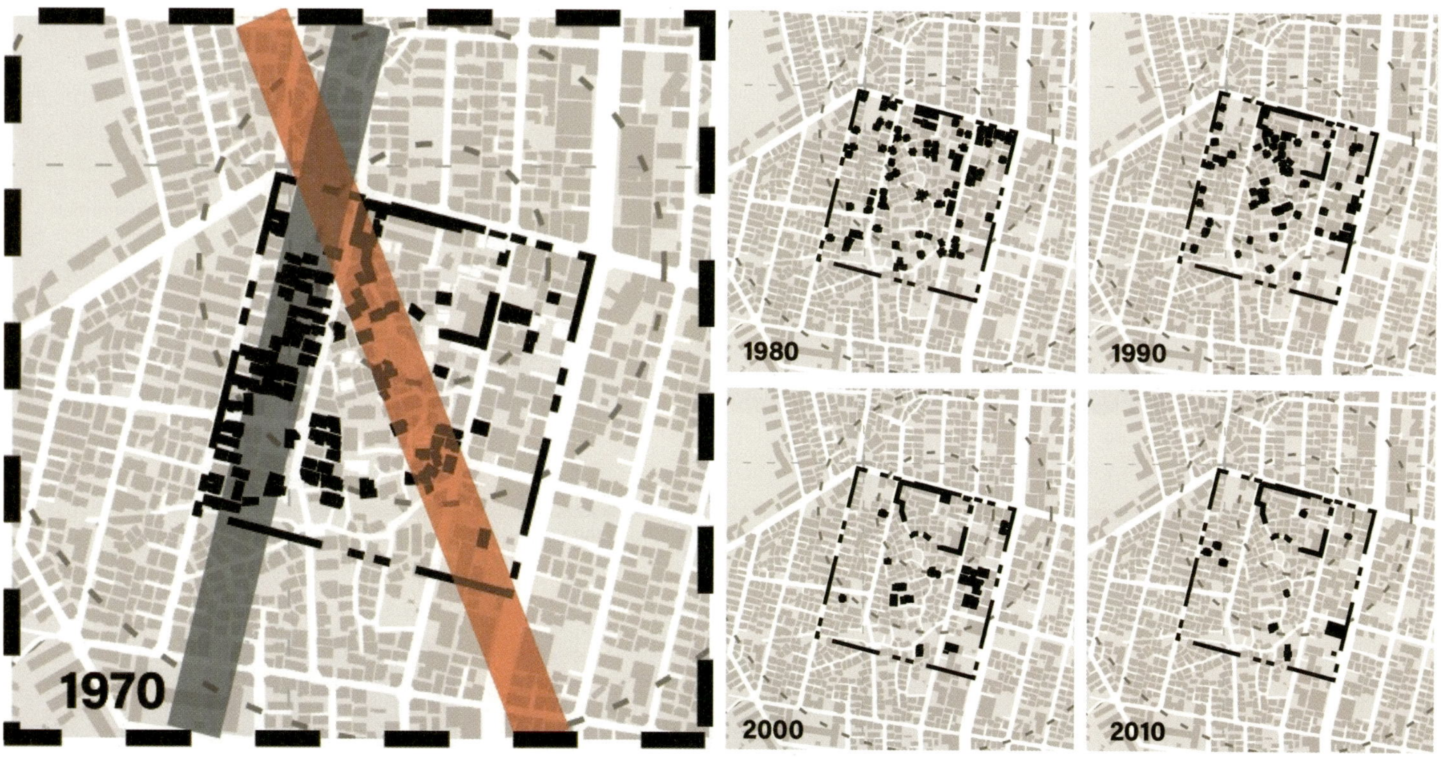

URBAN BLOCK FABRIC ANALYSIS

Grid Blocks

Vehicle-Oriented

Direct Movement

Fast-moving Space

High Degree of Integration

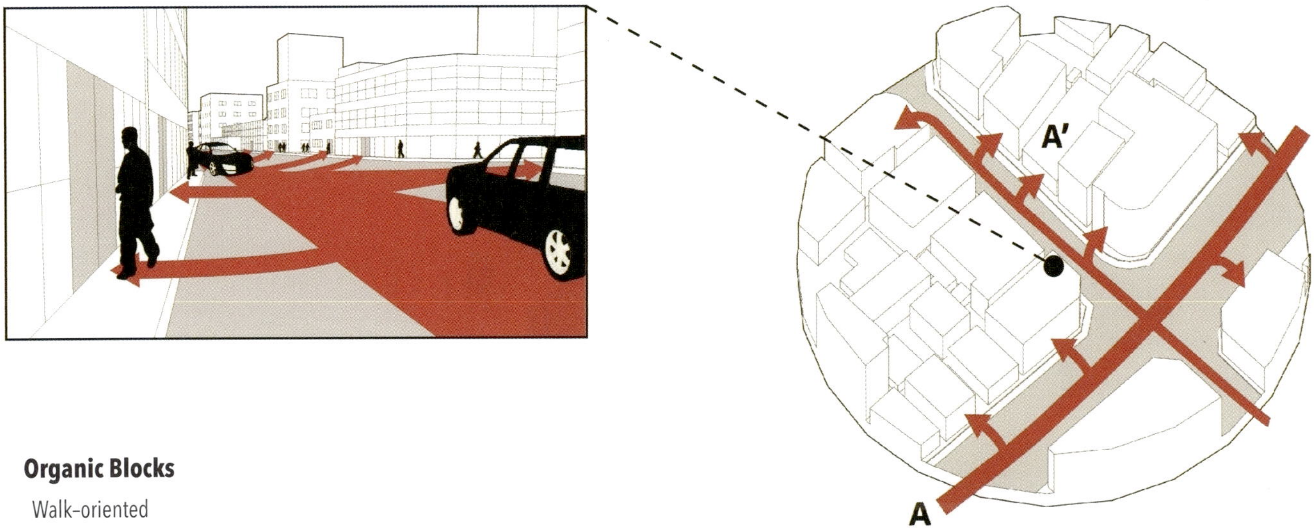

Organic Blocks

Walk-oriented

Indirect Movement

Slow-moving Space

High Degree of Spatial Depth

Influence on Natural Factors

Branch-like Structure

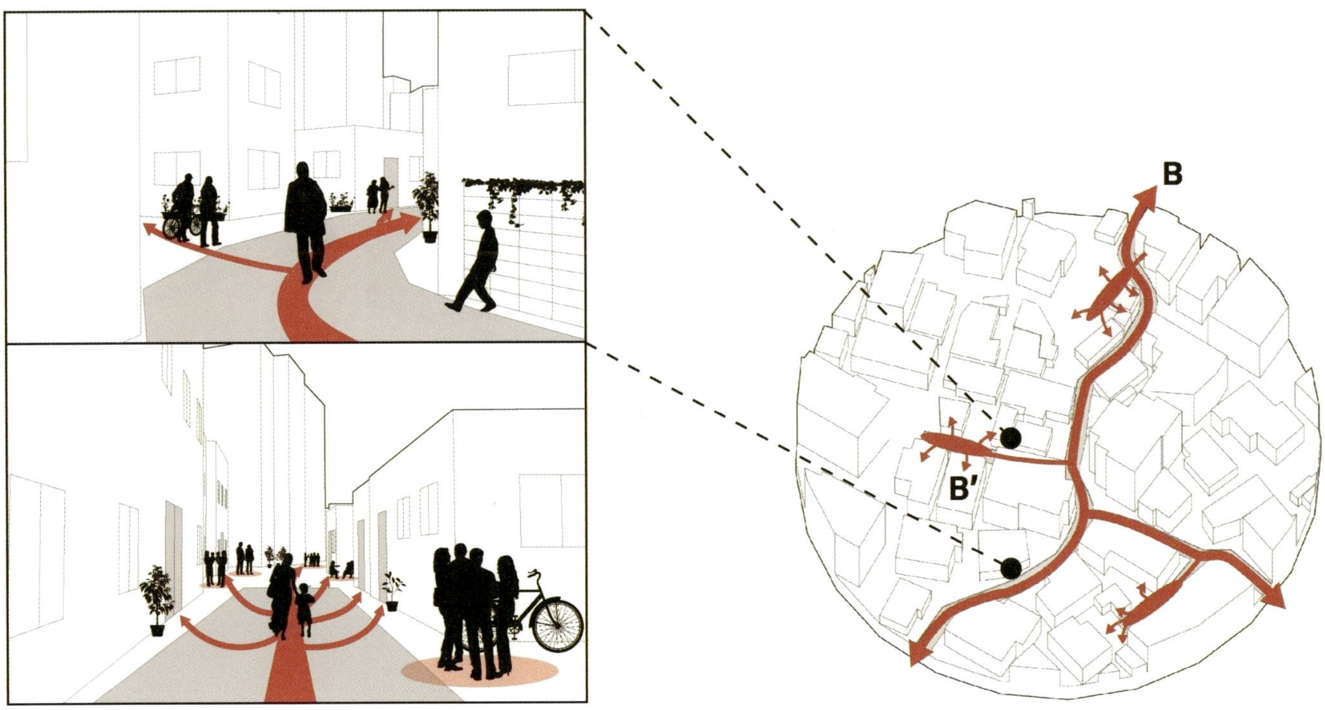

PERCENTAGE OF NUMBER OF HOUSEHOLD MEMBERS

RESIDENTIAL UNIT

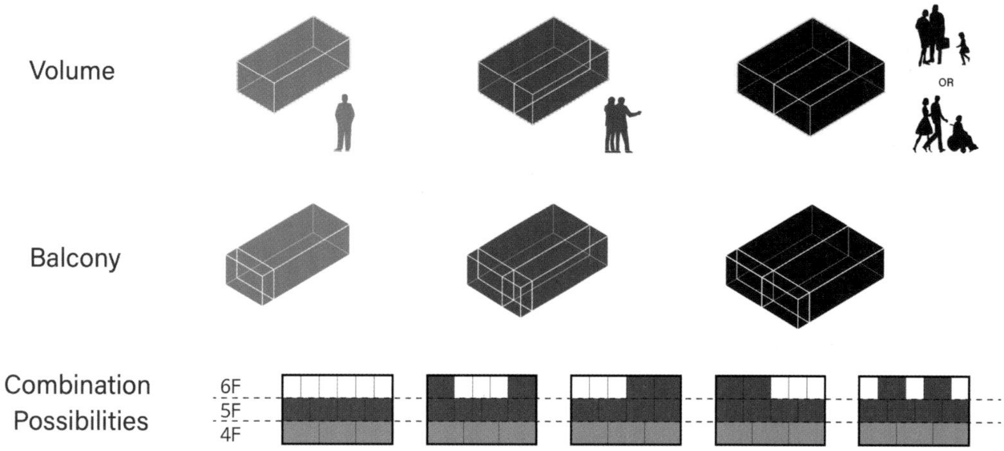

CORRIDOR PROCESS

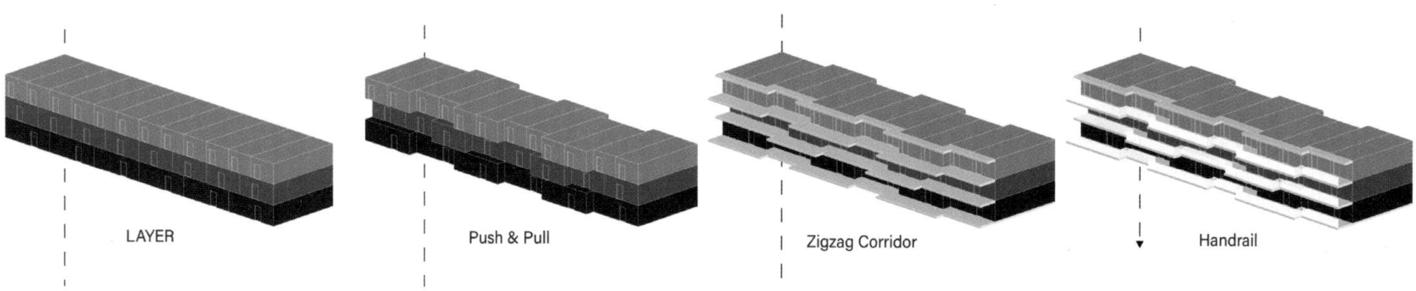

ARCHITECTURAL EQUIPMENT

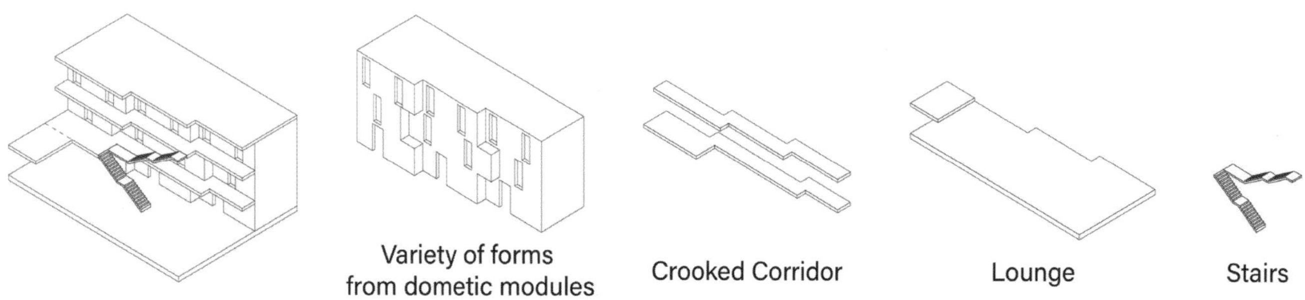

MASS PROCESS

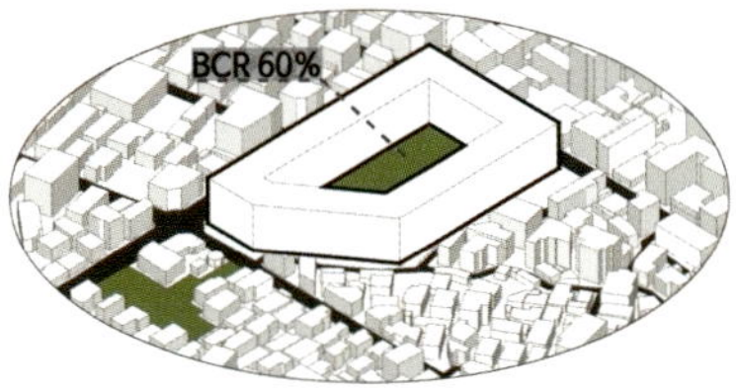

Boundary

Maximizing the Building Coverage Ratio
Vitalize as a commercial street
An enclosed interior space

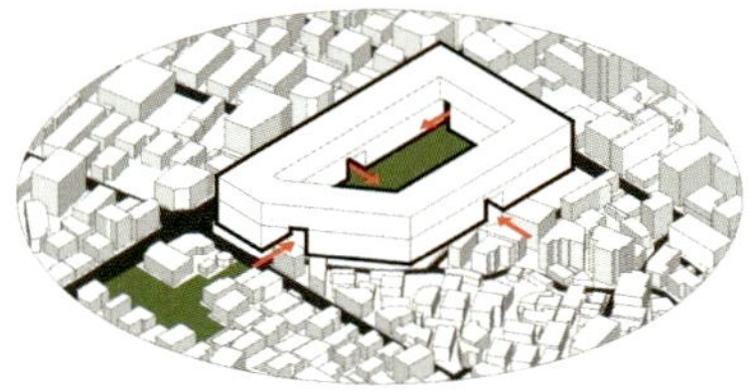

Opening

Connecting to Open Space in the City
Connecting to the Park
Internally oriented space

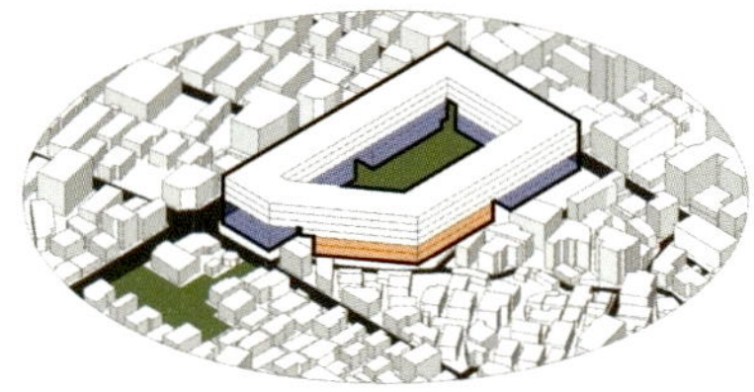

Layering

Low-rise Commercial Facilities
Low-rise Public Facilities for Community
Mid-level residential facilities

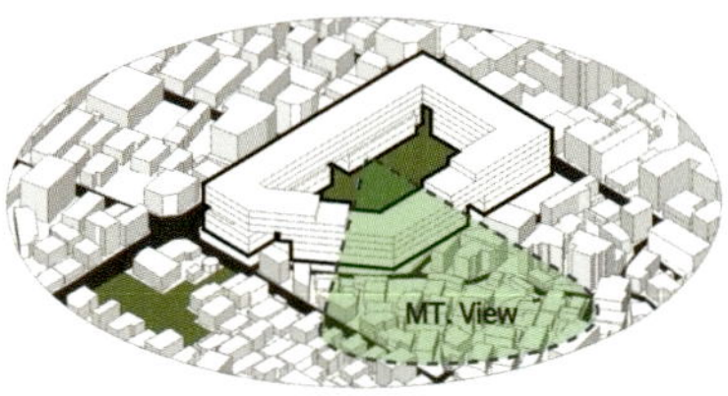

Scale

Scale for mountain views
Harmonization with Peripheral Buildings
Sunlight Increase to Open Space

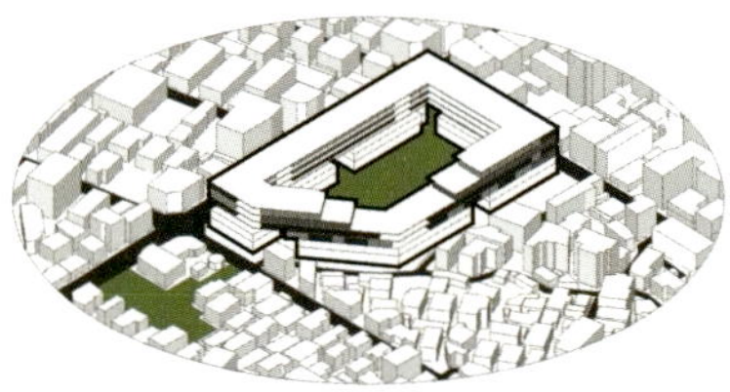

Domestic

Residential Unit Placement
Low-rise - housing for 3 to 4 people
High-rise - housing for 1 or 2 persons

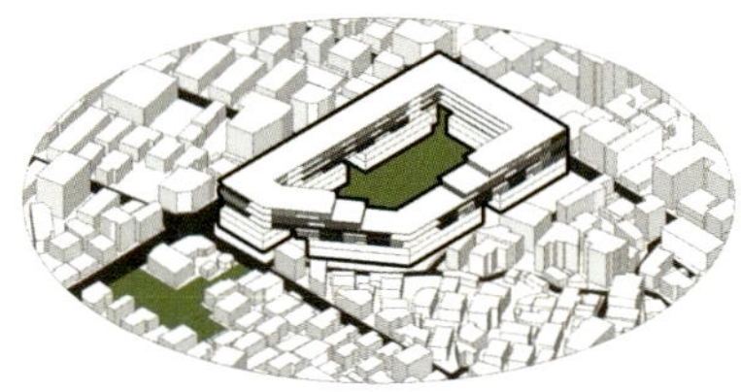

Unevenness

Reflection of Architectural Devices
the formation of an uneven corridor
Diversification of the Facade

MRMU PROGRAM

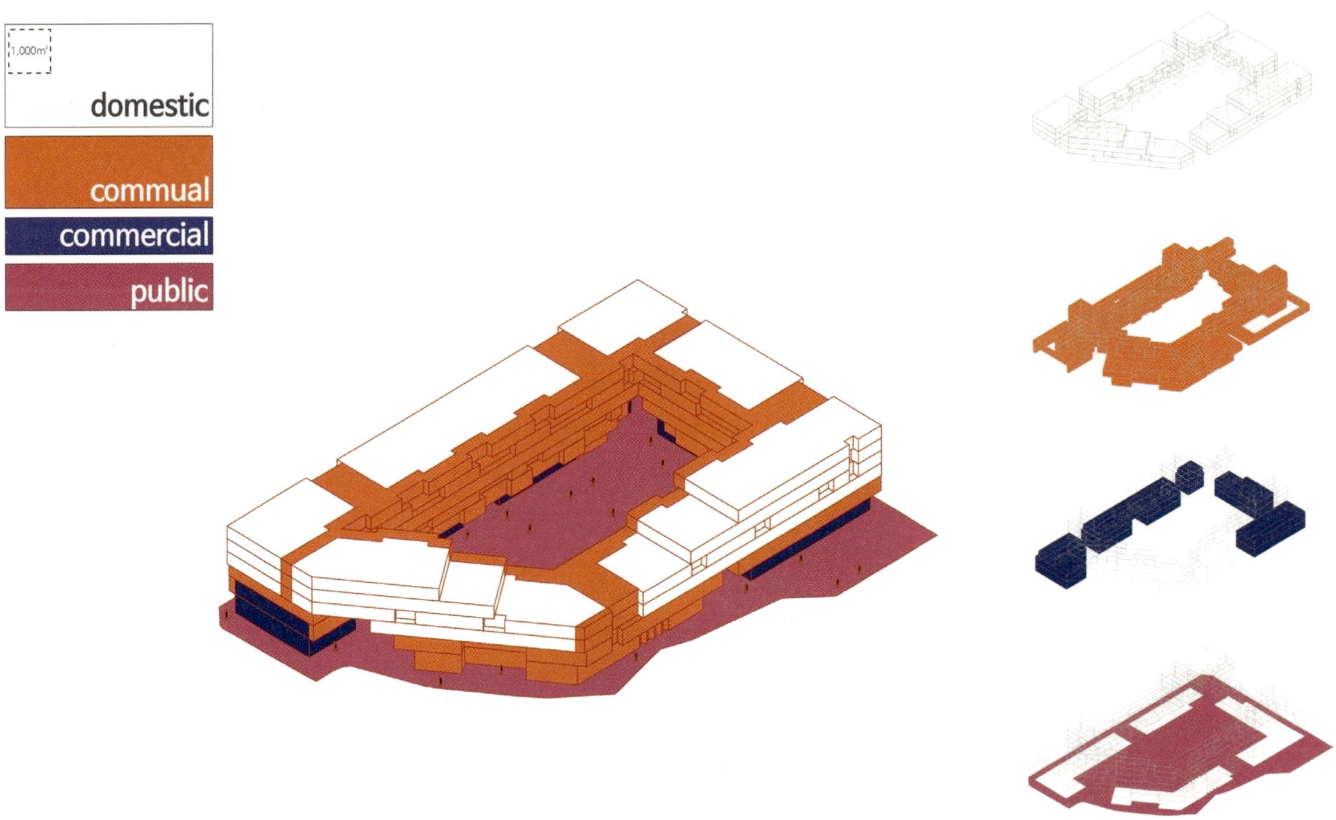

CORRIDOR EXPERIENCE SEQUENCE

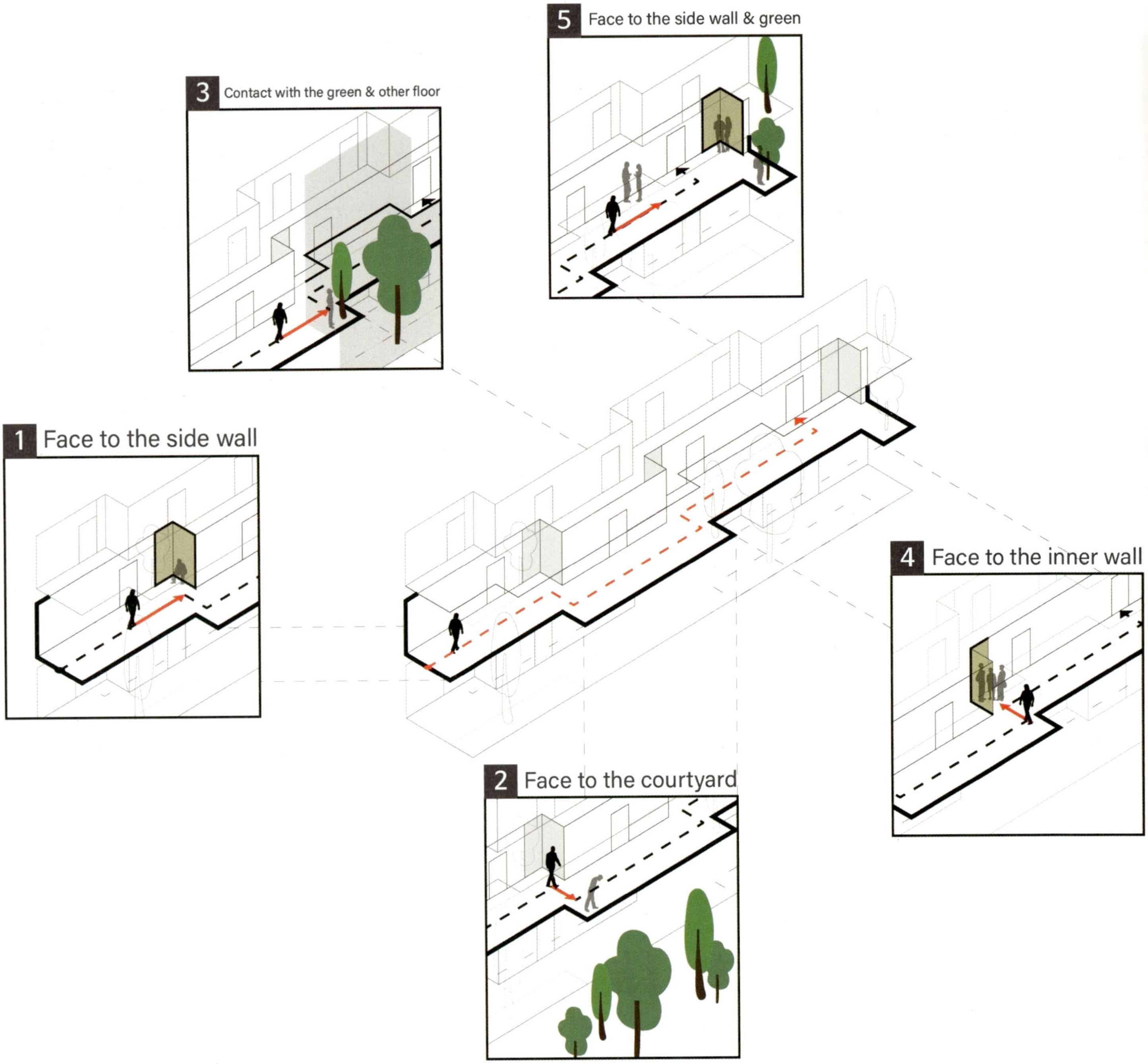

EQUIPMENT STUDY MODEL

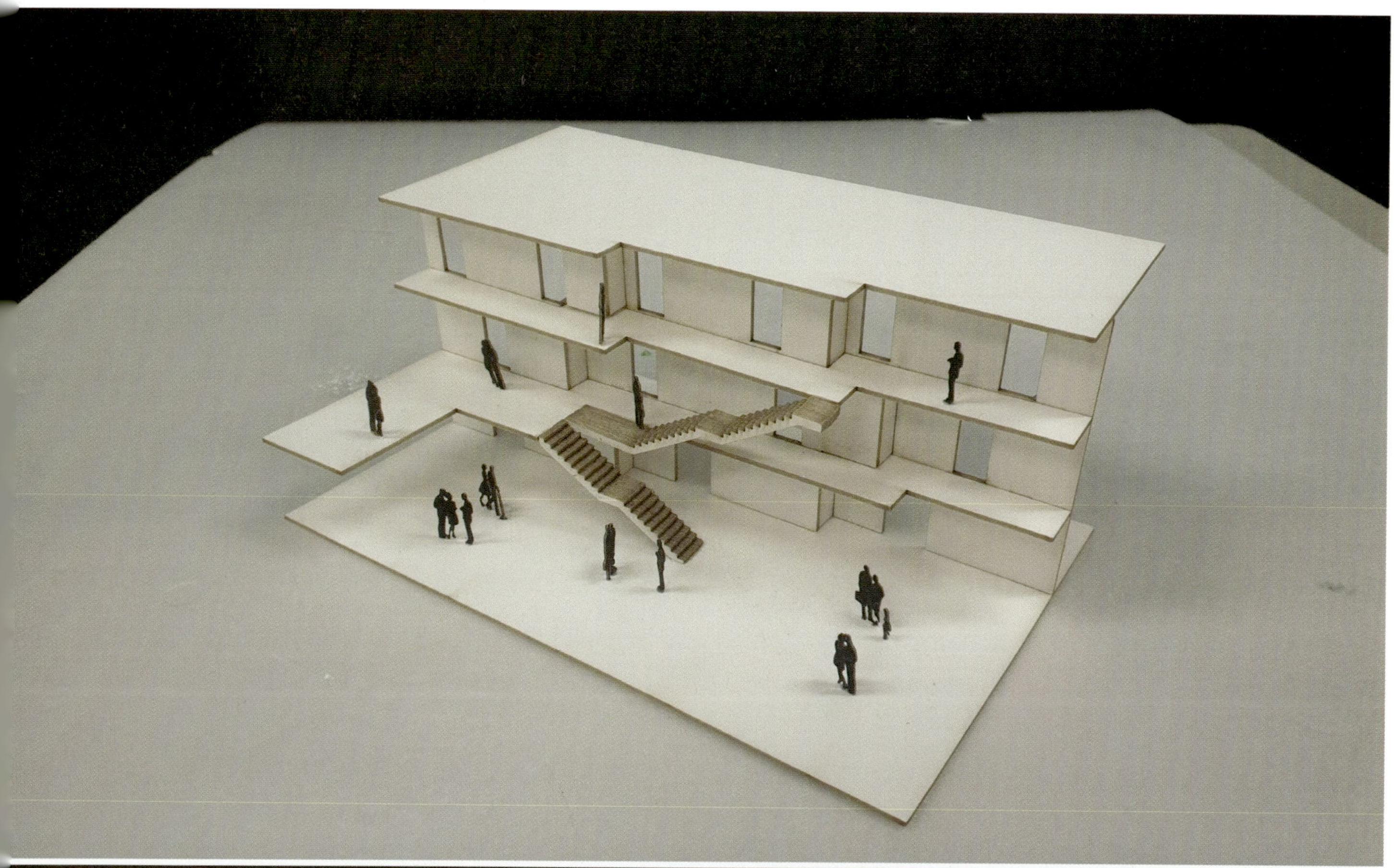

Main Entrance

Corridor

Corner of the Corridor

CORNER FUNITURE MODULE

CFM (S) – Shelf

CFM (M) – Bench for 2

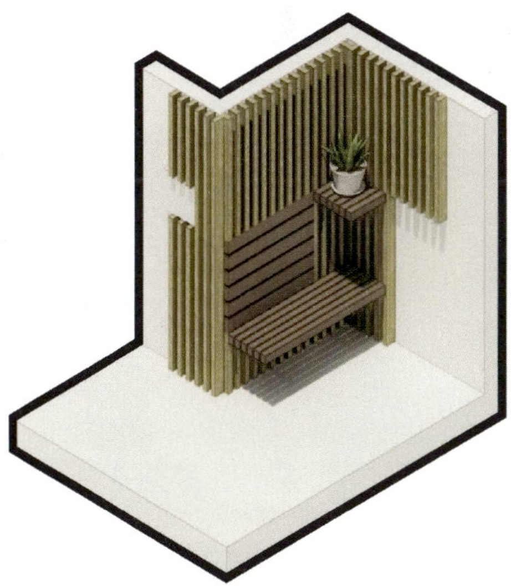

CFM (L) – Bench for 5

SECTION MODEL

SITE PLAN

PLAN

1F

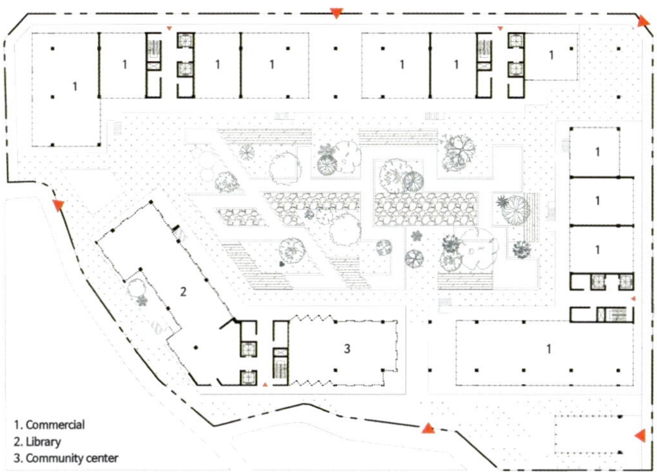

B1F

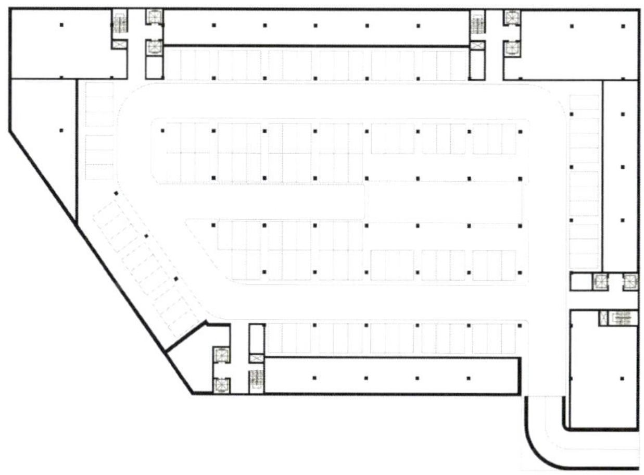

2F

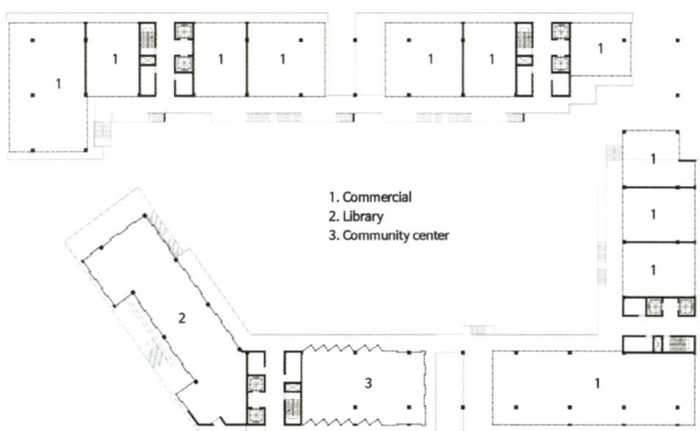

B2F

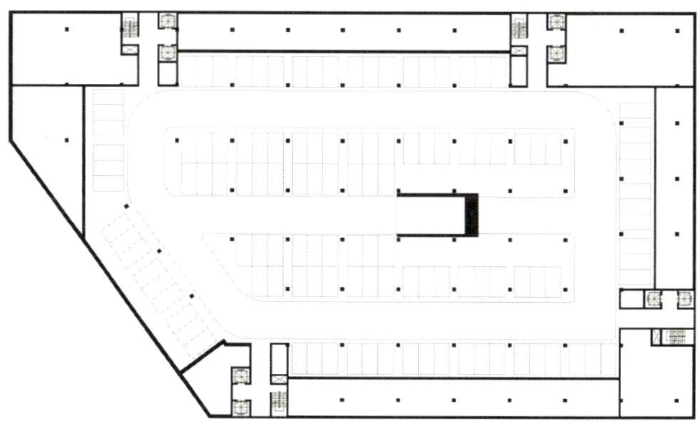

ELEVATION

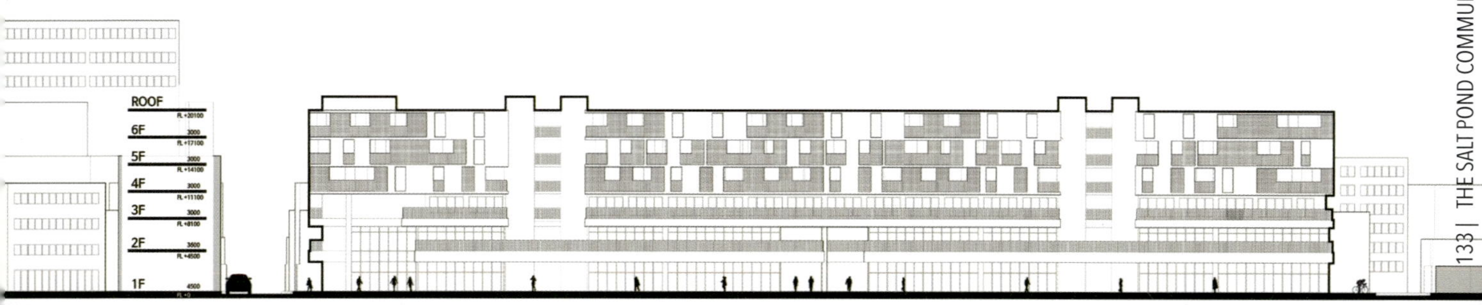

ORTH-EAST

3F

5F

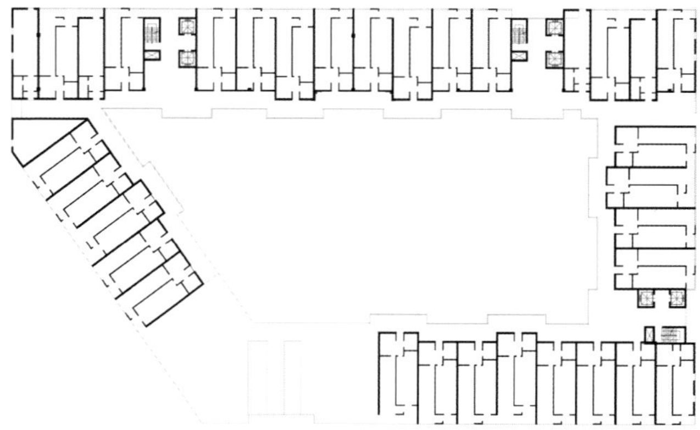

4F

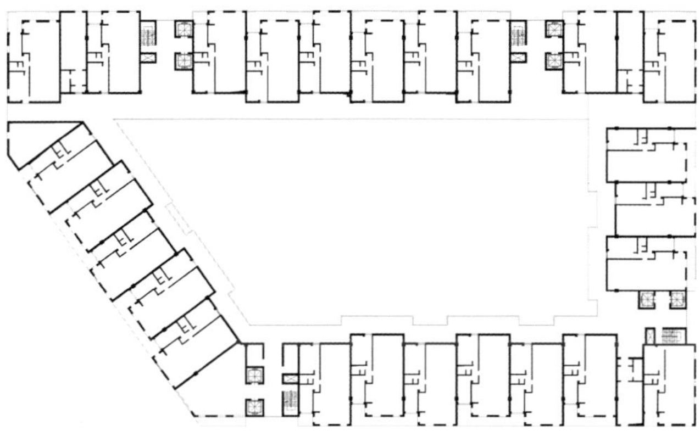

6F

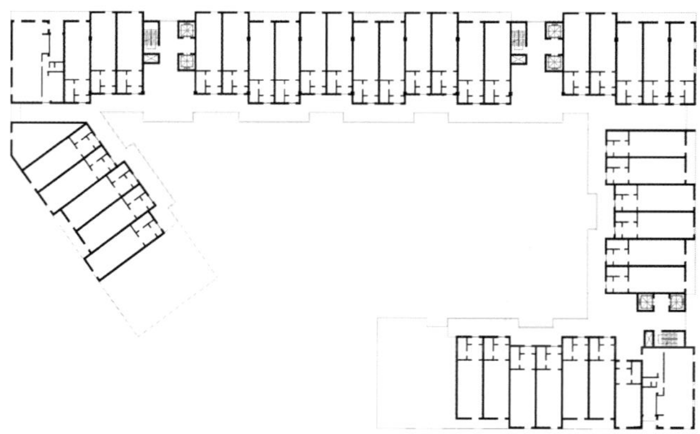

SECTION

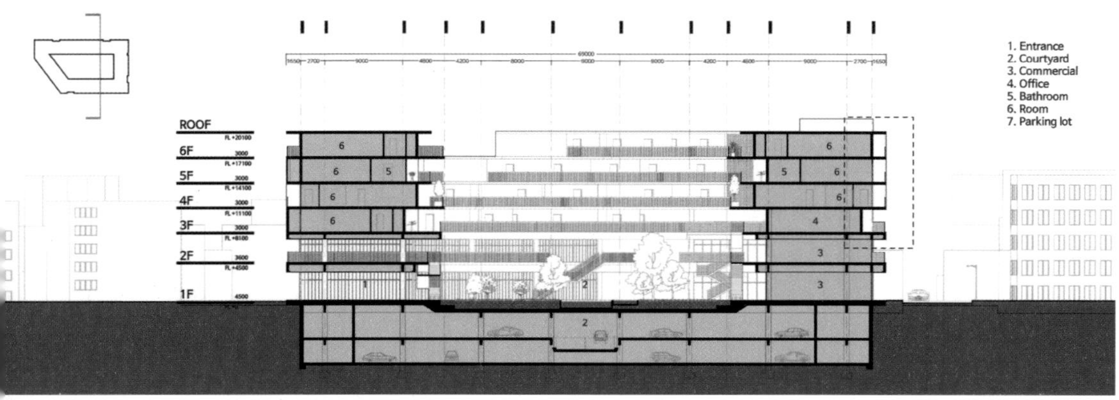

SECTION DETAIL

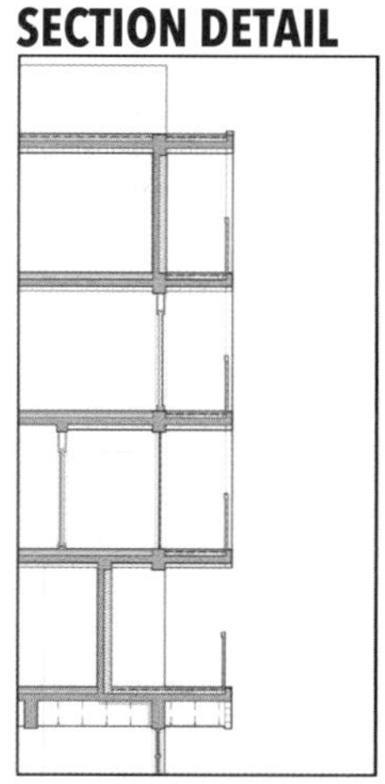

ELEVATION

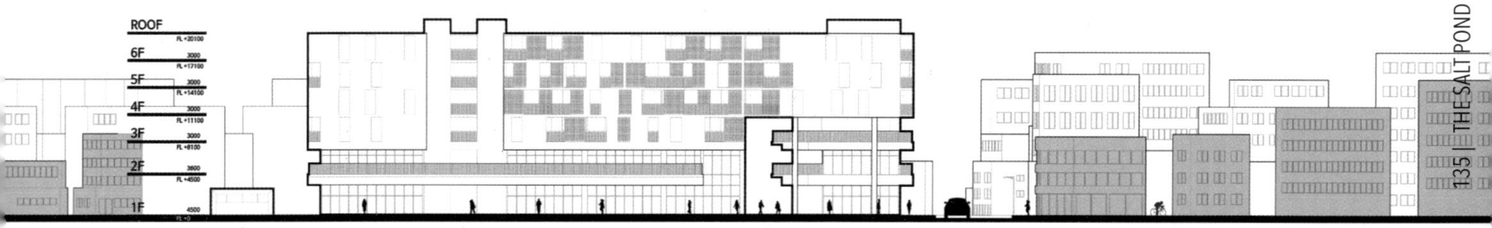

ST-SOUTH

JANGJEON-DONG, GEUMJEONG-GU, BUSAN

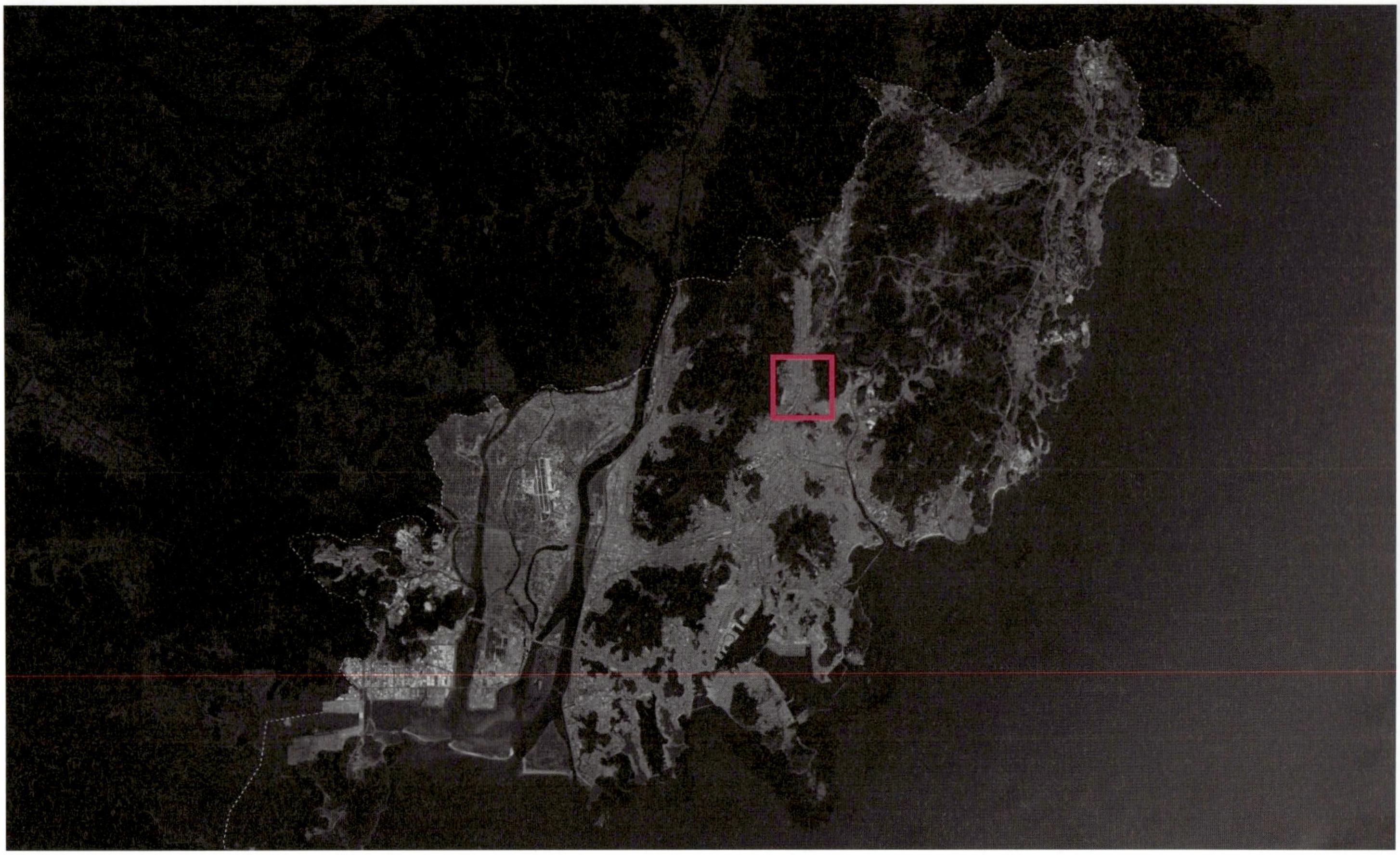

RESEARCH BY
HEONJU LEE
RYULHYEON KIM
JUHAE YU

Jangjeon-dong is a legal dong in Geumjeong-gu, Busan. It is divided into Jangjeon 1-2, administrative dong. Most of the residential areas were built against the backdrop of Geumjeongsan Mountain, and the public transportation route was convenient, so it met the optimal housing requirements. In particular, young cultural commercial districts are formed around the connection between the northern gate of Pusan National University and Jangjeon Station of the urban railway, and demand is high.

장전동(長箭洞)은 부산광역시 금정구의 법정동이다. 행정동 장전1~2동으로 나뉜다. 금정산을 배경으로 대부분 주거지가 조성되어 있고, 대중교통 노선이 편해 최적의 주거 요건을 갖추었다. 특히 부산대학교 북문과 도시철도 장전역의 연계를 중심으로 젊은 문화 상권이 형성되어 수요가 높다.

© Google Earth

Photo by Sam Balye on Unsplash

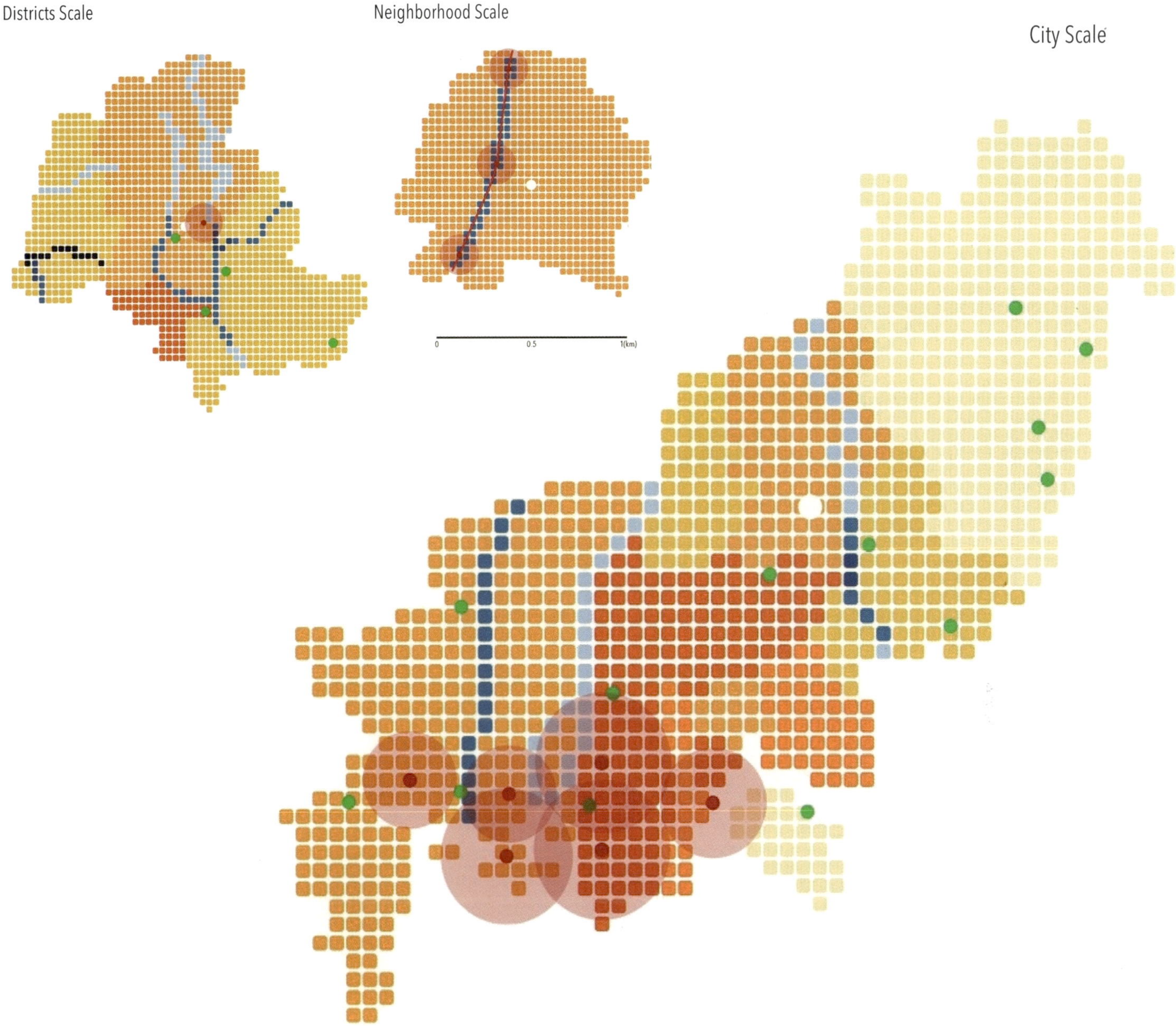

We tried to analyze pollution levels and explore the causes by focusing on the atmosphere and water. Air pollution levels used NO2 indicators from automobiles and factories. We thought this was the most urban element.

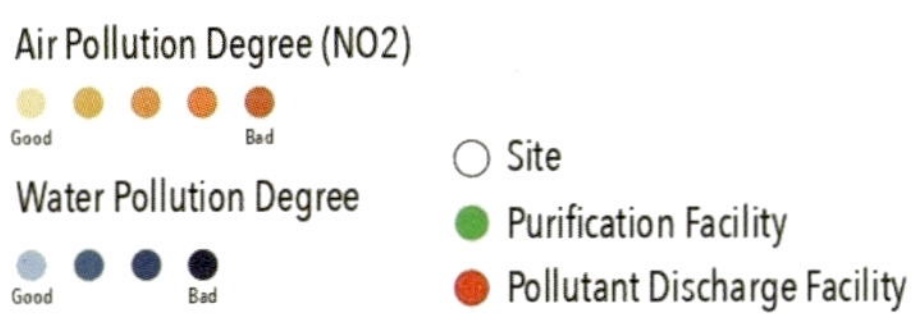

ENVIRONMENT

Mass Transit Scale

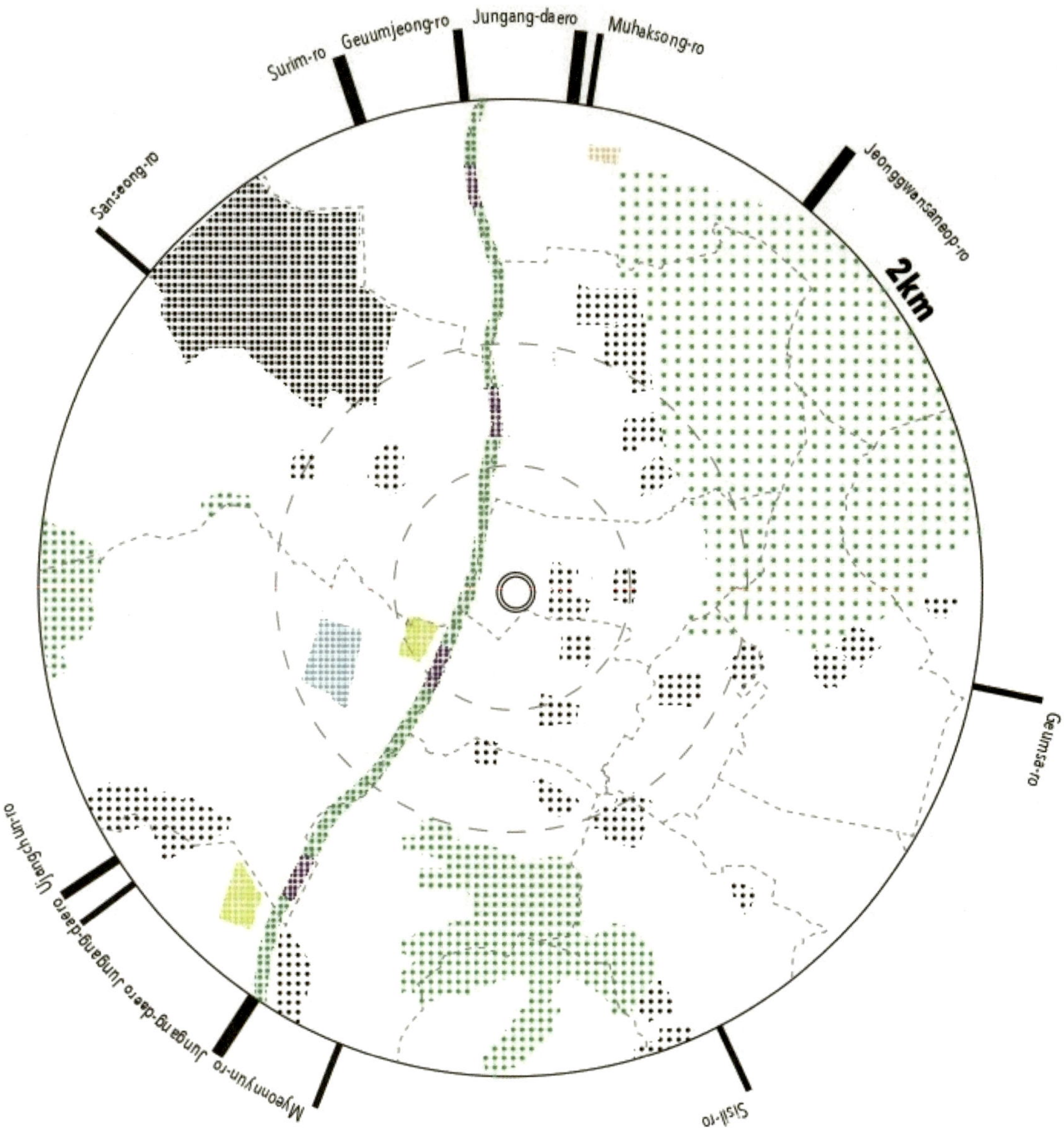

● Vegetation ● Traffic
● Education ● Library
● Medical ● Tourism
● Site ○ Supermarket
▬ Access to the road

The infrastructure around the site and insufficient infrastructure were explored in transportation, education, medical care, and vegetation. The type of facility was expressed through the color of the point, the density of the point, and the accessible road through the outline.

INFRASTRUCTURE

Walking Scale 1

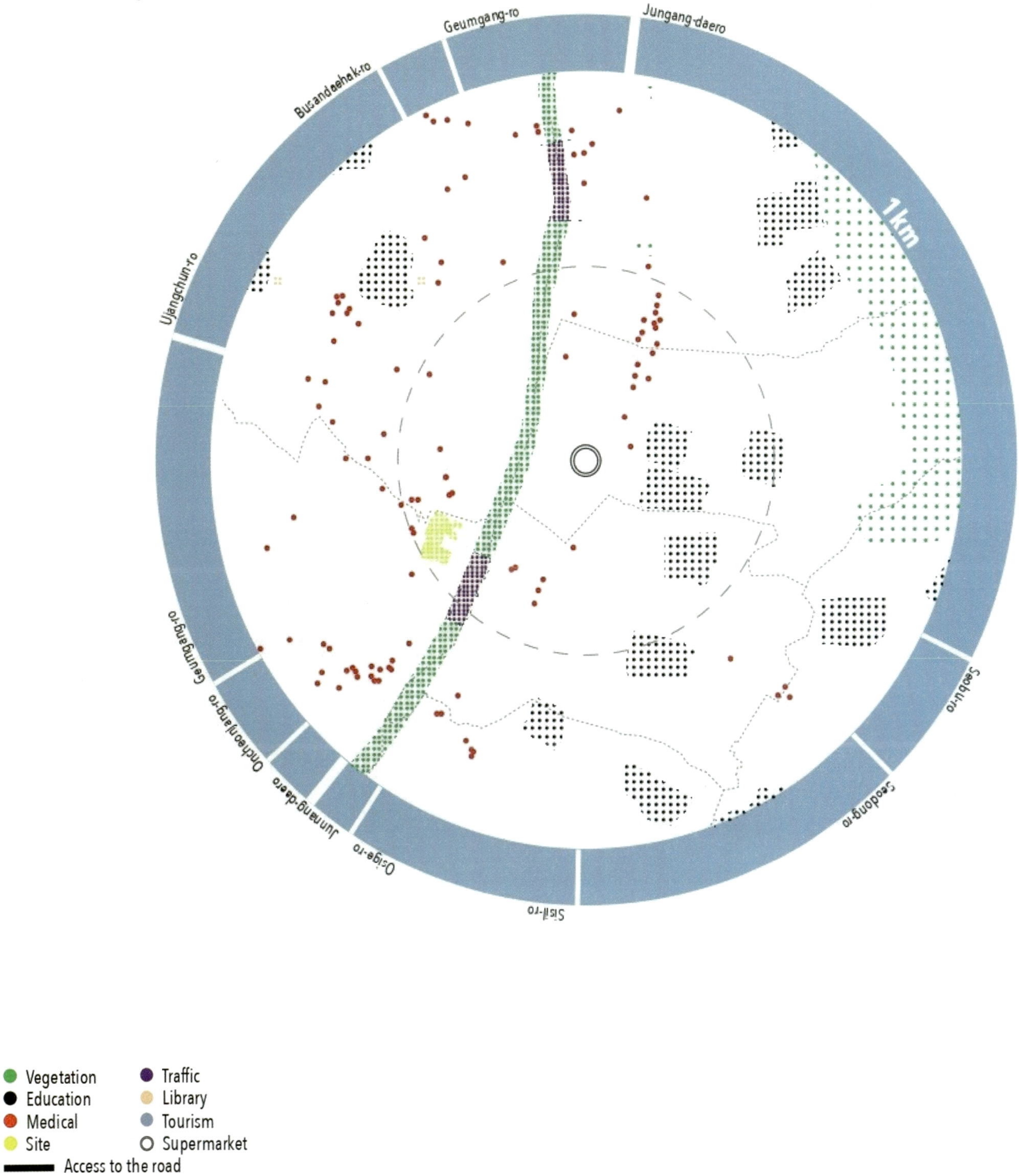

INFRASTRUCTURE

Walking Scale 2

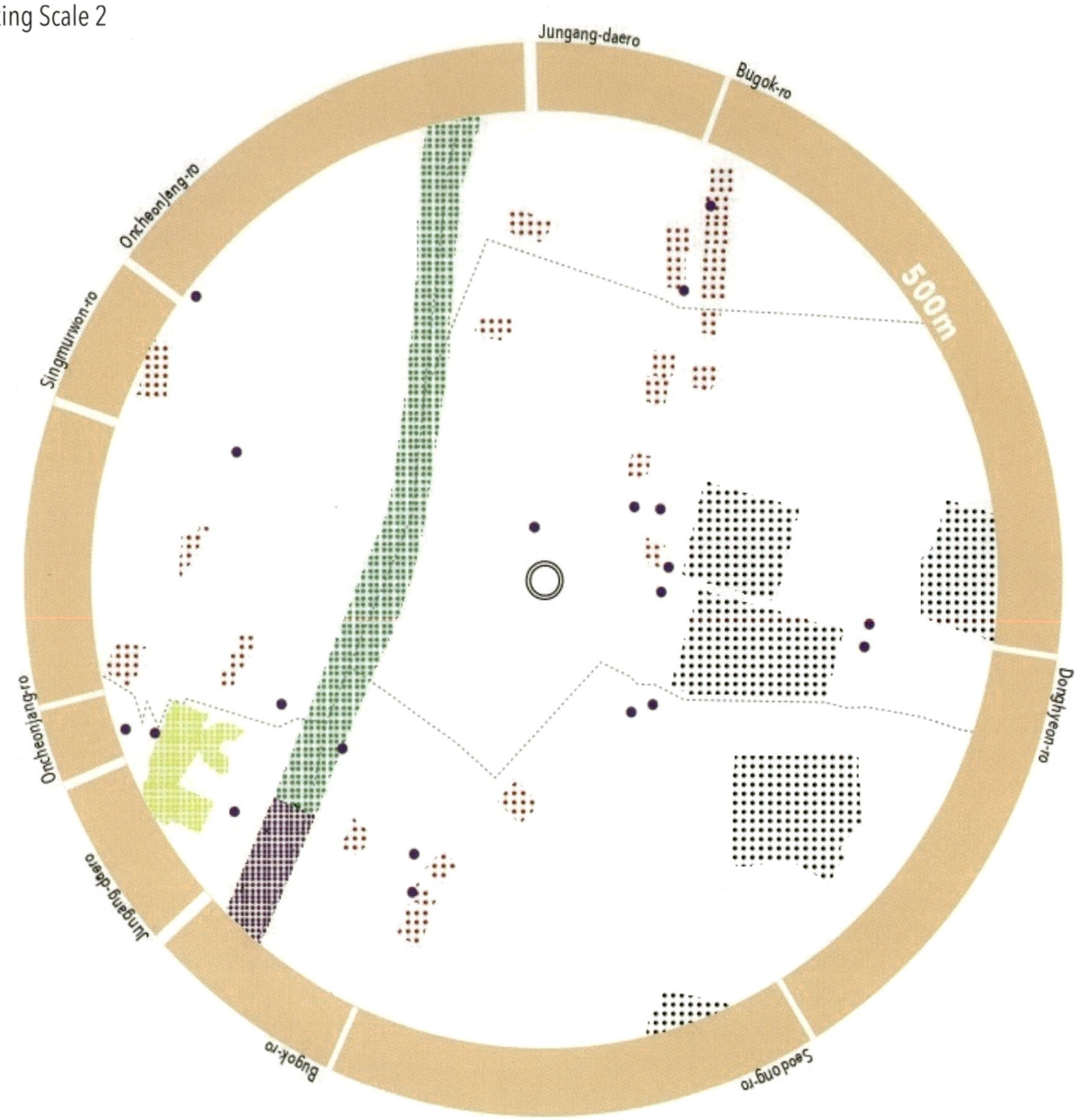

- Vegetation
- Education
- Medical
- Site
- Traffic
- Library
- Tourism
- Supermarket
- Access to the road

INFRASTRUCTURE

City Scale

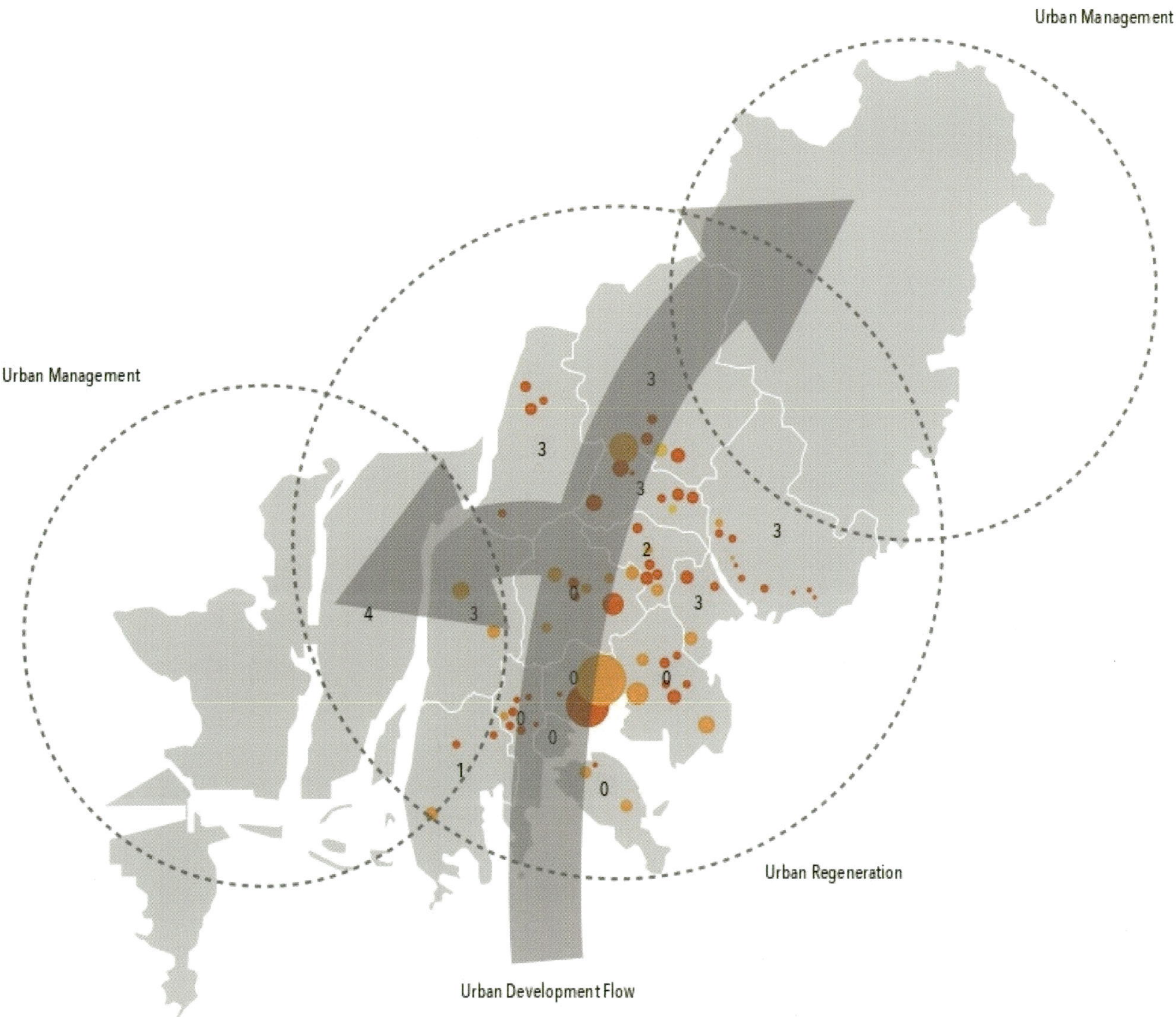

- Propulsion
- Progressing
- Completion

We explored the redevelopment flow from Busan to the site. The distance was expressed in the thickness of the line and the height of the building in a fan shape. The number of city scale is the order in which each region was formed, and through this, it was compared and analyzed with the status of redevelopment.

REDEVELOPMENT FLOW

Distircts Scale

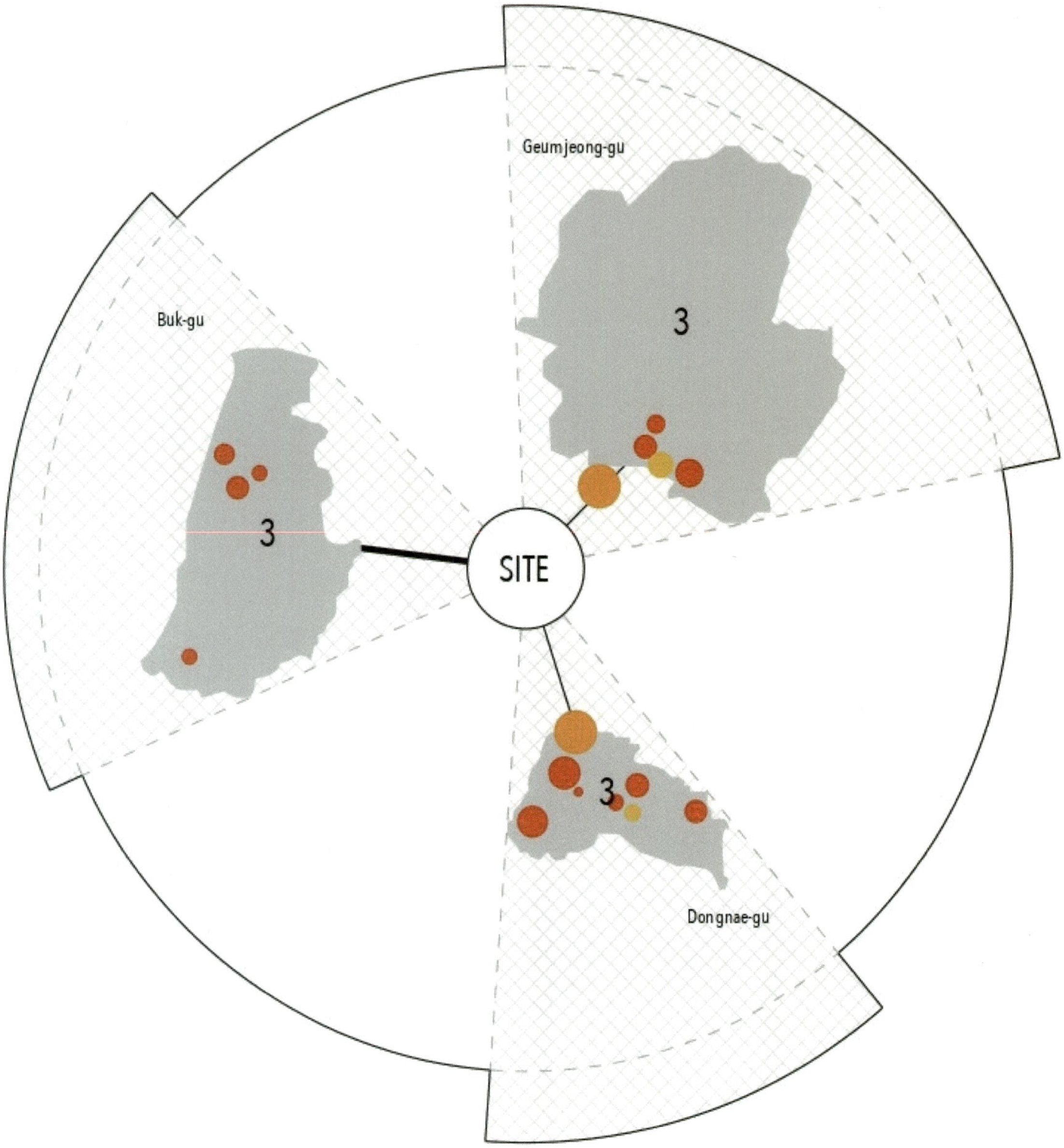

Propulsion
Progressing
Completion

REDEVELOPMENT FLOW

Neighborhood Scale

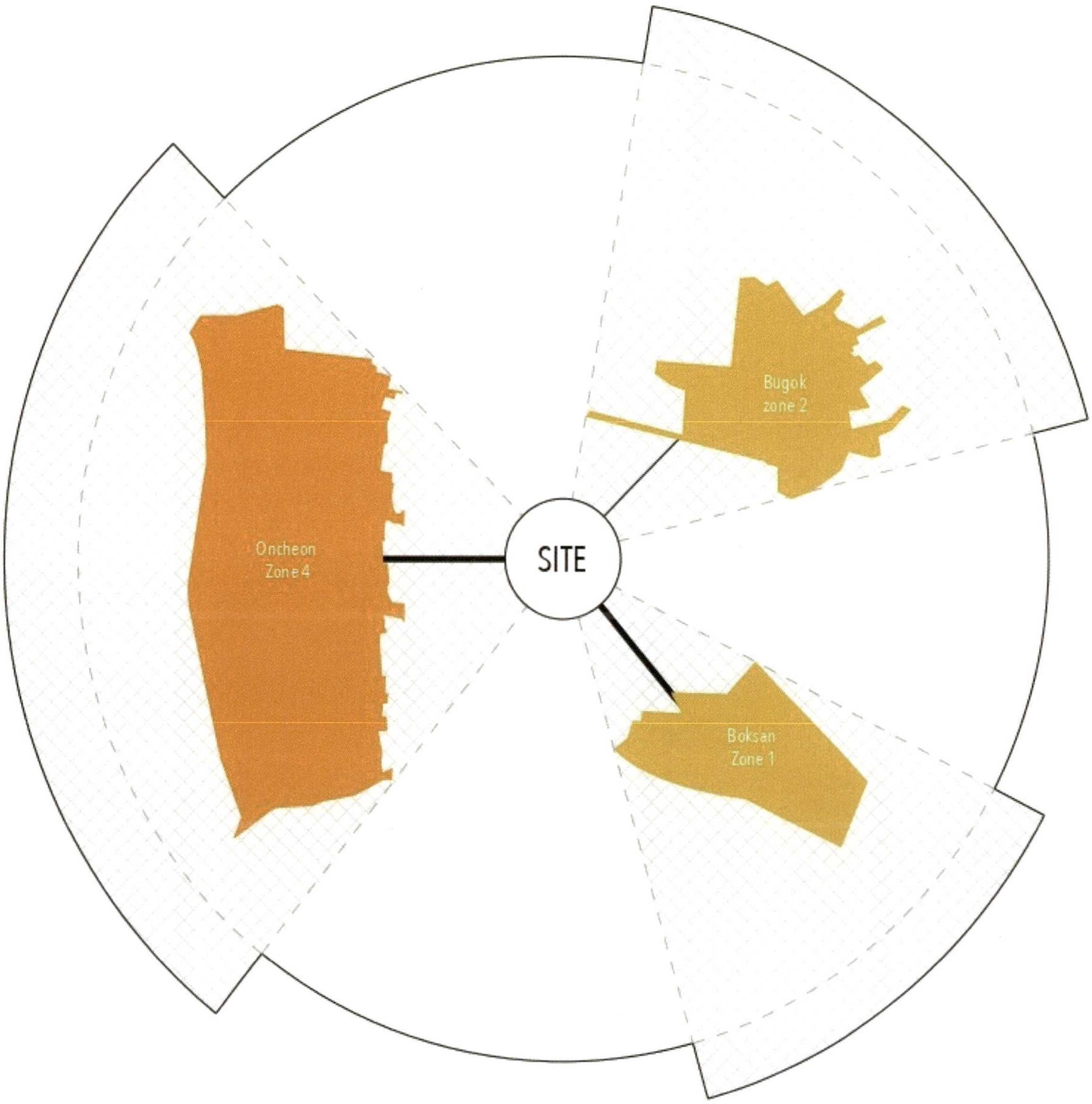

Propulsion
Progressing
Completion

REDEVELOPMENT FLOW

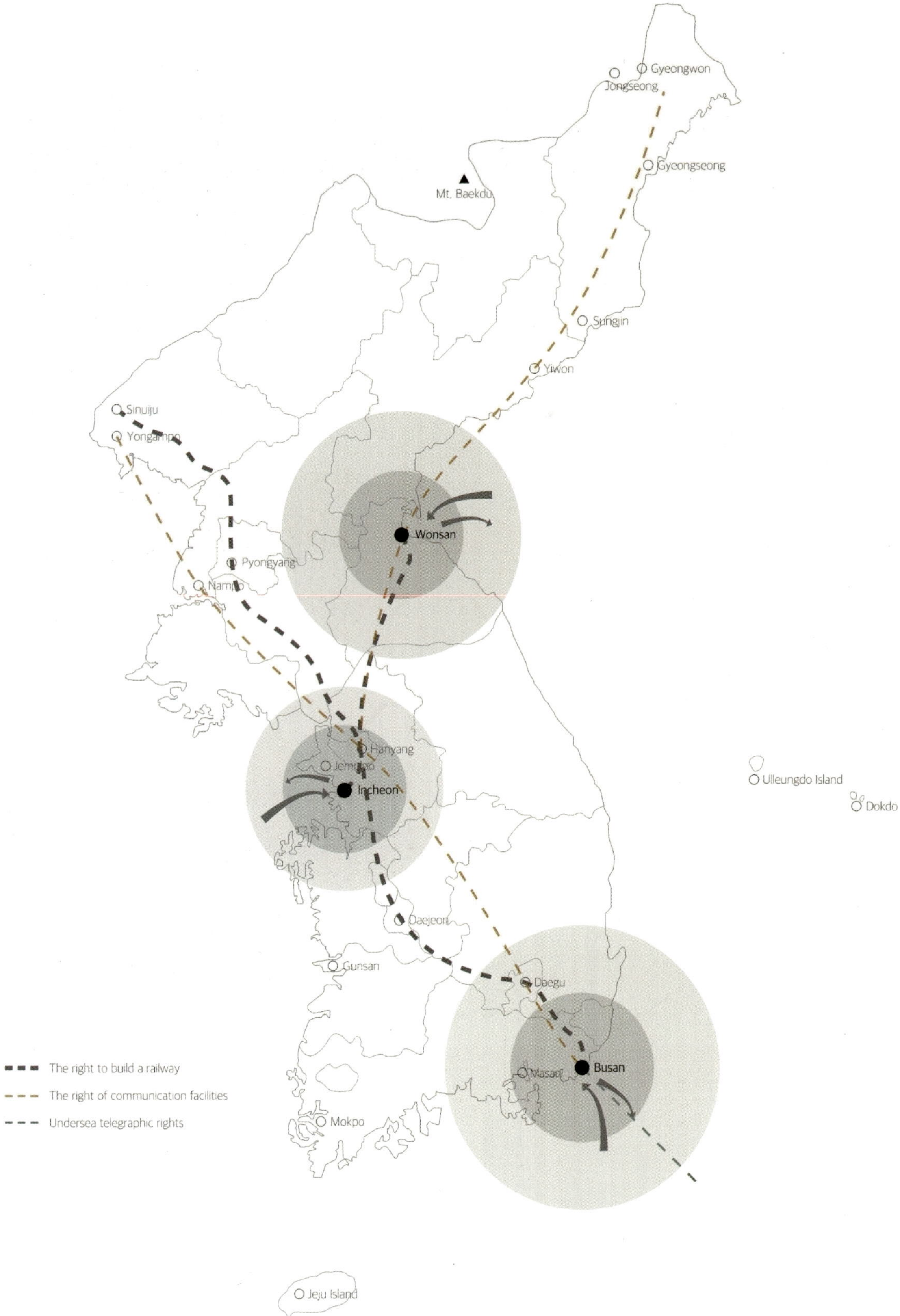

LOCAL IDNEITY

LOCAL IDNETITY

BUSINESS DISTRICT

BUSINESS DISTRICT

- F&B
- Healthcare facility
- Grocery store
- Shopping center
- Education center

BUSINESS DISTRICT

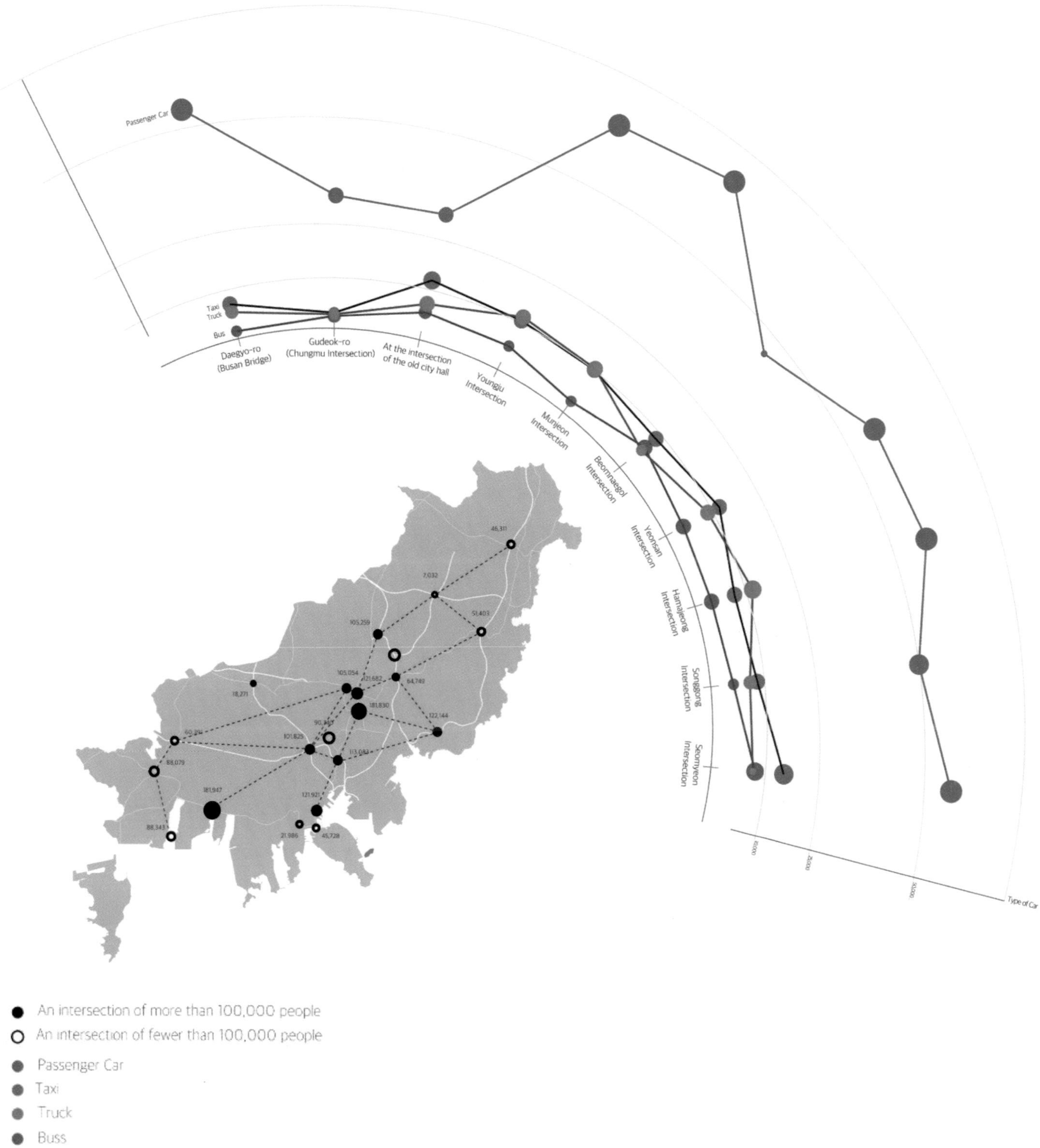

MASS TRANSIT

Region Scale

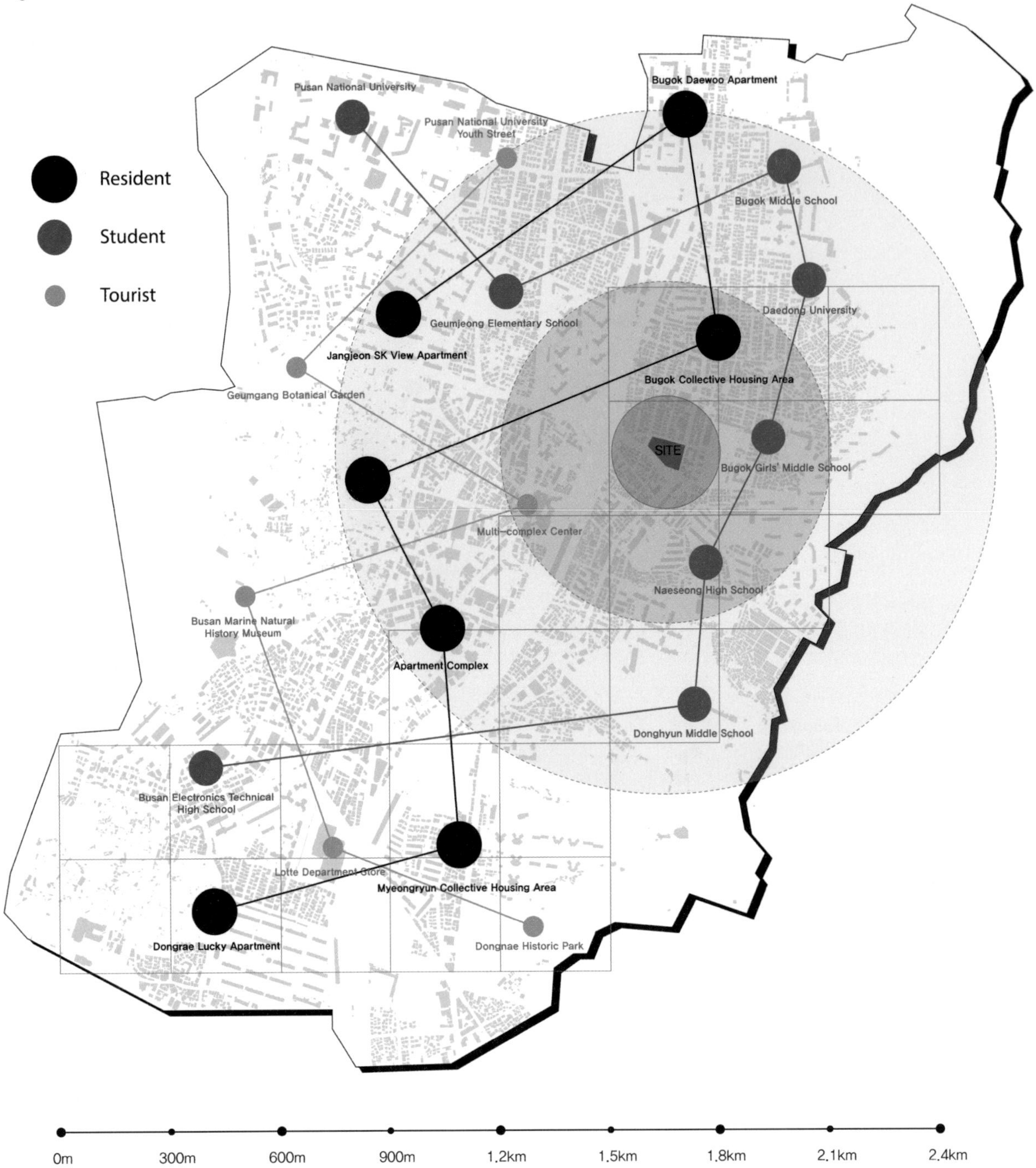

Since the necessary programs also vary depending on the given environment, the necessary programs are explored through population density and connectivity of various sizes.

DENSITY

Site Scale
SITE
Population
High
Low

Population Change
SITE
Population Change
Increase
0
Decrease

DENSITY

Region Scale

Public space is a social device that connects life and interacts with each other. It checks the location and connectivity of public spaces of various sizes, and explores appropriate conditions for a given site.

PUBLIC SPACE

PUBLIC SPACE

Region Scale

Noise degrades the quality of life. Investigate the degree and type of noise on the ground, place appropriate programs, and consider appropriate forms of mass.

NOISE

Site Scale

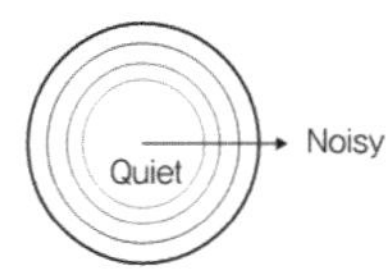

Direction

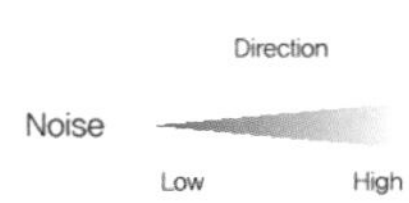

NOISE

FLOW TRANSITION

PROJECT BY

HEONJU LEE
RYULHYEON KIM
JUHAE YU

PROJECT INFORMATION

Location : 126-4, Dongmyeong-ro, Nam-gu, Busan, Republic of Korea
Use : Mid-Rise Mix-Use
Building area : 4,928sqm
Total Floor area : 21,124sqm
Building coverage ratio : 58%
Floor area ratio : 249%
Structure : SRC
Floor : B2 - 6F
Height : 20.1m
The number of parking spaces : 242
The number of households : 123

As the name suggests, Geumjeong-gu Hot Spring Market in Busan is an area formed around hot springs. After the Japanese colonial rule, as Japanese people flooded in, recreational facilities such as hot spring facilities and inns began to be installed in earnest for profit. In addition, the Busan tram connecting the city center where many Japanese lived and Dongnaecheon Stream opened in 1915, and until the 1980s, it was a representative bankruptcy center in Busan. However, as the terminal moved to other areas from the east of Oncheon-dong and commercial districts in nearby areas such as Pusan National University and Seomyeon developed, this area gradually declined, and now it has lost its past status and is stagnating.
The project site is a space where several axes intersect, and is a nodule point where various functions are concentrated and the flow slows down. It was easy to connect to other urban areas due to its good accessibility, and had excellent potential and potential for development, such as high floating population, distribution of various age groups, and the unique historicity of the land. However, the location of the land was not living properly due to the lack of space where people could gather or exchange, such as open space. We want to use an architectural device called OpenSpace, Guidance, and Louver to create a place where people may want to stay and to be the central axis that connects the flow of surrounding elements. By strengthening the role of meeting and contact, we can alleviate social phenomena such as individualization, neighborhood weakness, and community disintegration, and provide people with an experience of restoring locality by creating collective and shared memories.

부산광역시 금정구 온천장은 이름처럼 온천을 중심으로 형성된 지역이다. 일제강점 이후 일본인들이 밀려들어오면서 영리 목적으로 온천시설 확장과 여관 등 위락시설이 본격적으로 갖추어지기 시작했다. 또한 일본인들이 많이 살던 도심과 동래천을 잇는 부산전차가 1915년 개통해 1980년대까지 이곳은 부산의 대표적인 부도심지였다. 그러나 온천동의 동부터미널이 다른 지역으로 이전하고, 부산대와 서면 등 인근 지역의 상권이 발달하면서 상대적으로 이 지역은 점점 쇠퇴해 현재는 과거의 지위를 잃고 상권이 침체해 있다.
프로젝트 대지는 여러 축이 교차하는 공간으로 다양한 기능이 집중된 결절점이자 흐름이 느려지는 지점이다. 접근성이 좋아 다른 도심권과의 연결이 쉽고, 높은 유동 인구, 다양한 연령층의 분포, 대지 고유의 역사성 등 발전 가능성과 잠재력이 우수했다. 그러나 현재 오픈스페이스 등 사람들이 모이거나 교류가 일어날 수 있는 공간이 부족해 대지의 입지적 장점이 제대로 살지 못하고 있었다. 우리는 오픈스페이스, 가이던스, 루버라는 건축적 장치를 활용하여 사람들이 머물고 싶어 할 만한 장소를 만들고 주변 요소들의 흐름을 연결하는 중심축으로 만들고자 한다. 우리는 만남과 접촉의 역할을 강화함으로써 개인화, 이웃 약화, 공동체 해체 등의 사회 현상을 완화하고, 사람들에게 집단적 기억과 공동의 기억을 만들어 지역성을 회복하는 경험을 제공할 수 있다.

URBAN ISSUES

DENSITY

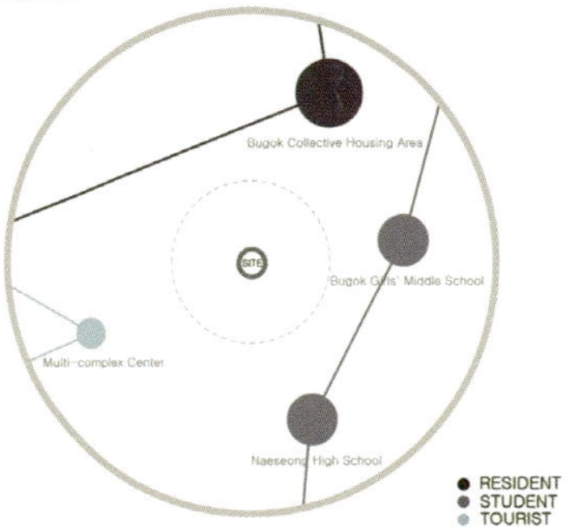

TRANSIT

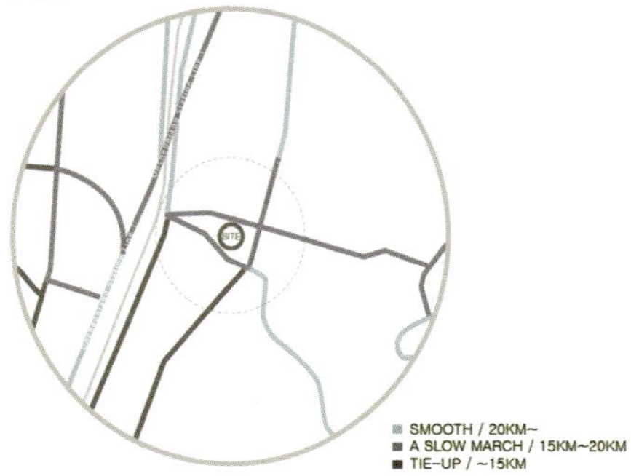

PUBLIC SPACE

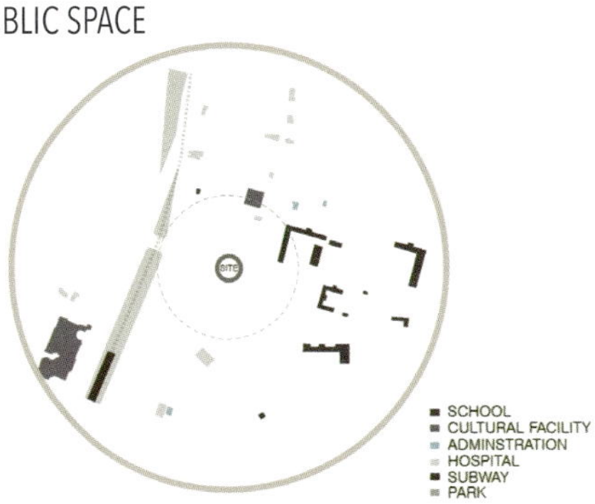

NOISE

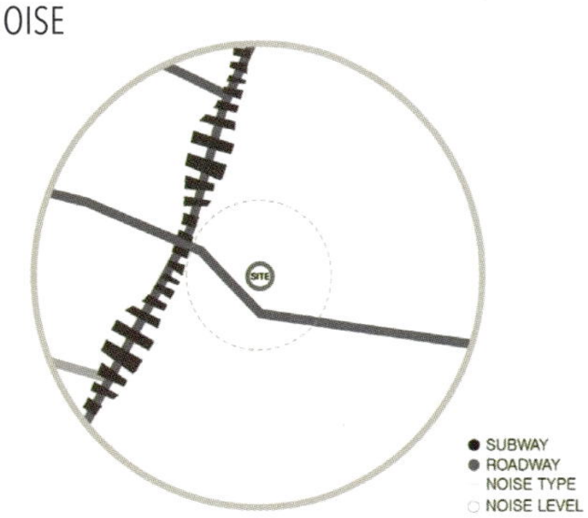

SITE ANALYSIS

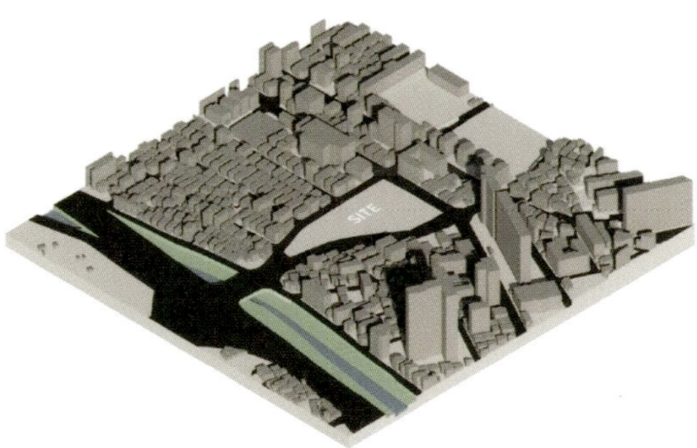

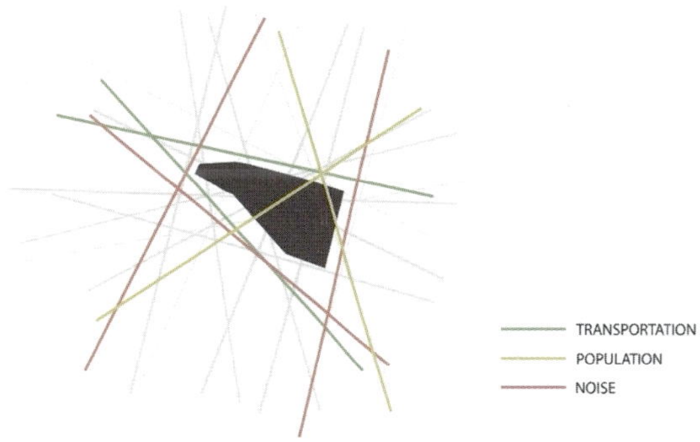

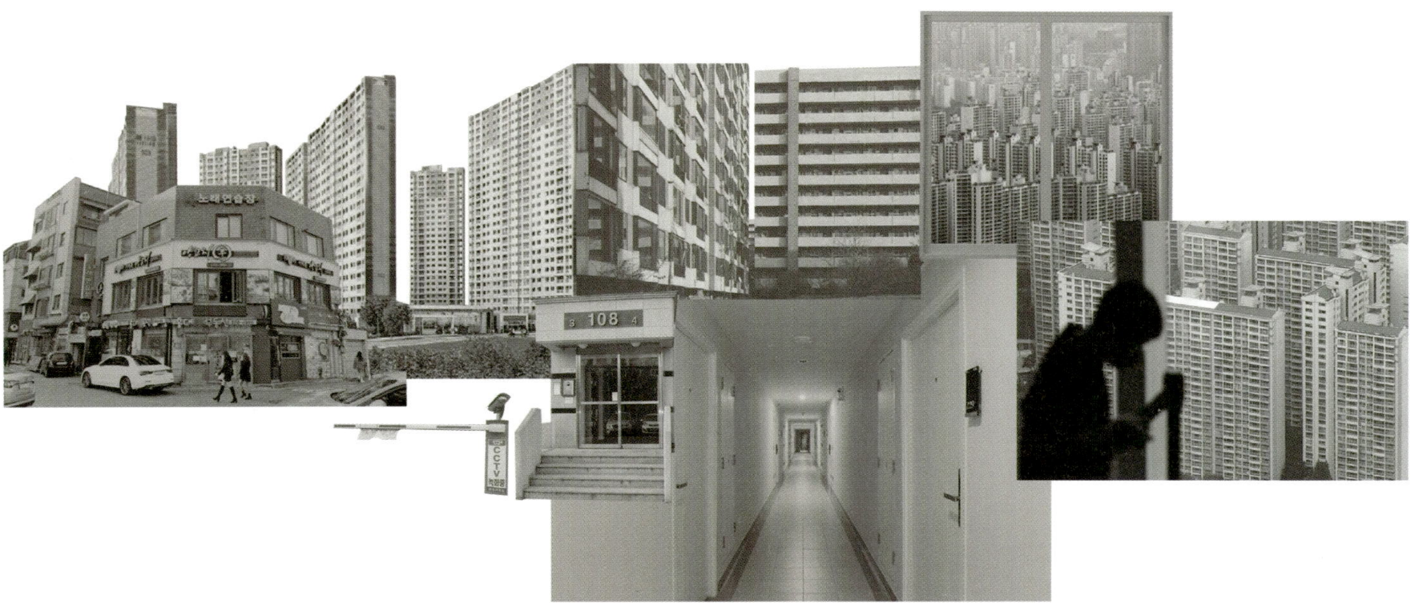

ARCHITECTURAL DEVICES

01 OPEN SPACE

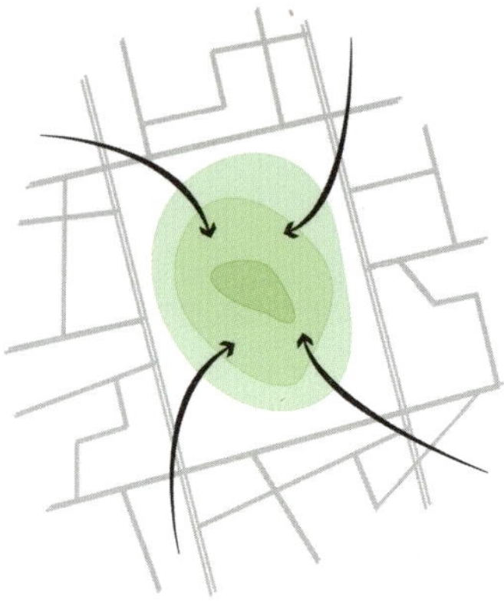

CENTRALITY

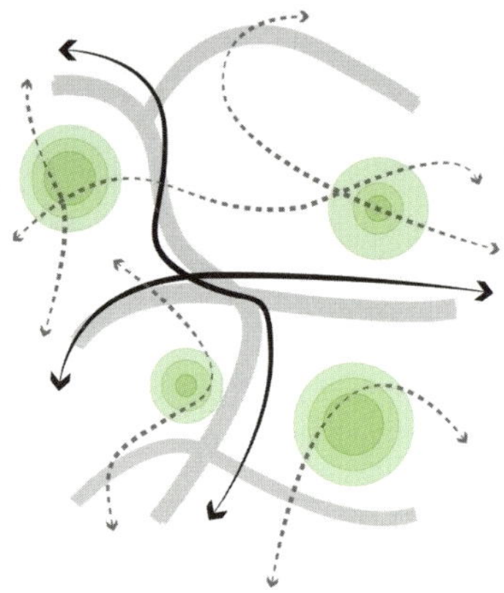

MULTI-CENTRALITY

RULES

01 Multi-Centrality (Interconnected)

02 The point at which the axes intersect

03 A personality that corresponds to the surrounding context

04 Surrounded by more than one curve

05 Within visual perception area unit (25 ~ 35m)

02 GUIDANCE

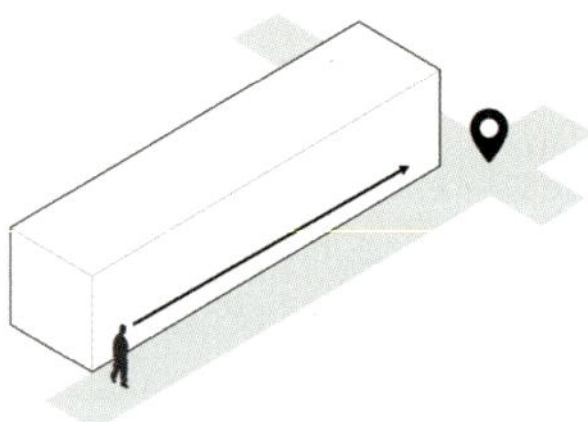

STRAIGHT
Predictable Movement

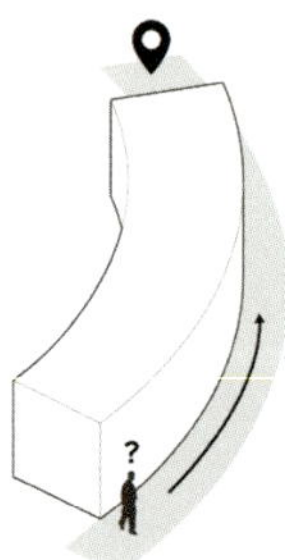

INDIRECT
Unpredictable Movements

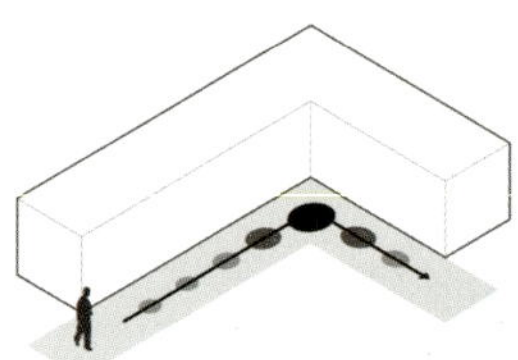

STRAIGHT
Uneven Concentration

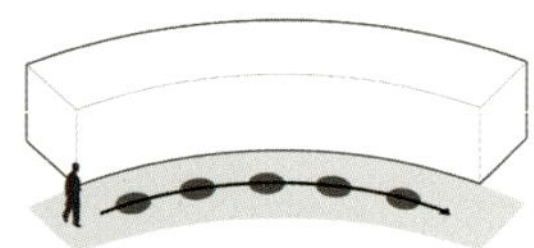

INDIRECT
Unalienated Commercial

03 LOUVER

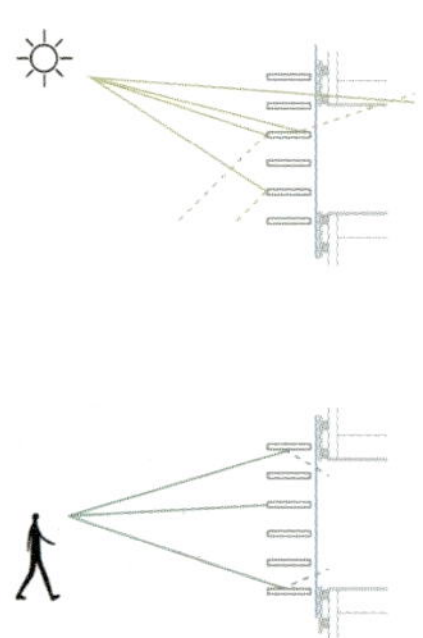

HORIZONTAL SHADING DEVICE
Horizontal shading devices protect from sun at higt angles and opposite to the wall to be shaded such as south sides.

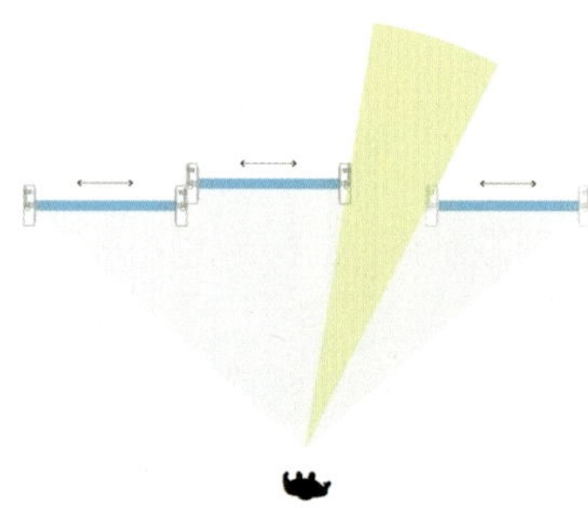

SELF-REGULATING LOUVERS
Privacy is selectively protected through a sliding louver that the user can autonomously control.

APPLICATION

01 OPEN SPACE

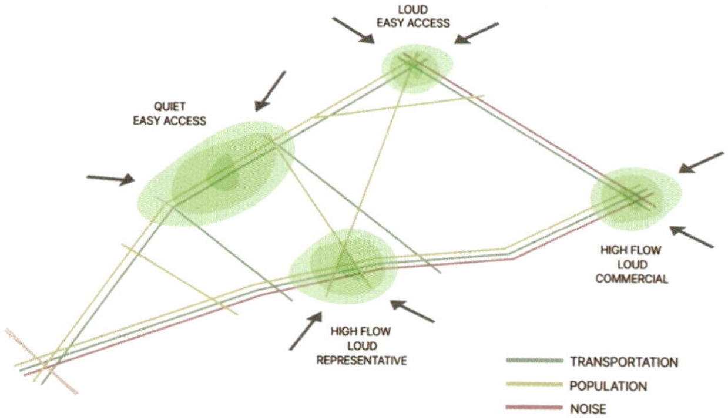

MULTI-CENTRALITY
Freely connect the axes by placing the open space at the point where the axes intersect.

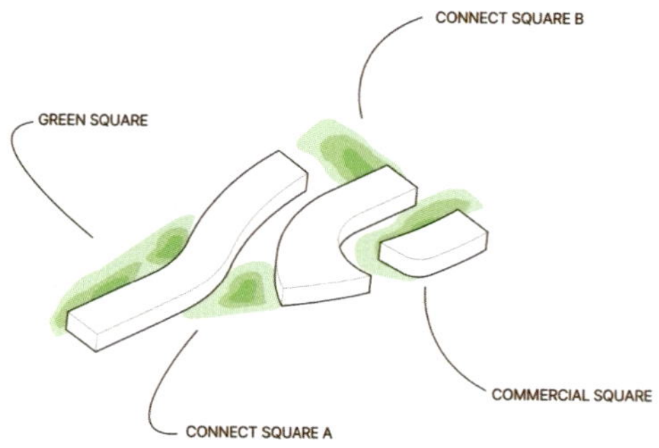

CHARACTERIZATION
Through the nature of the intersection passing through the open space different characteristics were given for each open space.

02 GUIDANCE

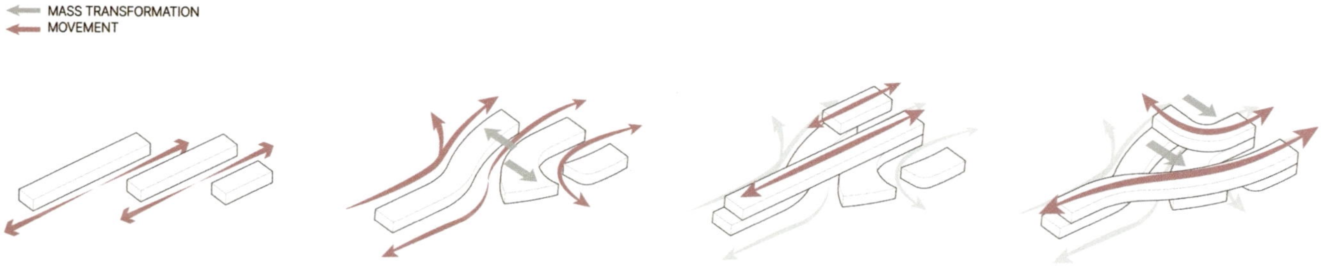

FRAGMENTARY ROUTE
Constant and Typical Movement

MULTIPLE ROUTE
Irregular and Eclectic Movement

FRAGMENTARY ROUTE
Constant and Typical Movement

MULTIPLE ROUTE
Irregular and Eclectic Movement

03 LOUVER

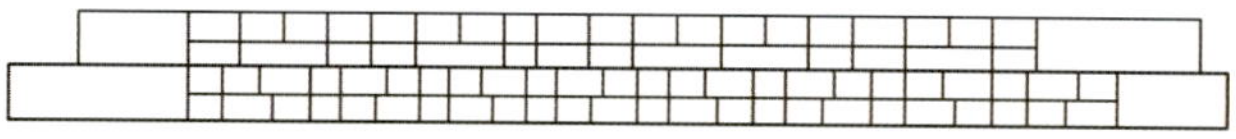

DOMESTIC FLOOR
Domestics were placed on the mass of the 3rd to 6th floors. And the surface is formed according to the various residential units.

BASIC LOUVER DEPLOYMENT
Solar radiation control, which is the role of a basic louver, was performed through a louver.

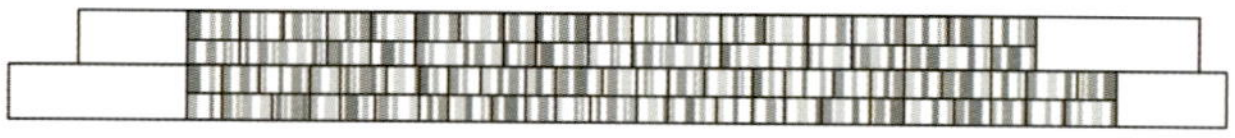

SELF-REGULATING LOUVER
Various facades are formed through sliding louver that can be autonomously adjusted for each unit

MASS PROCESS

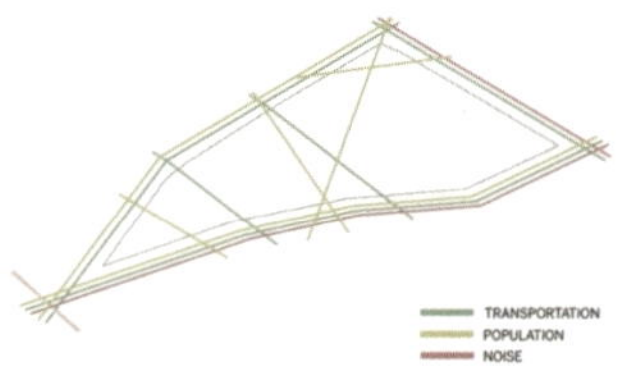

AXIS
Intersections were identified by classifying three axes passing through the site.

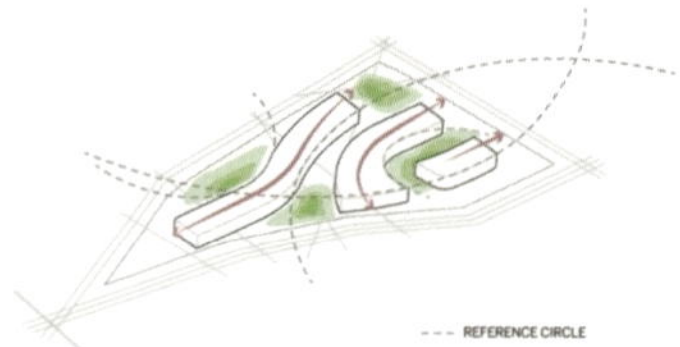

CONNECTION
An open space was arranged at the intersection of the axes and the flow was connected using a curved mass.

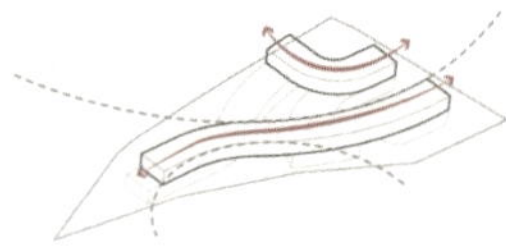

EXTEND
An axis extending from the end of the site in various directions is represented.

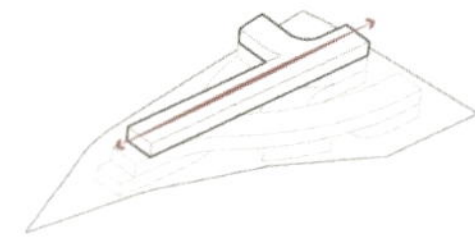

EMPHASIS
It maximizes contrast with curves by emphasizing the axis of a straight line starting at the end of the site.

PROGRAM DIAGRAM

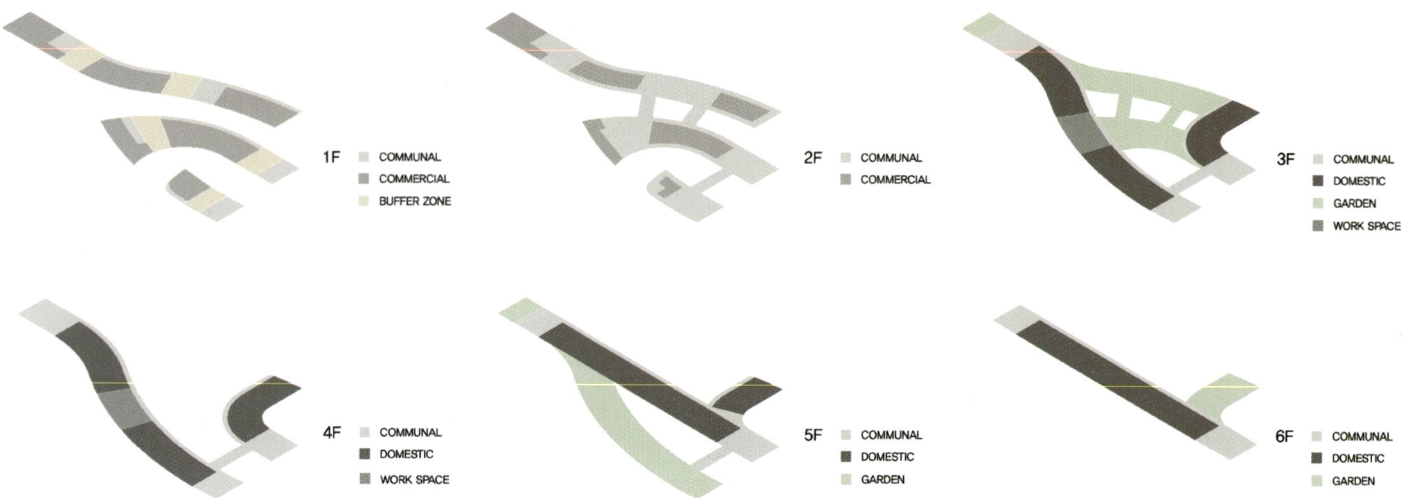

MRMU PROGRAM

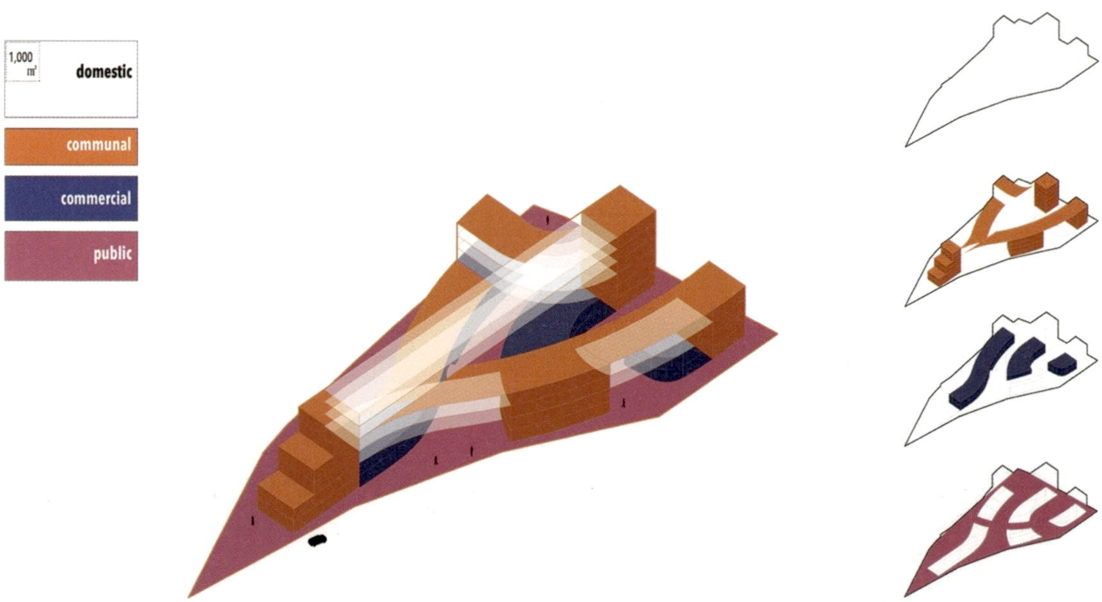

SEQUENCE COLLAGE

CONCEPT COLLAGE

A cross-sectional collage that expresses the porosity of a residential complex with various age groups. The cross-section of the mass represents the characteristics of the diversity and high density of the residential unit, and the side of the mass represents the continuity of commercial facilities revealed in the surrounding land.

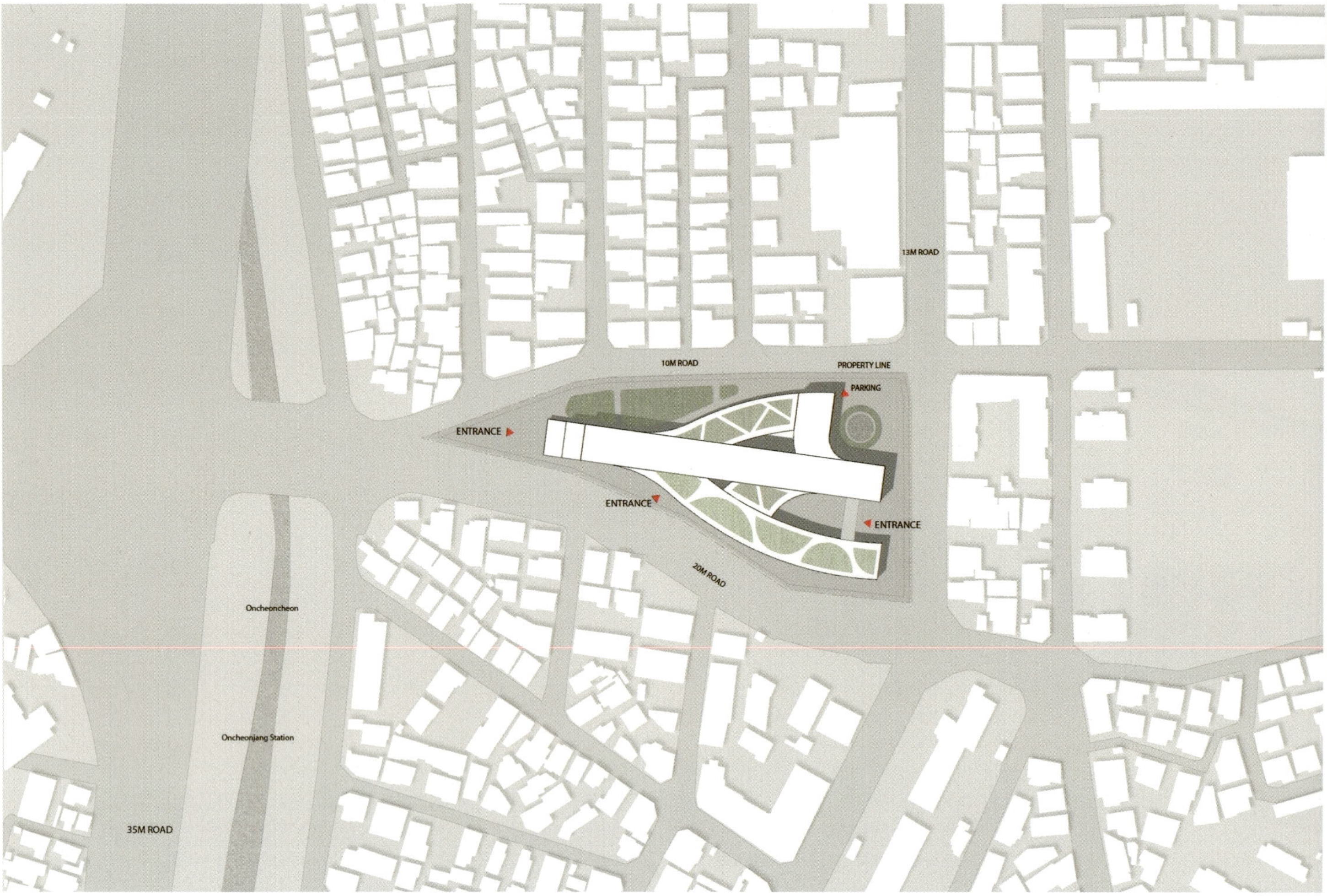

ELEVATION_NORTH & EAST

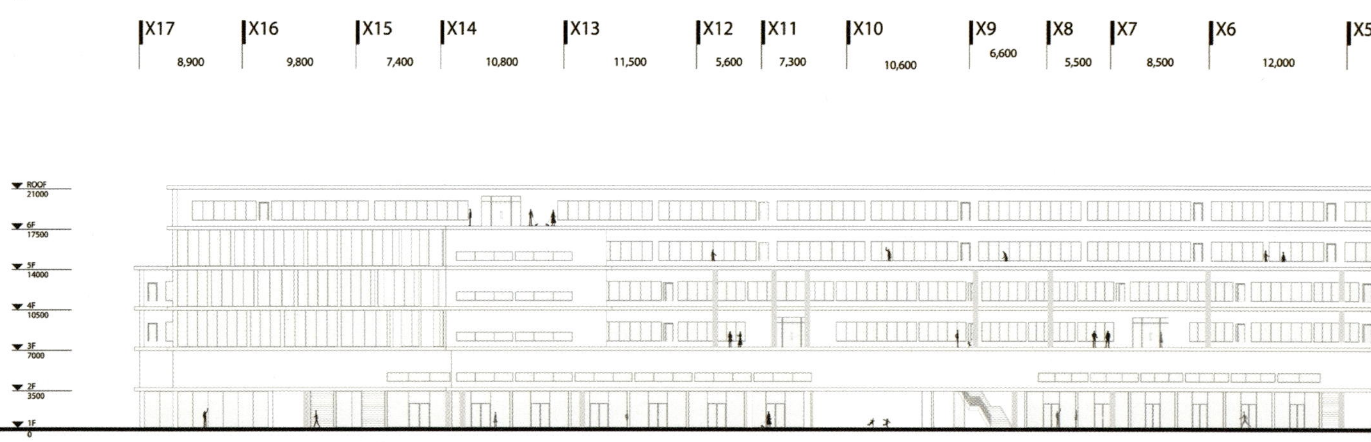

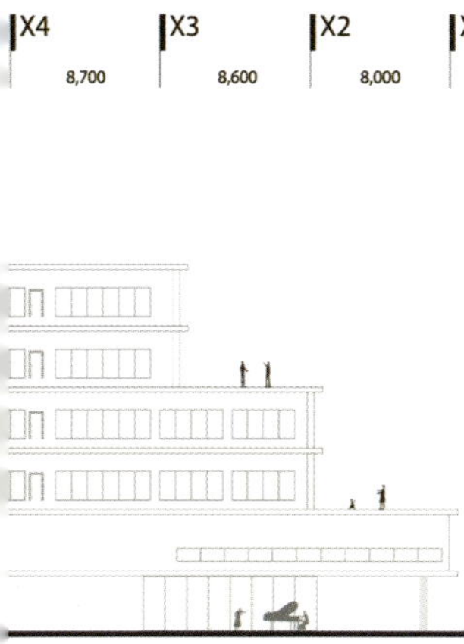
X4
X3
X2
X1
8,700
8,600
8,000

Y1
Y2
Y3
Y4
Y5
Y6
Y7
Y8
Y9
Y10
Y11
8,000
6,300
7,500
4,000
6,000
6,900
4,100
10,800
4,300
12,300

1. COMMERCIAL 2. GREEN SQUARE 3. CONNECT SQUARE

4. COMMERCIAL SQUARE 5. BUFFER SPACE 6. PARKING

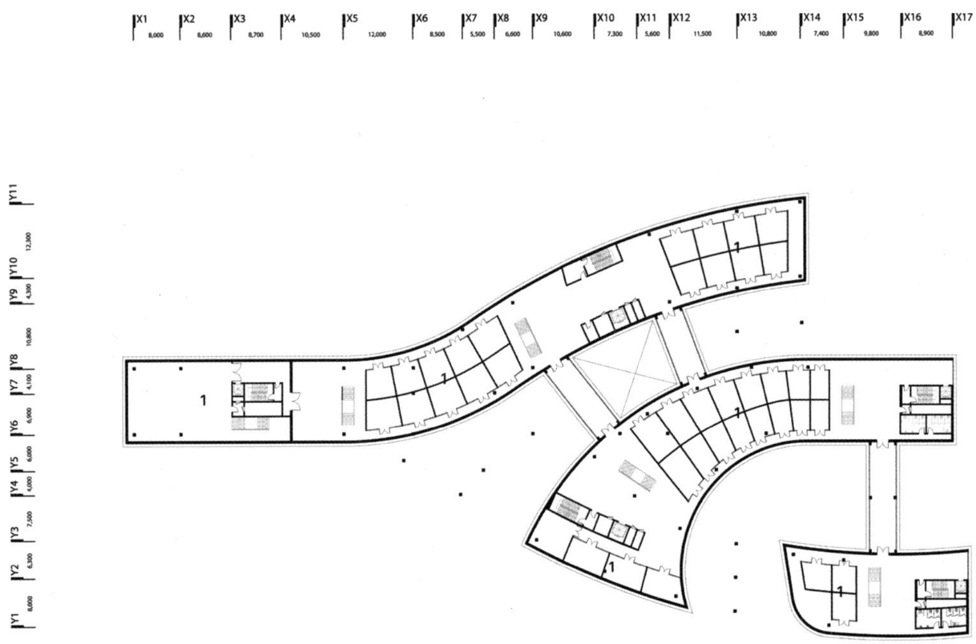

1. COMMERCIAL

2F

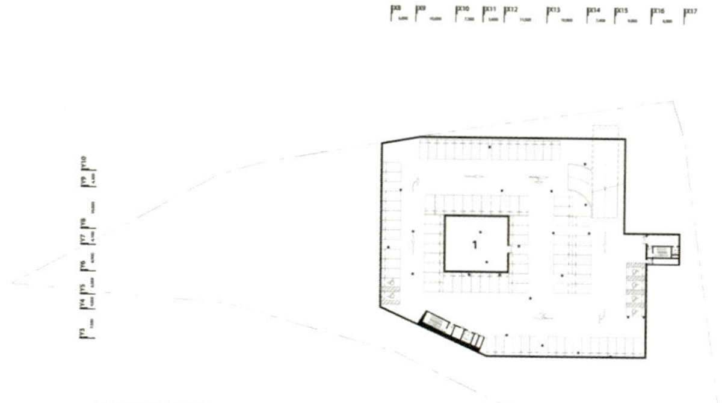

B1F

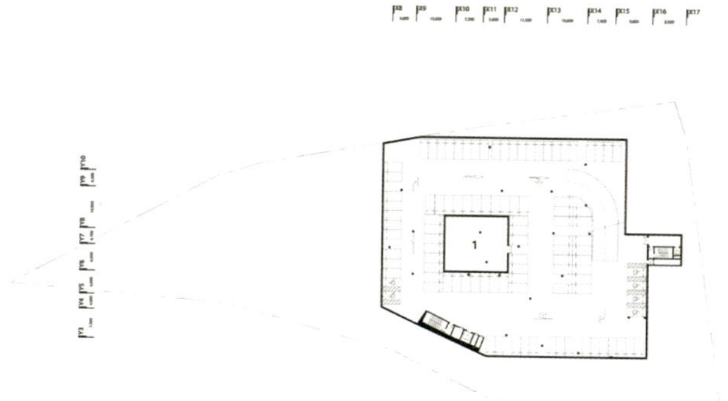

B2F

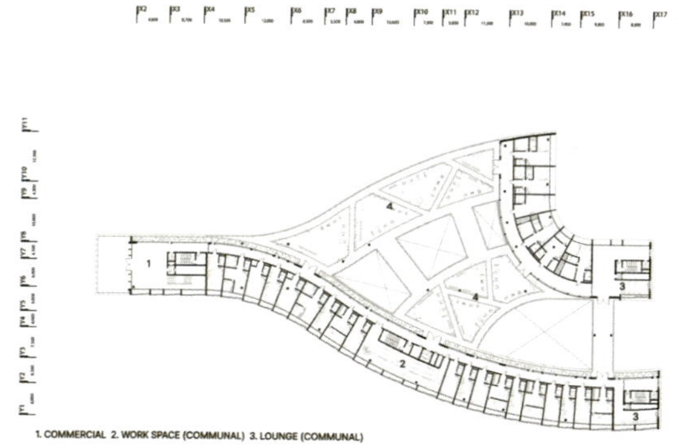

3F

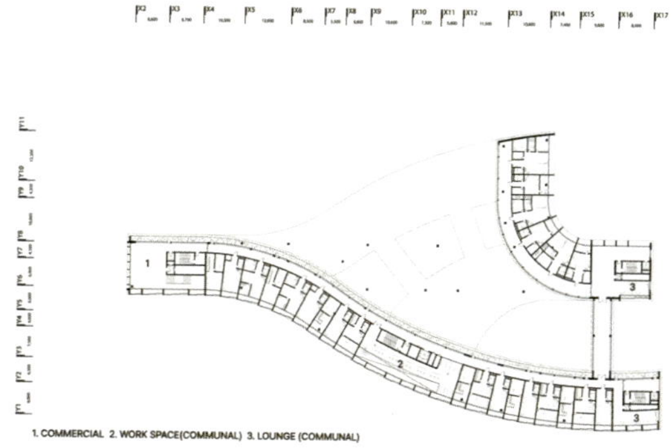

4F

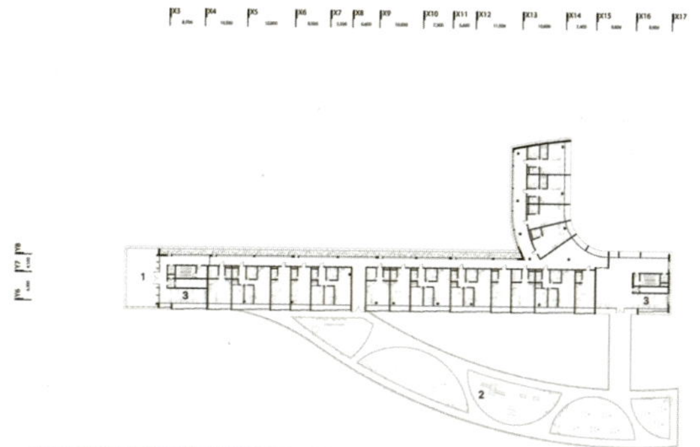

5F

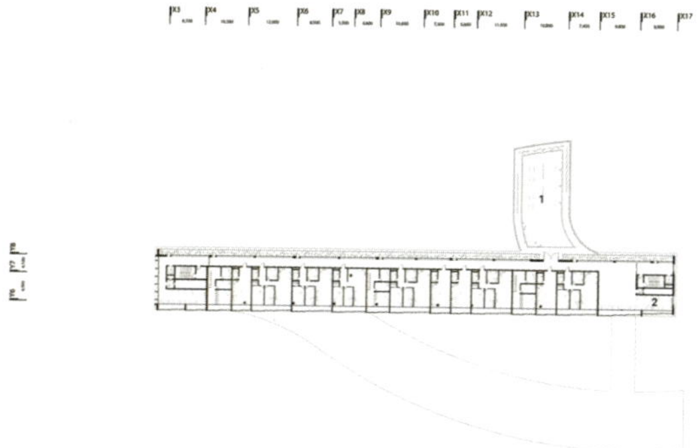

6F

SECTION

A-A' SECTION

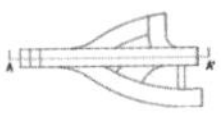

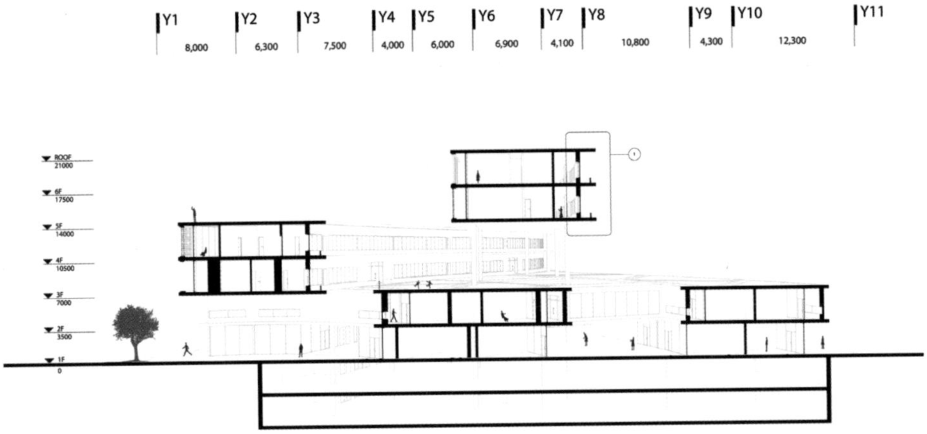

B-B' SECTION

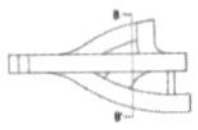

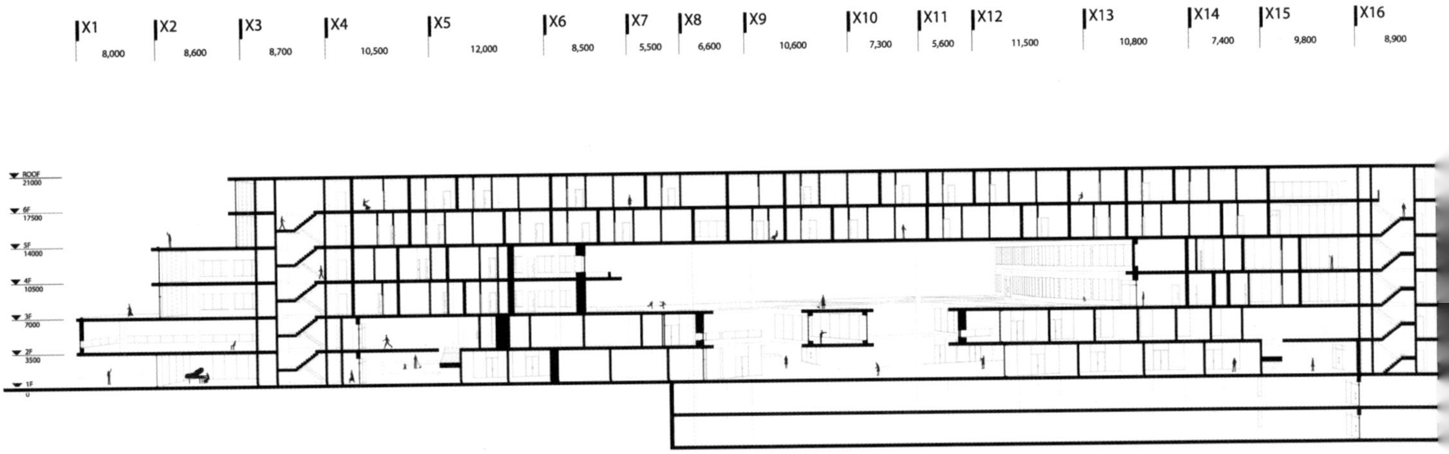

ELEVATION_SOUTH & WEST

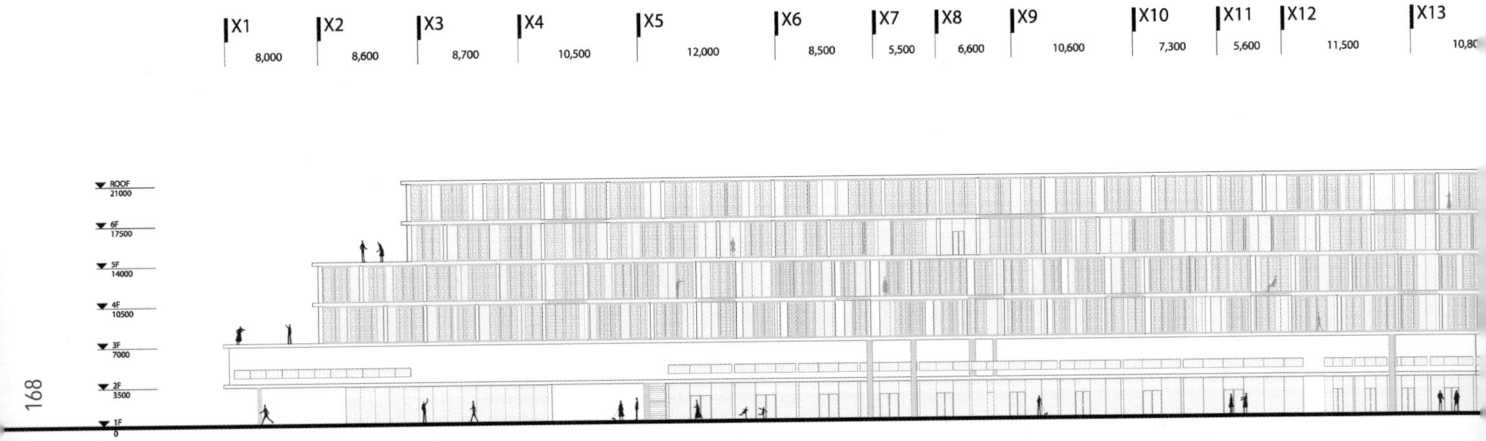

SECTION DETAIL

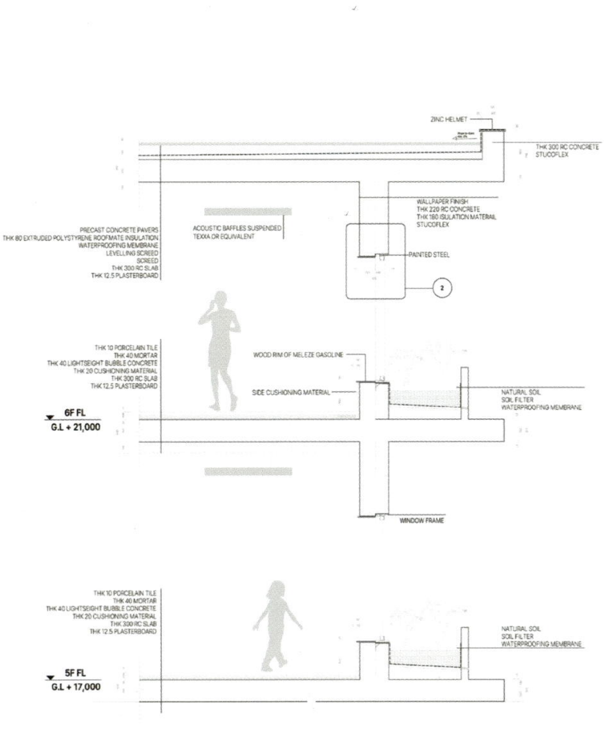

① CORRIDOR SECTION DETAIL
SCALE 1:50

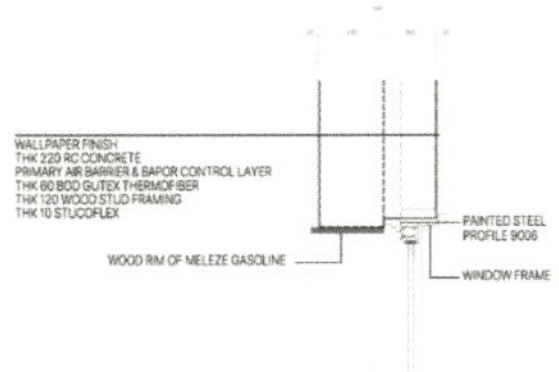

② PARTIAL SECTION DETAIL
SCALE 1:100

ELEVATION EXPLANATION (SOUTE & WEST)

A cross-sectional collage that expresses the porosity of a residential complex with various age groups. The cross-section of the mass represents the characteristics of the diversity and high density of the residential unit, and the side of the mass represents the continuity of commercial facilities revealed in the surrounding land.
A common area for communication and harmony of residents was placed between each residential floor, and a plaza was placed in the open space of the shopping district on the first floor to connect the flow and attract inflow from the outside. The square has different characteristics along the axis of the earth and serves to buffer the flow transition between the outer space and the inner space.

건물의 주 진입로인 서측 입면과 남측 입면이다. 특히 서측 입면은 여러 흐름이 모여 대지 안으로 들어오는 영역이므로 하단부를 필로티로 계획하고, 대지에서 서측 조망이 갖는 이점을 살리기 위해 비교적 큰 규모의 상업시설을 배치했다. 남측 입면 하단부는 보행자의 동선을 자연스럽게 유도하고자 곡면 벽을 활용해 오픈스페이스를 계획하였다. 상단부 주거 입면은 각 매스가 가진 축의 성격(유닛)을 드러내고자 유닛별로 프레임이 돌출되게 디자인하였다.

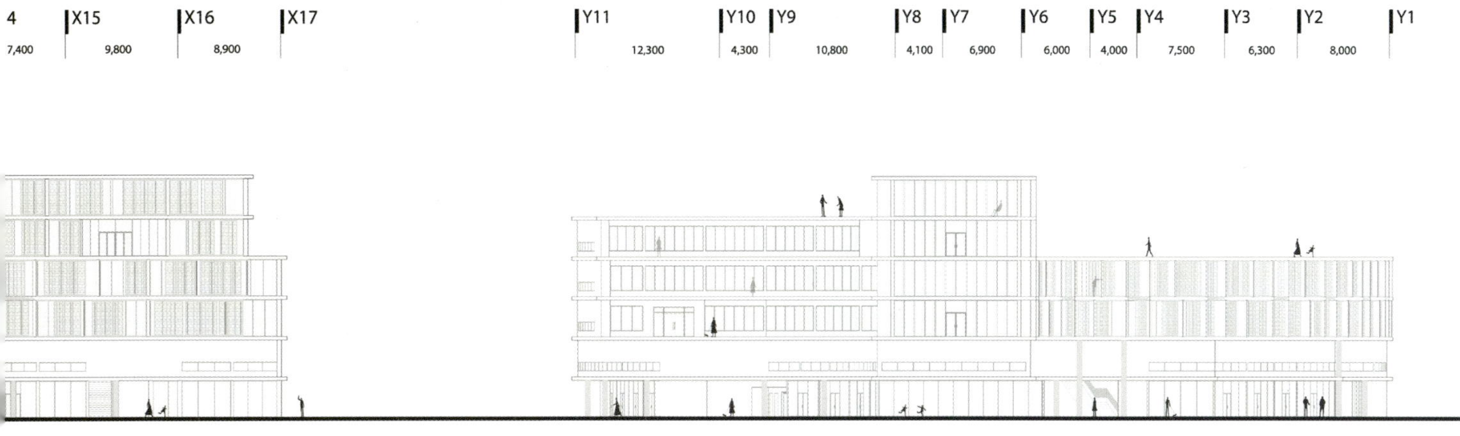

YEONGSEON-DONG, YEONGDO-GU, BUSAN

RESEARCH BY
MYEUNGBIN KIM
UNGHUI JUNG
SANGJEEN MOON
GYUSIK SON

Yeongseon-dong is the only dong in Yeongdo-gu, Busan, which consists only of islands among all autonomous districts in Korea. It is divided into administrative dongs Yeongseon 1-2. The entire area is flat, and most of the commercial facilities of Yeongdo-gu are concentrated, making it the busiest area. White Yeoul Culture Village and Bongnaesan Mountain are located, where the site of the village formed first in Yeongdo-gu remains.

영선동(瀛仙洞)은 대한민국의 전체 자치구 중 유일하게 섬으로만 구성된 부산광역시 영도구(影島區)의 동이다. 행정동 영선1~2동으로 나뉜다. 지역 전체가 평지이며, 영도구의 상업시설 대부분이 몰려 가장 번화한 지역이다. 영도구에서 가장 먼저 형성된 마을의 터가 남은 관광지 흰여울문화마을과 봉래산 자락이 소재한다.

Busan

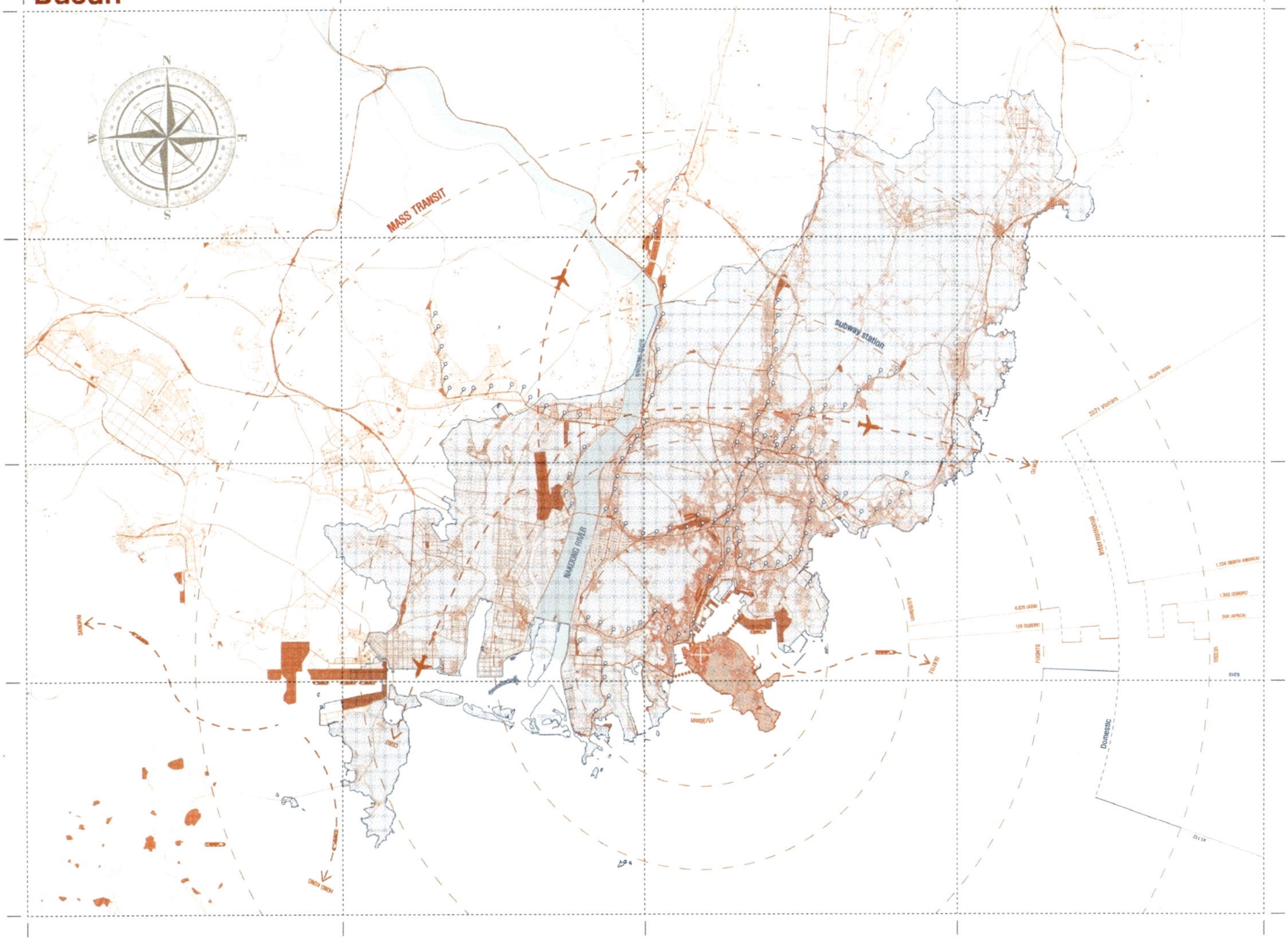

MASS TRANSIT, WATER, VISITOR & TOURISM

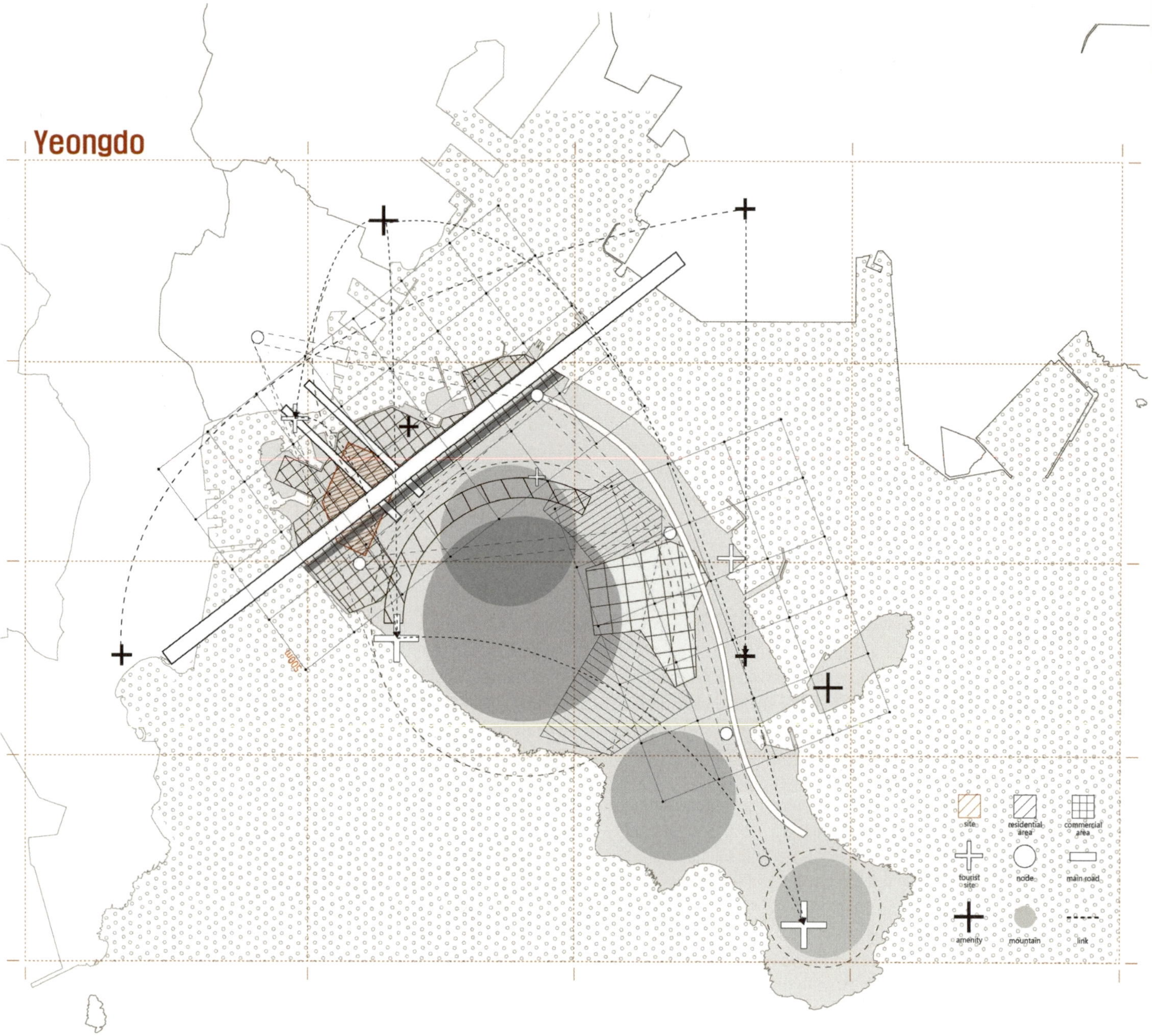

MASS TRANSIT, WATER, VISITOR & TOURISM

Yeongseon 1-dong

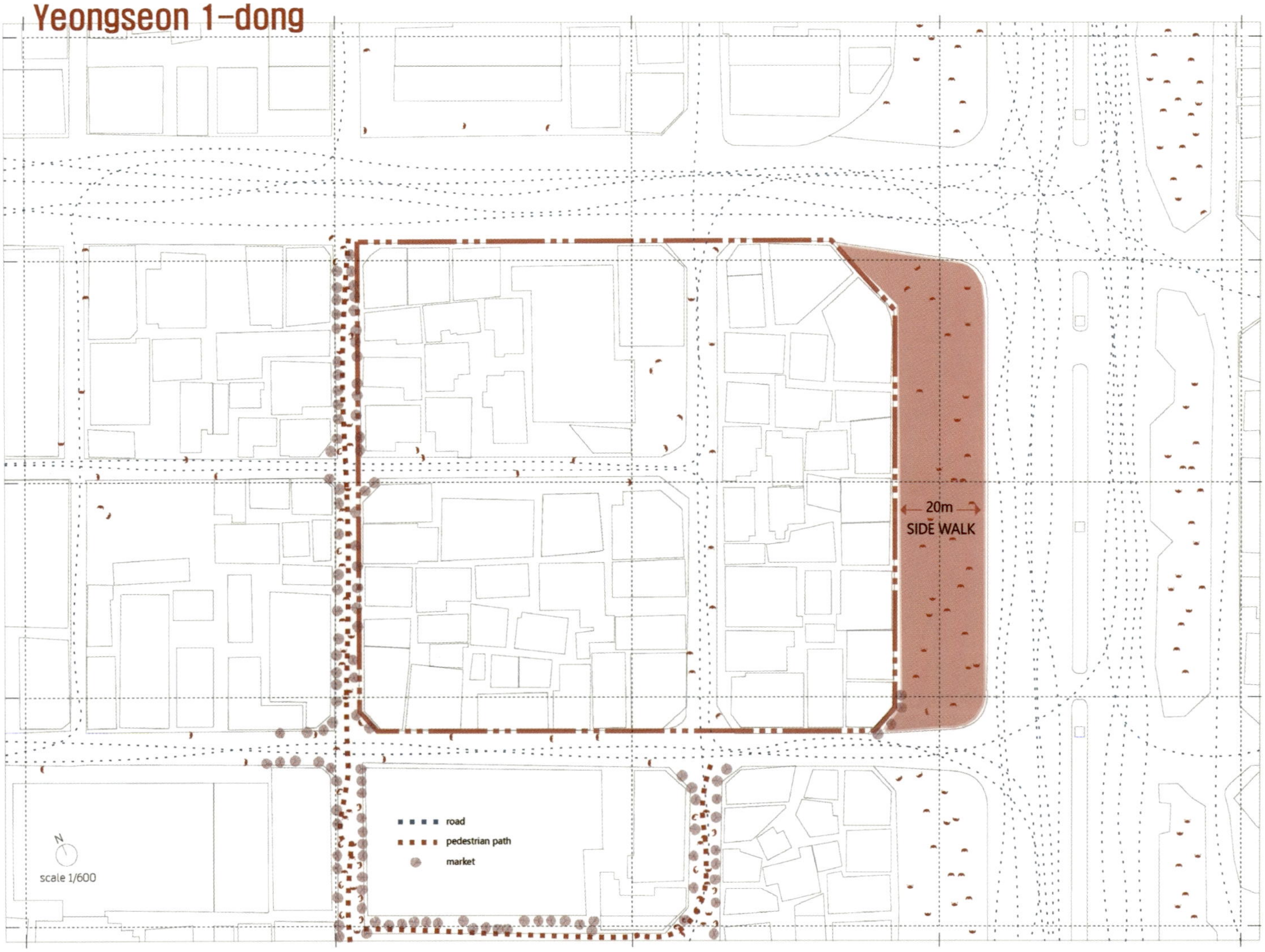

MASS TRANSIT, WATER, VISITOR & TOURISM

SCALE : CITY

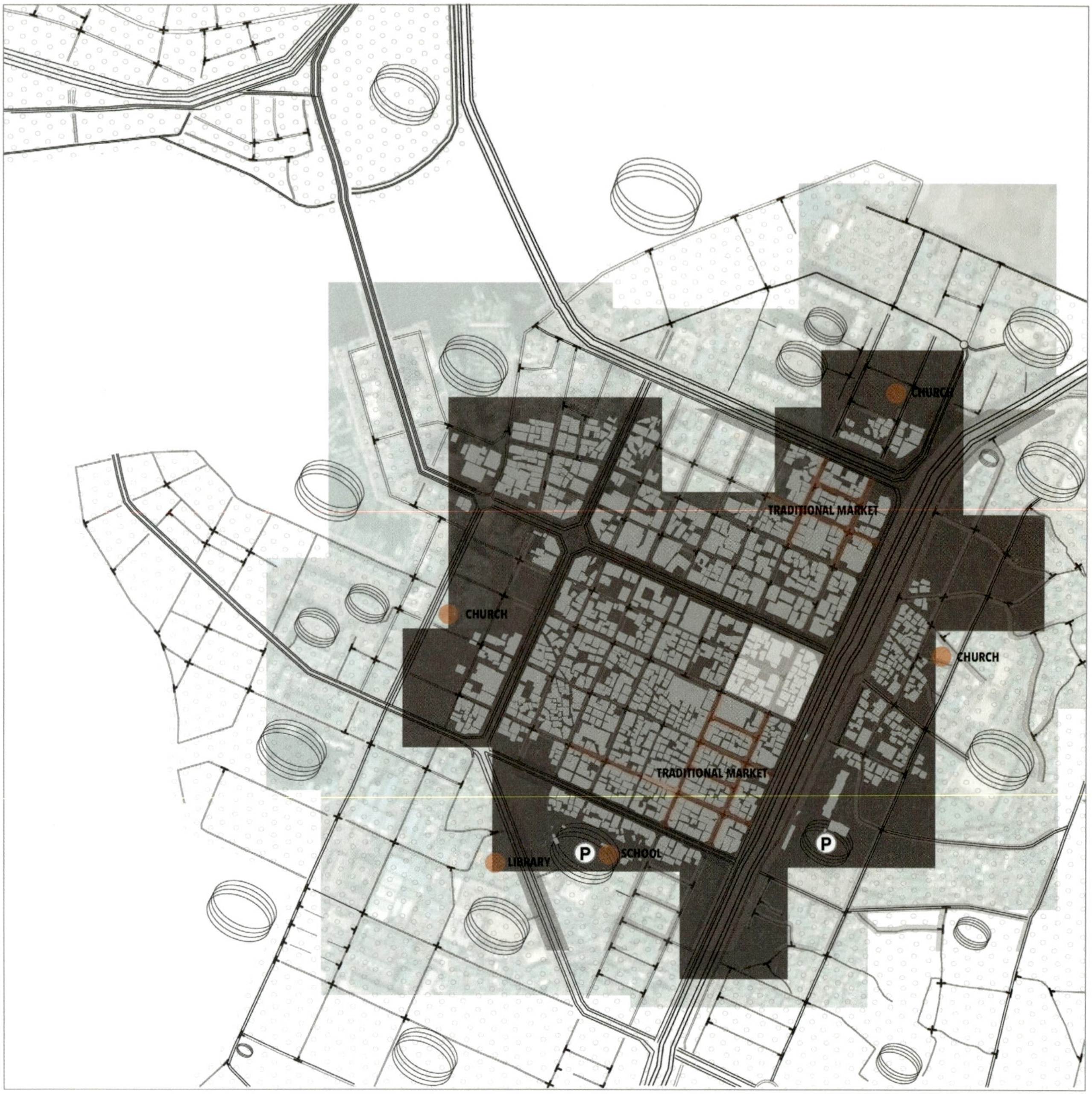

BUILDING
PUBLIC SPACE
ROAD (PUBLIC SPACE)
INTERSECTION
PARKING LOT (PUBLIC SPACE)
SITE

How many of the spaces classified as PUBLIC SPACE in Yeongdo are actually for people? There is a traditional market around the site, and people's communication is active around it. In addition, there were churches, libraries, and schools. But above all, the space occupied will be the way we walk in our daily lives.

PUBLIC SPACE

SCALE : NEIGHBORHOOD

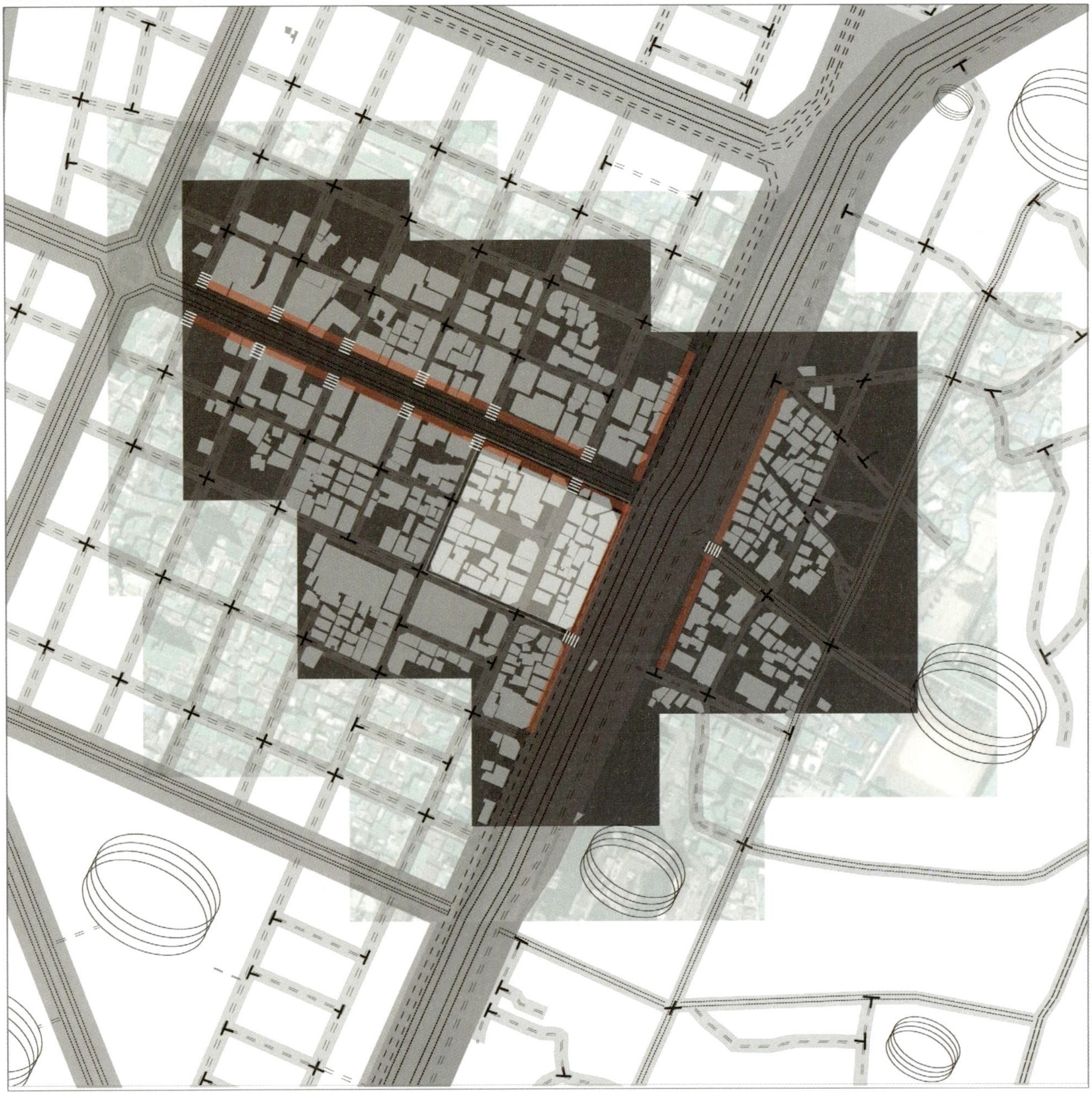

BUILDING
SIDE WALK
ROAD (PUBLIC SPACE)
INTERSECTION
PARKING LOT (PUBLIC SPACE)
SITE

PUBLIC SPACE

SCALE : HUMAN BODY SCALE

PUBLIC SPACE

SCALE : CITY

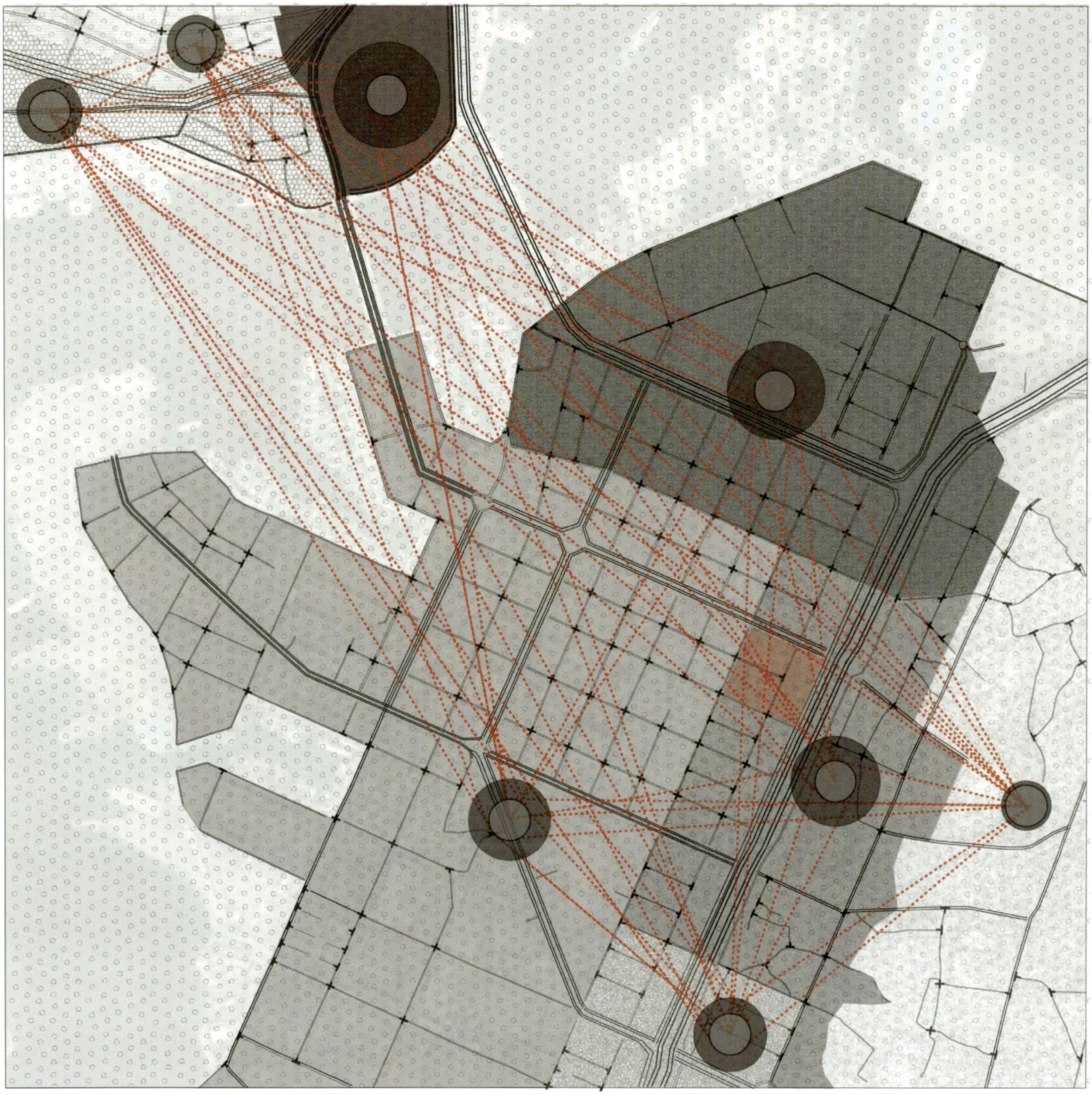

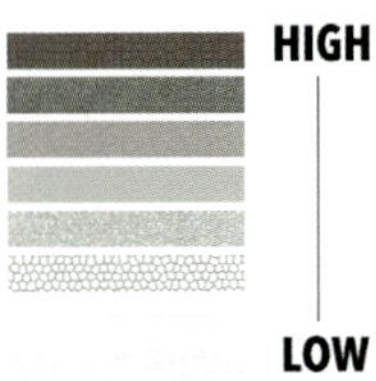

The rate of population increase and decrease and the attractiveness of the region were analyzed by combining them. Areas with high population inflow were selected as attractive, and areas with high outflows were selected as low attractive, and the possibility of inflow and outflow of surrounding areas was expressed by dotted lines. The size of the node point edge of each zone is the relative attractiveness index of the region.

DENSITY

SCALE : NEIGHBORHOOD

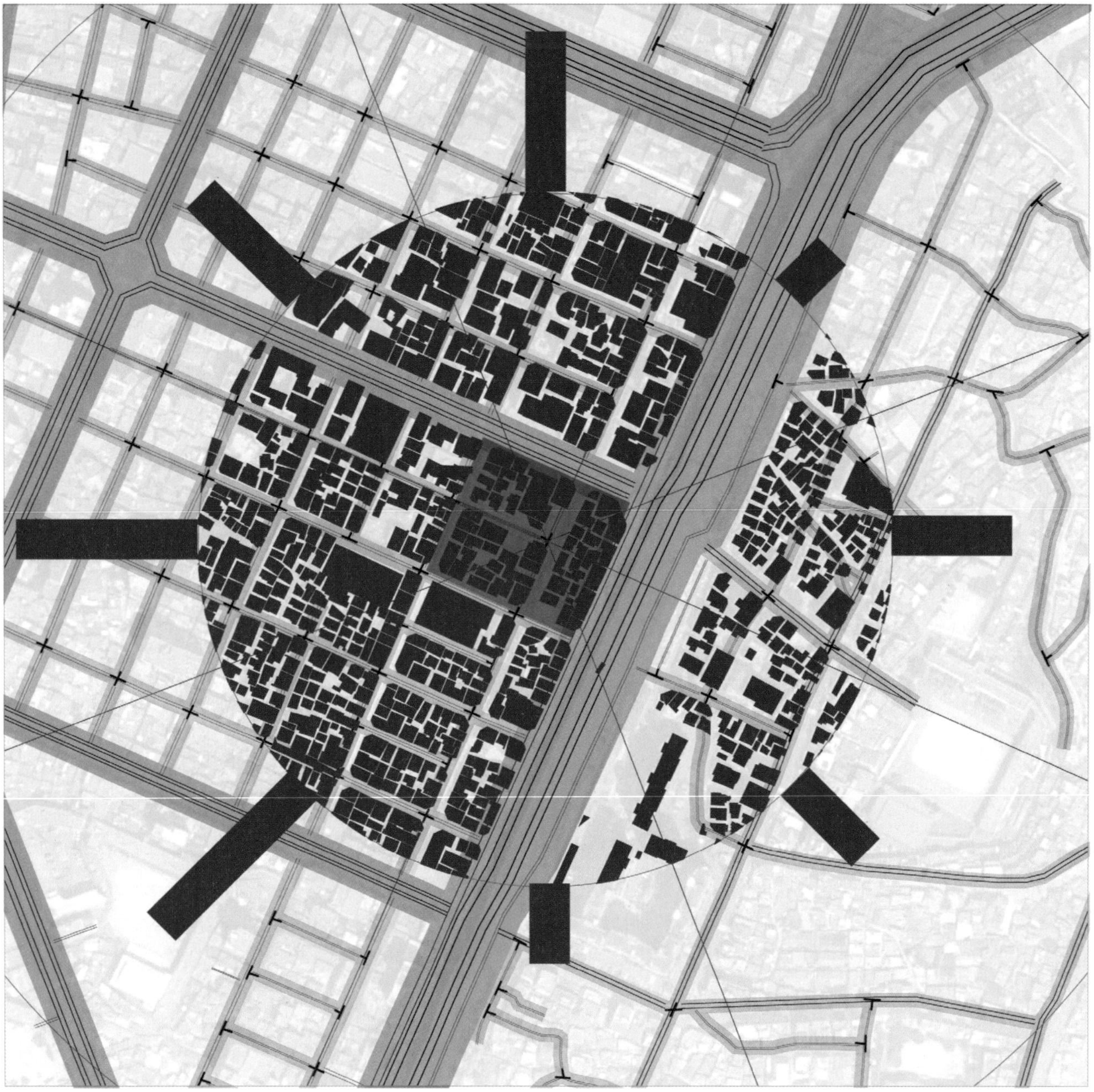

SOLID

DENSITY

SCALE : SITE

PARKING LOT
SIDEWALK
BUILDING

DENSITY

SCALE : CITY

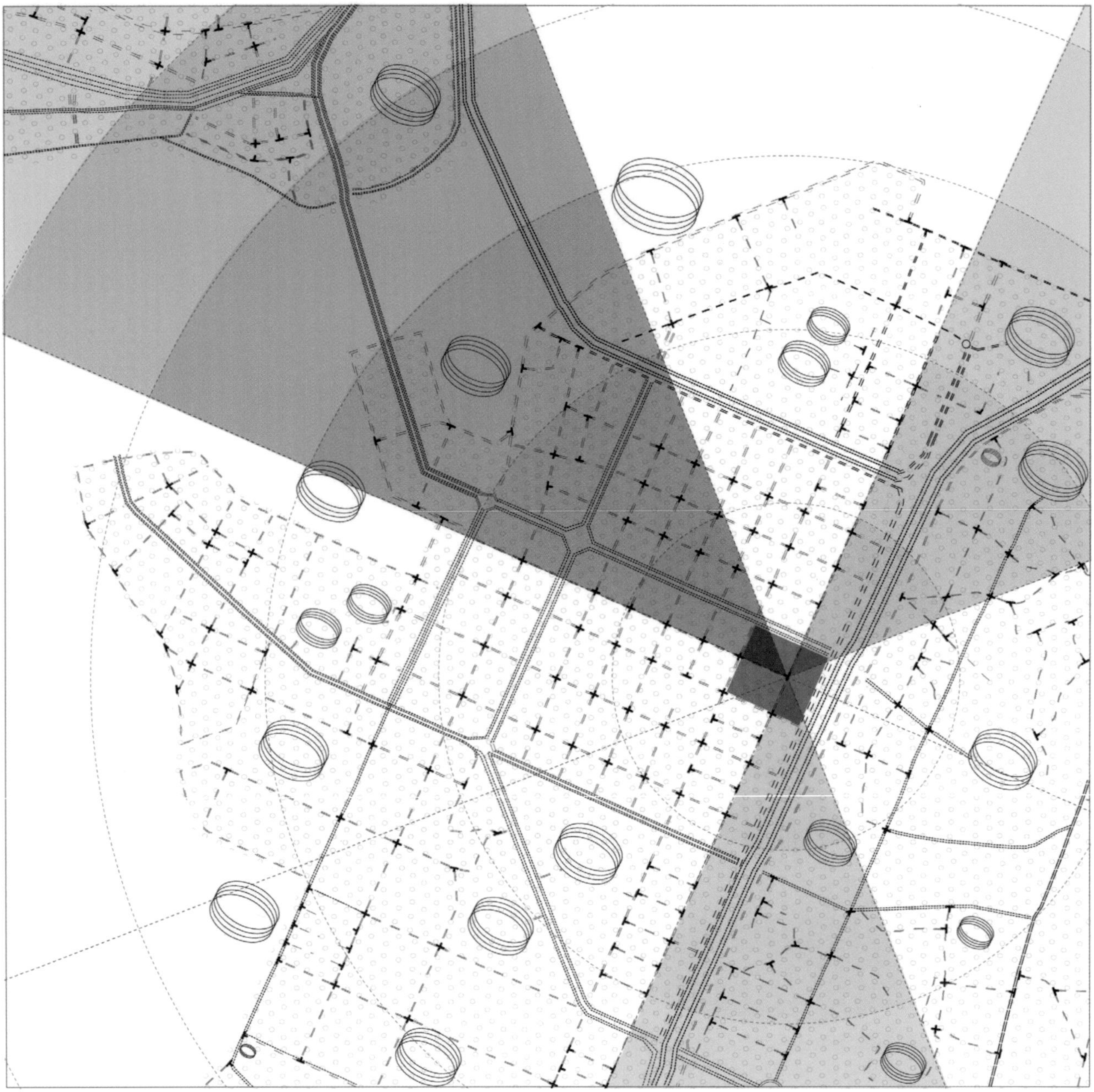

EXPRESSWAY
MAIN ROAD
SMALL ROAD
INTERSECTION
PARKING LOT
SITE

Unlike in the past, when the only way to get to the land was by boat you can now go to the city in about 10 minutes by car. It expressed the circulation flow around Yeongdo Bridge, the center. It expressed the circulation of roads with fast speed and heavy traffic and the opposite, and in detail, it showed parking lots and deck spaces where intersections and circulation paths of roads are concentrated.

CIRCULATION

SCALE : NEIGHBORHOOD

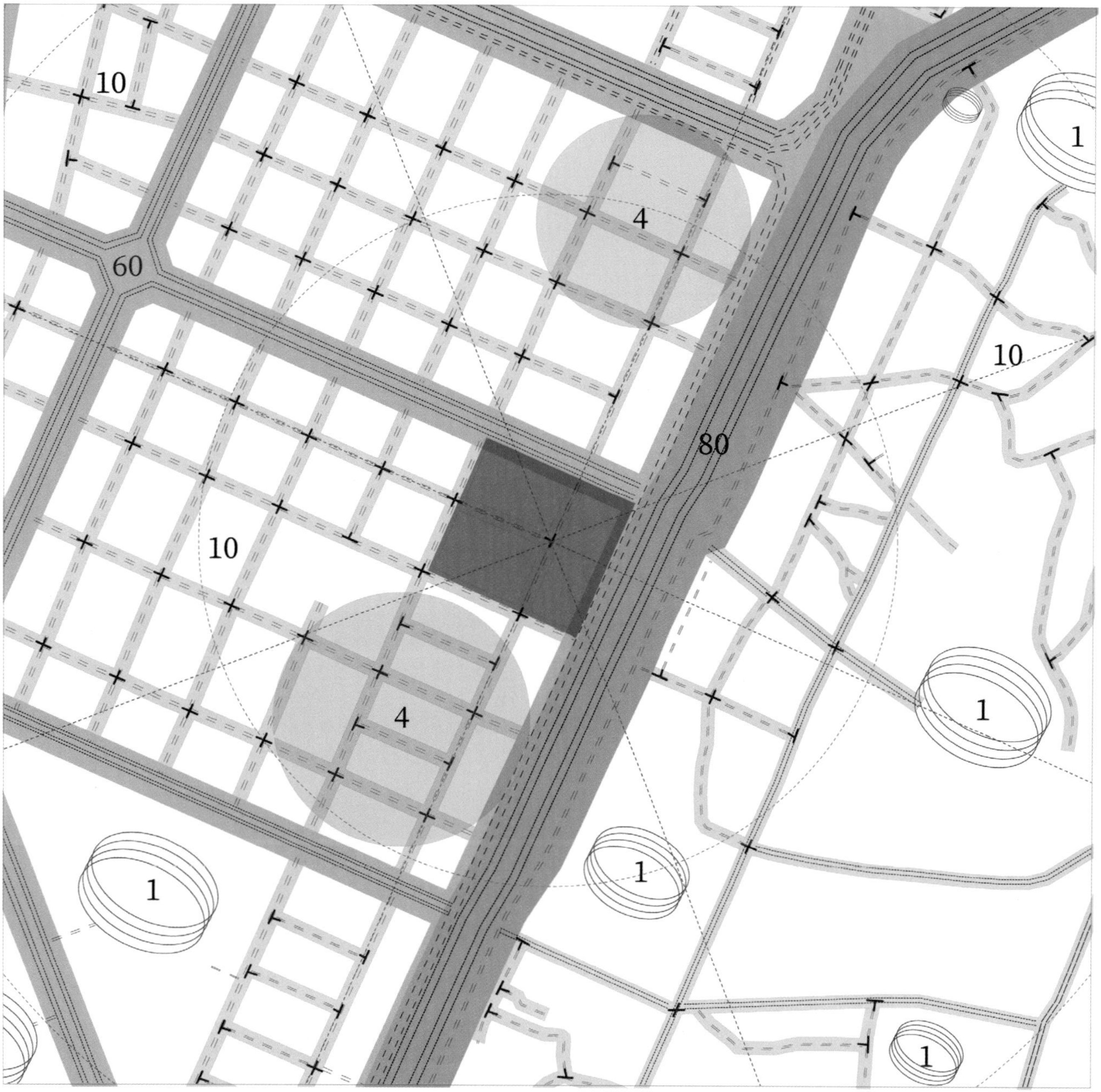

EXPRESSWAY
MAIN ROAD
SMALL ROAD
INTERSECTION
PARKING LOT
SITE

CIRCULATION

SCALE : HUMAN BODY SCALE

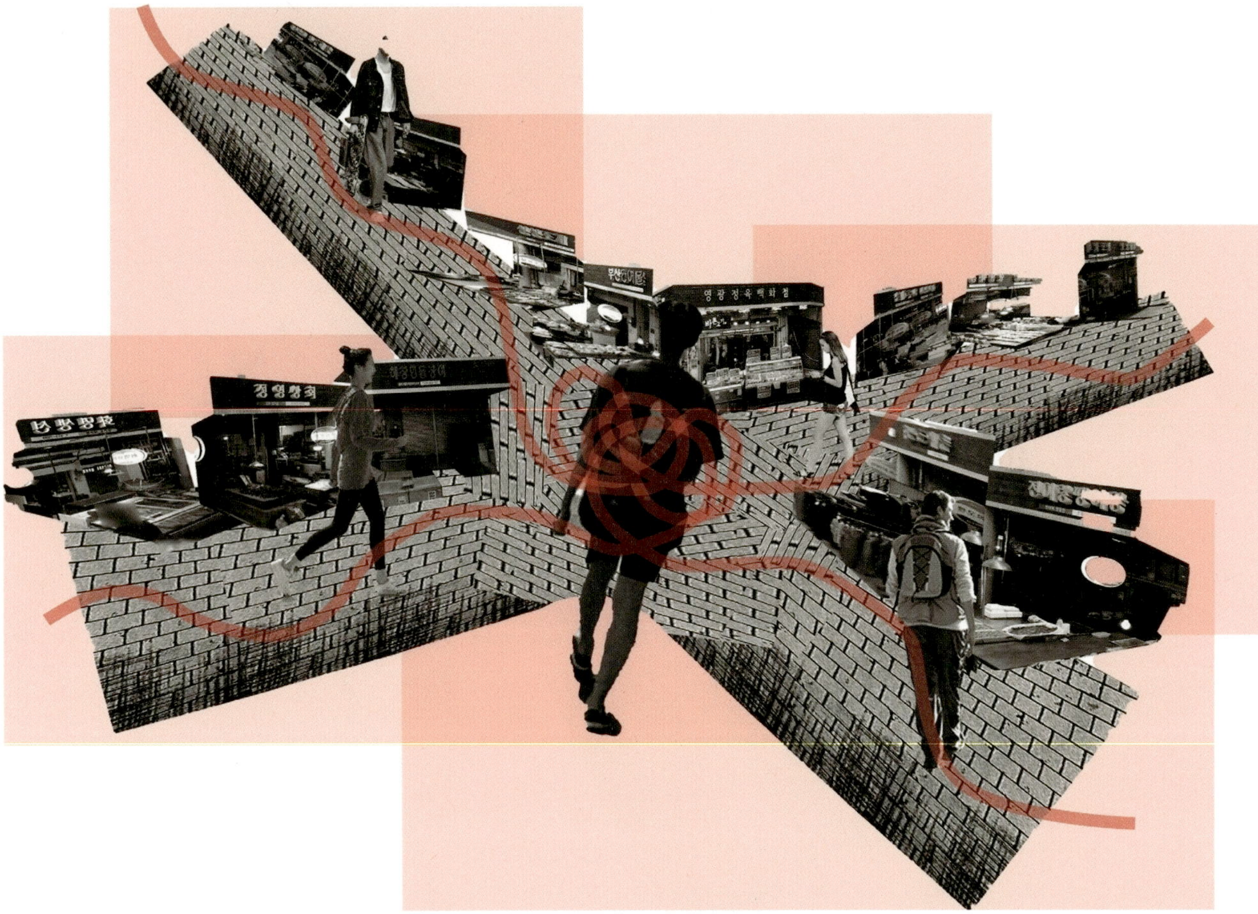

CIRCULATION

: Maintenanace project &
Development Project
: Street managenent

URBAN PLAN

: Bongnae tunnel
: Urban railway
: Bongnae mountain Environment
: City (located in Dongsam)

URBAN PLAN

HEIGHT AND PURPOSE RELATIONSHIP

NEIGHBORHOOD SCALE

In order to examine the form of urban structure, the characteristics of urban blocks were studied. Apartments and school buildings independently occupy urban blocks, and small blocks gathered in residential areas.

MORPHOLOGY

RESIDENTIAL INFRASTRUCTURE

CITY SCALE

It is a place with the best residential infrastructure in Yeongdo-gu. However, while there are many welfare facilities for the elderly, cultural facilities such as movie theaters are insufficient.

INFRASTRUCTURE

PEDESTRIAN PATH ENVIRONMENT

NEIGHBORHOOD SCALE

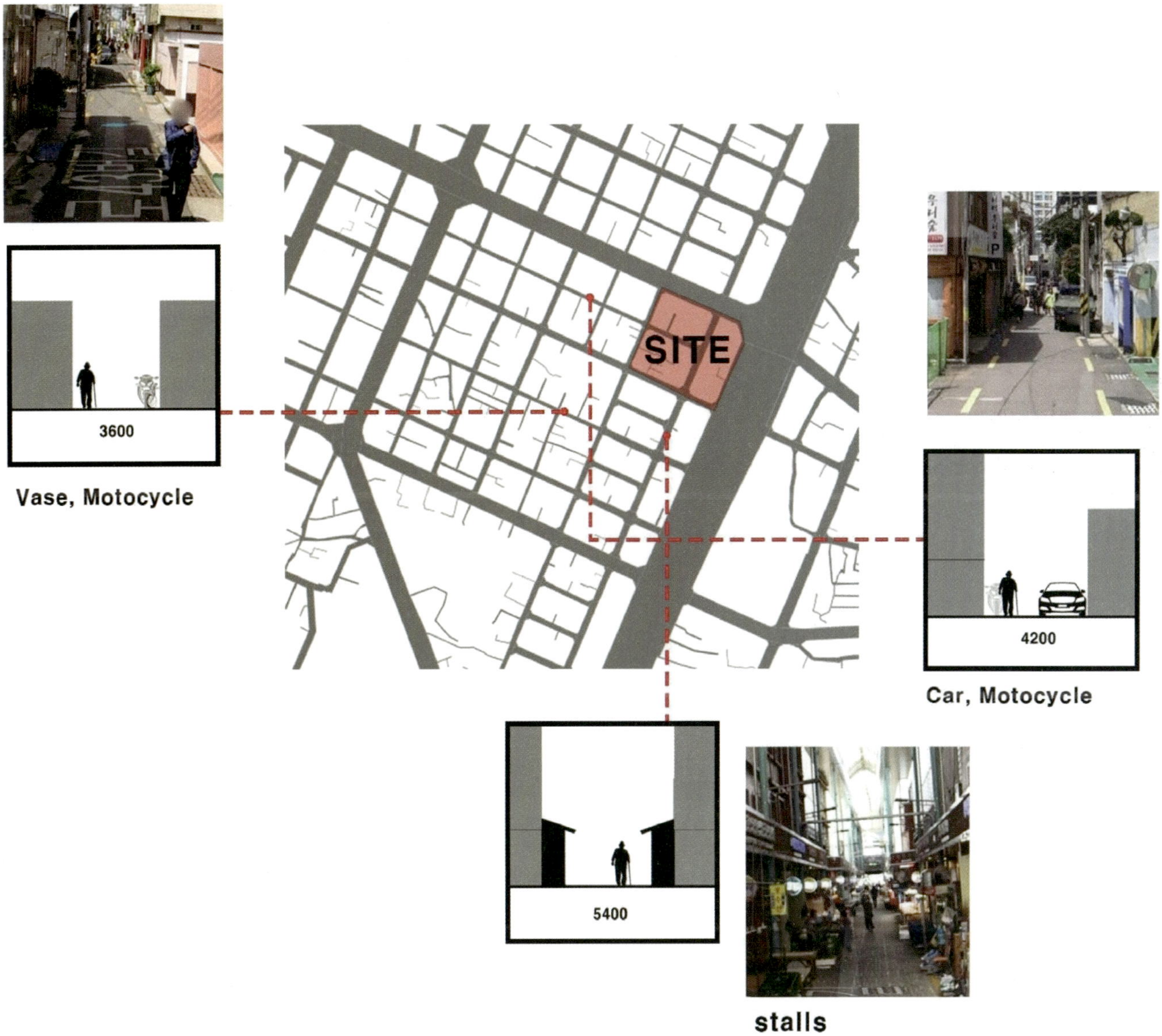

There are trails and street markets around the site. Cars and motorcycles are parked on the road, and a narrow alleyway of 5.5m wide with a cart is mainly formed.

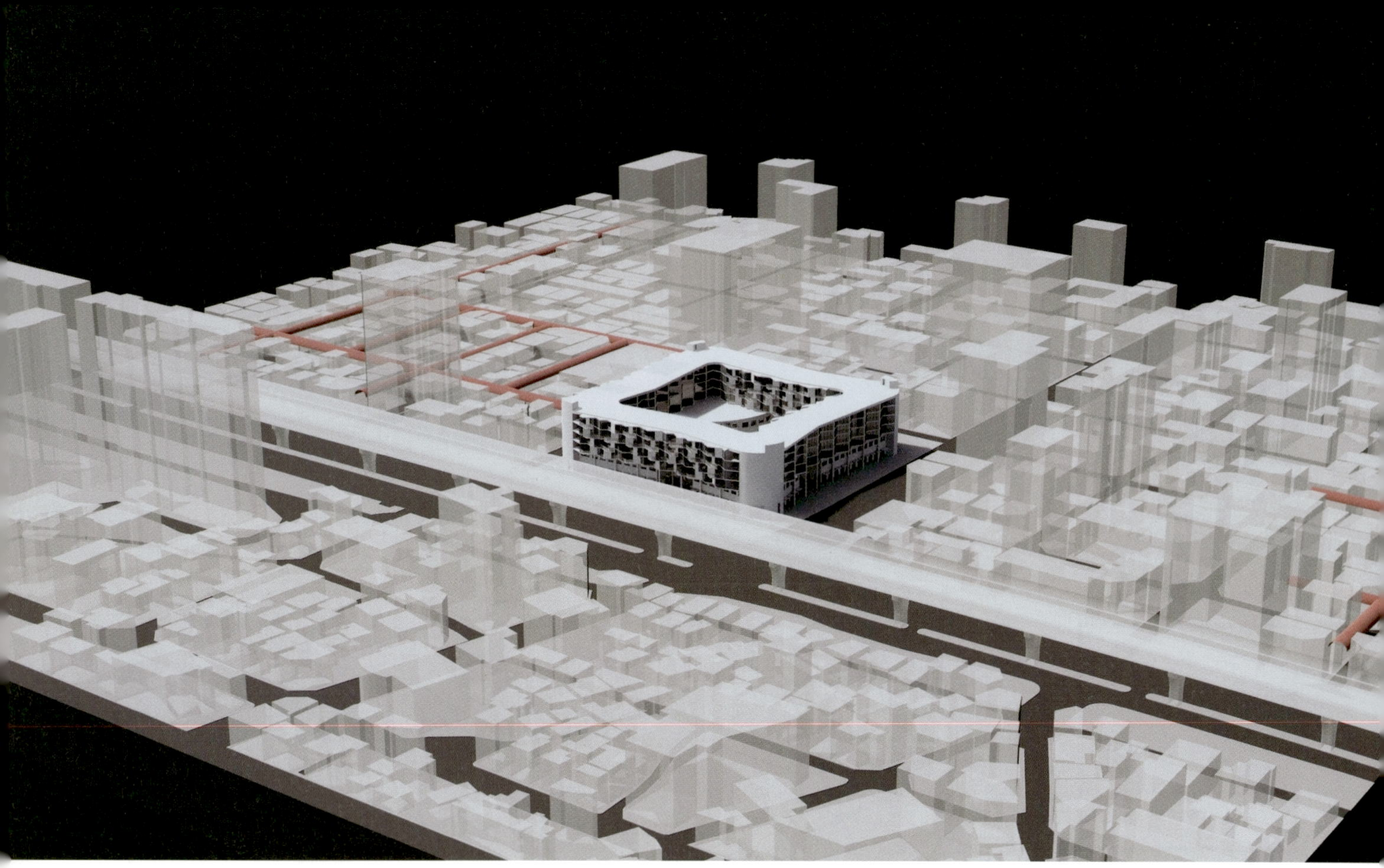

ALLEY MARKET

PROJECT BY
MYEUNGBIN KIM
UNGHUI JUNG

PROJECT INFORMATION
Location : 1110, Taejong-ro, Yeongdo-gu, Busan, Republic of Korea
Use : Mid-Rise Mix-Use
Building area : 10,000sqm
Total Floor area : 37,000sqm
Building coverage ratio : 58%
Floor area ratio : 249%
Structure : SRC
Floor : B1 - 6F
Height : 22m

Our project began with the stage of analyzing existing spaces around a site with publicity, aiming to create public spaces that can communicate with neighbors within the site. On the vertical axis of the site, Namhang Market and Bongnae Market were facing, and on the horizontal axis, commercial and residential building complexes and large boulevard with high bridges were located. In the case of the vertical axis, the flow around the two markets was the most active, and a community between certain generations centered on the elderly occurred. However, the program and spatial structure were uniform and limited for various generations in the region to gather and communicate, so connection and openness were insufficient. In the case of the horizontal axis, small and low buildings were densely built with high density without public space. In addition, the overwhelming overpass cut off areas and sites beyond the boulevard, resulting in a lack of comfort and accessibility.
Based on this analysis, we want to create a cultural hub space that solves generational disconnection and weakening of neighborhood ties by connecting the flow of traditional markets within the site and forming a square at the intersection of the vertical and horizontal axes.

우리의 프로젝트는 대지 내에 이웃과 함께 소통할 수 있는 공공공간의 창출을 목표로, 공공성을 지닌 대지 주변의 기존 공간들을 분석하는 단계부터 시작되었다. 대지의 세로축에는 남항시장과 봉래시장이 면해 있었고, 가로축에는 상업용 및 주거용 건물단지와 높은 다리가 있는 큰 대로가 자리 잡고 있었다. 세로축의 경우 두 시장 주변의 유동 흐름이 가장 활발하고 고령자 위주 특정 세대 간의 커뮤니티가 발생했다. 그러나 지역의 다양한 세대가 모여 소통하기에는 프로그램과 공간 구조가 일률적이고 제한적이어서 연계성과 개방성이 부족했다. 가로축의 경우 작고 낮은 건물들이 공공공간 없이 높은 밀도로 빽빽하게 들어서 있었다. 또한 압도적인 규모의 고가도로는 대로 너머의 지역과 대지를 단절시키고 있어 쾌적성과 접근성이 부족했다.
이러한 분석을 토대로 우리는 강한 공공성을 가진 세로축의 관입을 통하여 대지 내에서 전통시장의 흐름을 연결시키고 세로축과 가로축의 교차부에 광장을 형성해 접근성, 연계성, 개방성, 쾌적성을 높인 공공성을 가진 공간을 창출하여 세대간 단절과 이웃 유대 약화를 해결하는 문화 허브 공간을 만들고자 한다.

SITE ANALYSIS

CONTEXT / LOCATION

Yeongdo-gu, Busan, once boomed due to the revival of the shipbuilding industry, but now after the industrial downturn, young people have fled inland looking for jobs, resulting in an aging population and intensifying disconnection between generations.
MRMU, which will be newly built in areas where social problems such as lonely death arise due to weakening relations with neighbors, was important to play a public role in helping restore relations with neighbors as a cultural hub space for communities that can communicate with various generations.

부산광역시 영도구는 한때 조선업의 부흥으로 호황을 이루었지만, 산업 침체기 이후 현재는 젊은 사람들이 내륙으로 일자리를 찾아 빠져나가면서 고령화가 발생하고 세대 간 단절이 심화되었다. 이웃과의 관계 약화로 고독사와 같은 사회문제가 대두되는 지역에서 새롭게 지어지게 될 MRMU는 다양한 세대와 상호 소통 가능한 지역 사회의 문화 허브 공간으로서 이웃과의 관계 회복을 보조할 공공적 역할을 행하는 것이 중요했다.

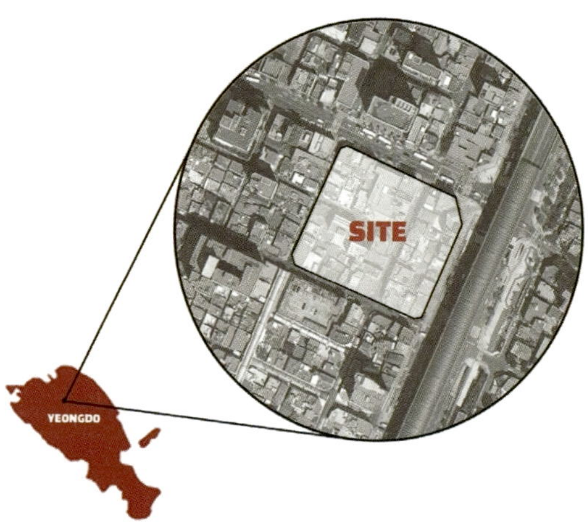

A FACTOR OF INTEREST

TRADITIONAL MARKET

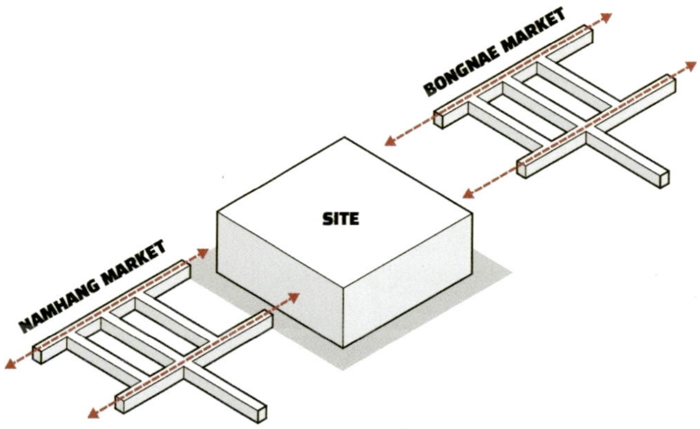

Our site is located between Namhang Market and Bongnae Market, the largest traditional market in Yeongdo. And these two traditional markets are the most public spaces nearby.

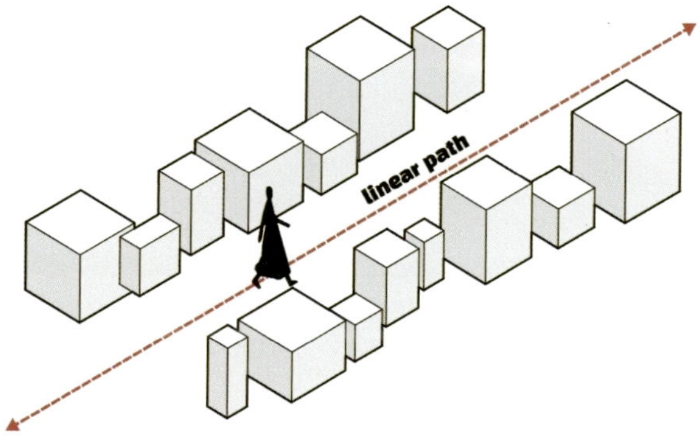

We tried to find a new solution to publicity by conducting a spatial analysis of the existing traditional market in order to establish a close relationship with the site. In addition, it was found that most of the existing traditional markets are spaces with linear paths.

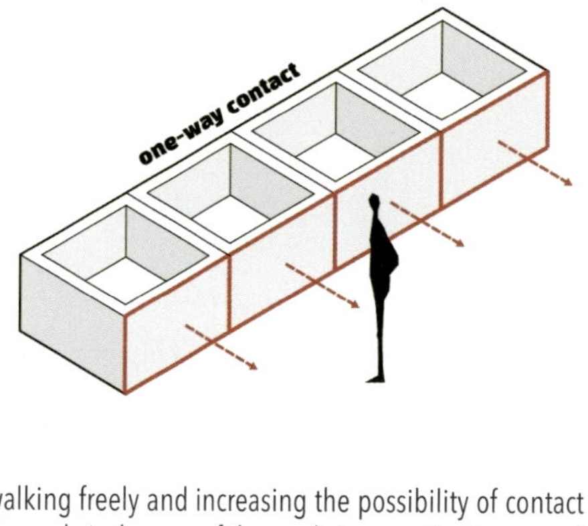

Although walking freely and increasing the possibility of contact between objects and people is the core of the market space, the linear market allows only two directions of gaze and movement, and the stand and the customer face only one side.

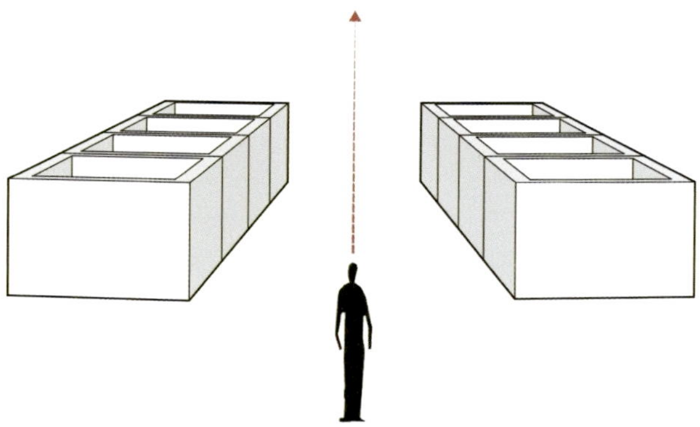

This linear space is interpreted as a space of speed and efficiency for the purpose of passing through, which can be said to be a space structure in which staying, meeting, and communication are impossible.

STATEMENT

YEONGDO TIMELINE

1934

Yeongdo Bridge was the first Dogaegyo Bridge established in Korea and developed through this Yeongdo Bridge by interacting with the land.

1601

Yeongdo is originally a place name shortened from Jeolyeongdo Island. The meaning of Jeolyeongdo Island may be a metaphorical name for an island with a broken shadow, that is. Sandoksan Mountain, which stopped visiting.

1960

Due to its regional geographical characteristics. Yeongdo was once greatly revived as a shipbuilding industry. but now it has declined due to the industrial downturn.

2008

Currently, Yeongdo is an area where young people leave and aging intensifies, and social problems such as the lonely death of the elderly are emerging.

DESIGN CONCEPT

Alley market – The Realization of Traditional Market
with the Characteristics of Alleyways

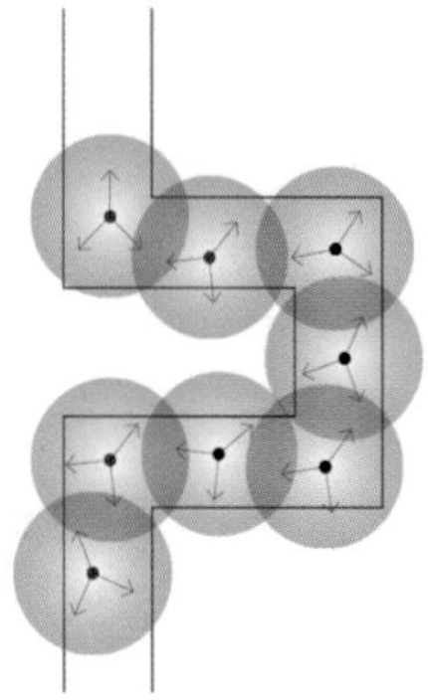

Various Directions

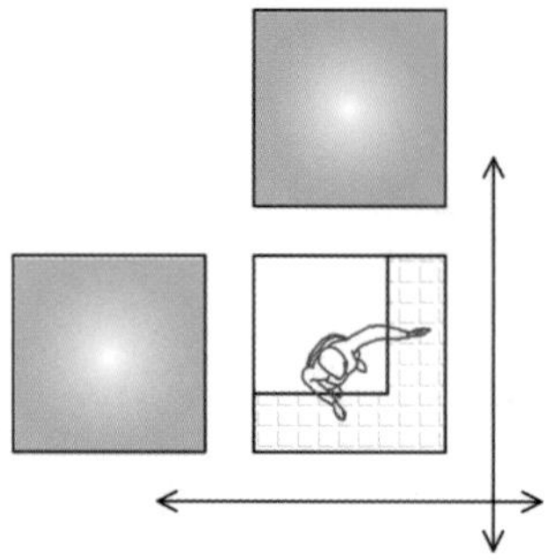

Multi-faceted Store

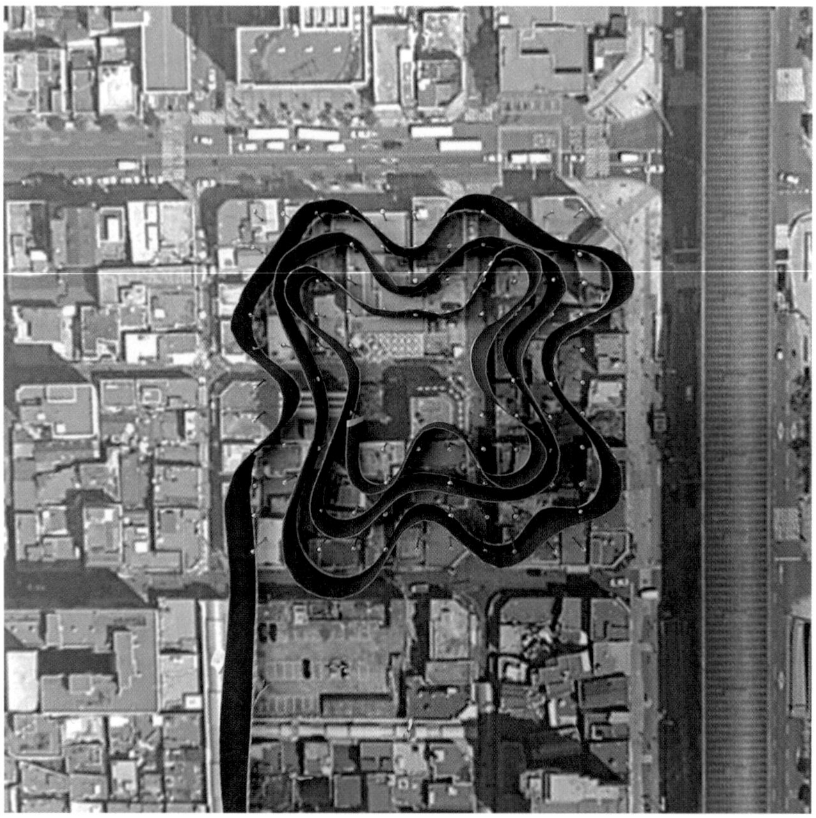

We propose a new cultural hub space that connects the flow of the existing traditional market to the site but restores the disconnectivity of the existing site by increasing the opportunity for people to stay and communicate through the implementation of spaces with the characteristics of alleys rather than linear spaces.

기존 전통시장의 흐름을 현장으로 연결하되 선형 공간이 아닌 골목길의 특성을 가진 공간 구현을 통해 사람들이 머물며 소통할 수 있는 기회를 늘려 기존 현장의 단절성을 회복하는 새로운 문화허브 공간을 제안한다.

ALLEY MARKET

The market, which was created with the motif of alleys, actively communicated with each other by increasing unexpected encounters with people using ordinary spaces and winding roads.

골목길을 모티브로 만든 시장은 평상공간, 구불구불한 길을 이용해 사람들과의 예기치 못한 만남을 늘림으로써 서로 소통이 적극적으로 일어나도록 하였다.

PLAZA, MARKET

The central square and surrounding market allow people to come out to the square to eat, talk, and also enjoy shopping.

중앙의 광장과 그를 둘러싼 시장 공간은 사람들이 나와 음식을 먹거나 이야기를 나눌 수 있고, 또한 쇼핑도 즐길 수 있도록 한다.

DOMESTIC: PYEONGSANG SPACE

It is a space located at four corners where the movement lines intersect. When you come up through the core, you can see the space where people can sit and communicate, the void space inside the building, and the hallway connected to the outside air at a glance.

동선이 교차하는 4개의 모서리에 위치한 공간. 코어를 통해 올라오면 골목길의 평상처럼 사람들이 앉아 소통할 수 있는 공간과 건물 내부의 보이드 공간, 그리고 외기와 연결된 복도가 한눈에 들어온다.

DOMESTIC CORRIDOR

The corridor connected to each house becomes a space where communication with people takes place actively like an alleyway.

각자의 집으로 이어지는 복도는 골목길처럼 사람들과의 소통이 적극적으로 일어나는 공간이 된다.

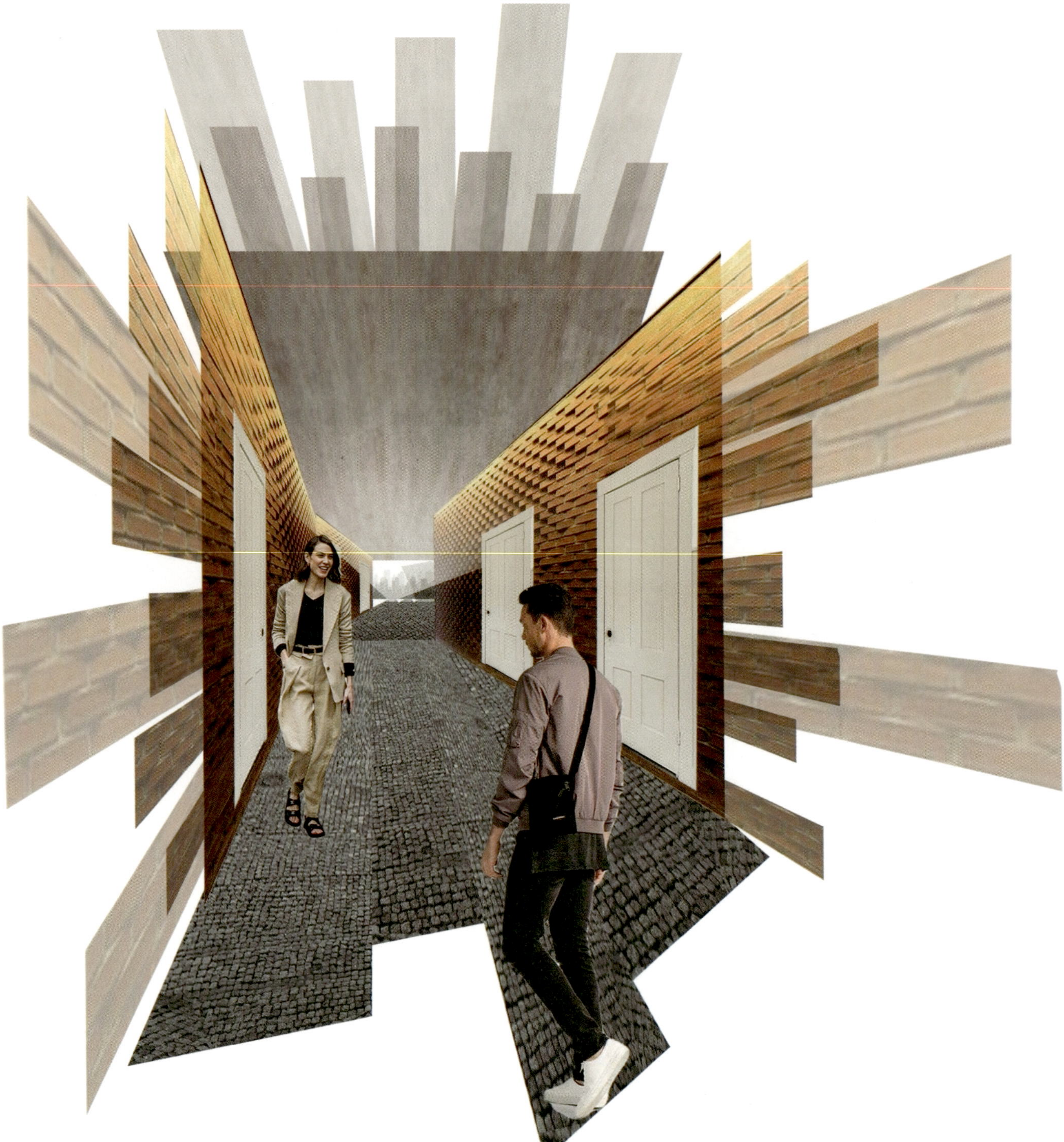

AN ARCHITECTURAL DEVICE FOR THE ALLEY MARKET

1. WINDING PATH

The winding movements of alleys and their characteristics

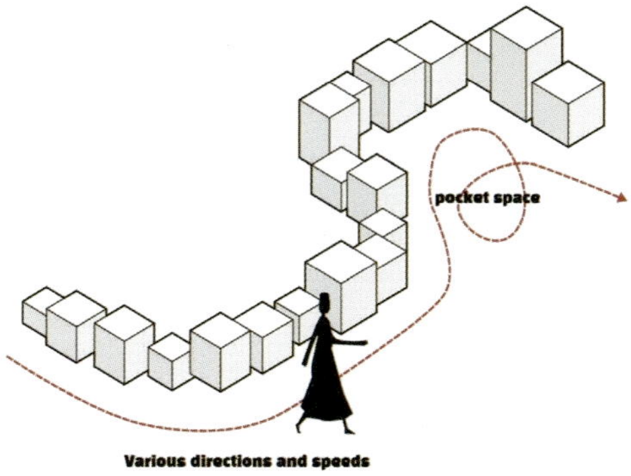

Rule

About three to four stores are set to a range where people can visually recognize and interact at the same time, and the movement line is bent based on this range.

2. PYEONGSANG SPACE

The Pyeongsang Space of alleys and their characteristics

Rule

Formed in a space where two or more axes intersect and multiple directions converge. Sequence of entering each destination after passing through the normal space.

APPLY - Commercial

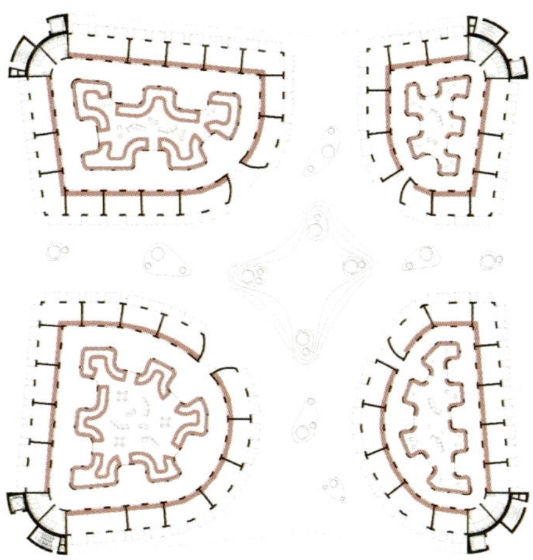

APPLY - Commercial

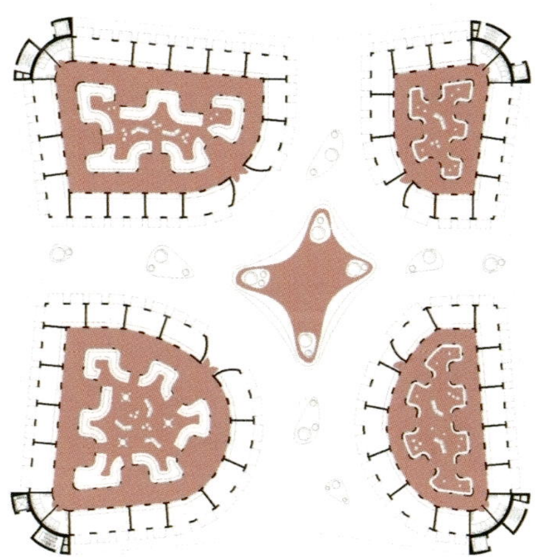

APPLY - Domestic

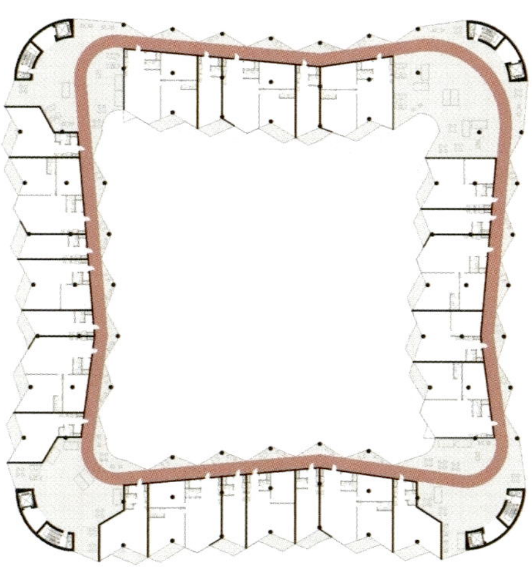

APPLY - Domestic

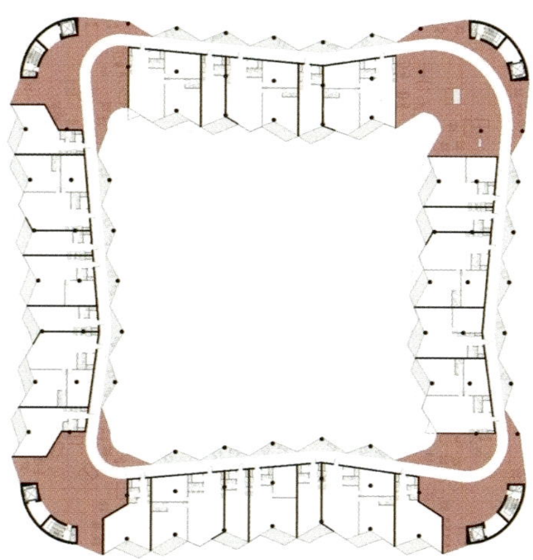

MASS PROCESS

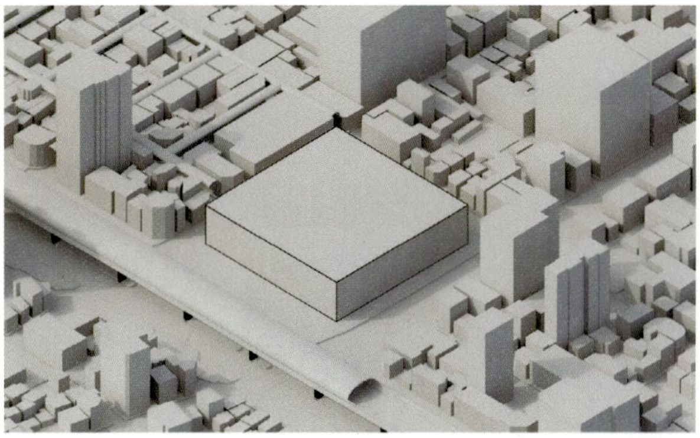

BUILDING BLOCK

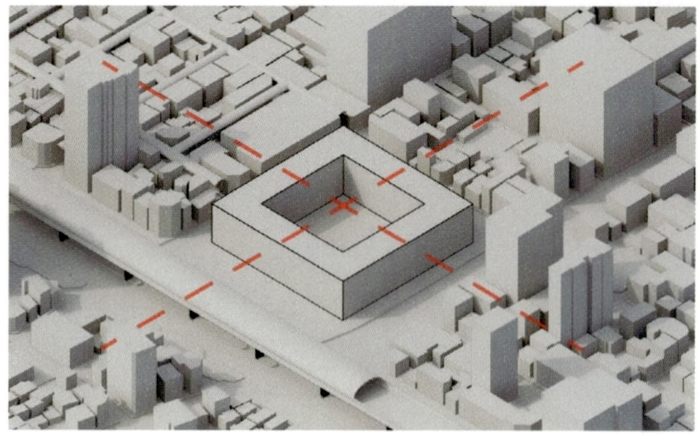

INSERT
PUBLIC SPACE

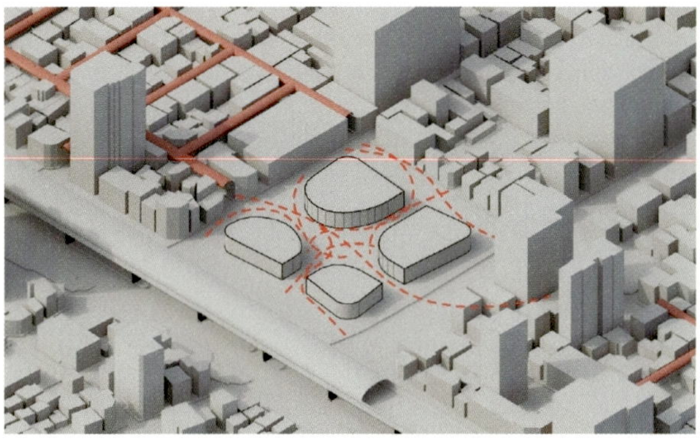

CONNECTING MARKET
ALLEY MARKET

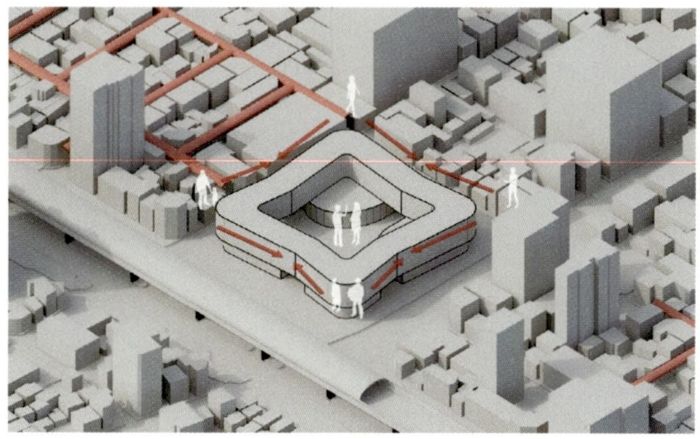

INSERT
AXIS INFLOW

DOMESTIC MASS PROCESS

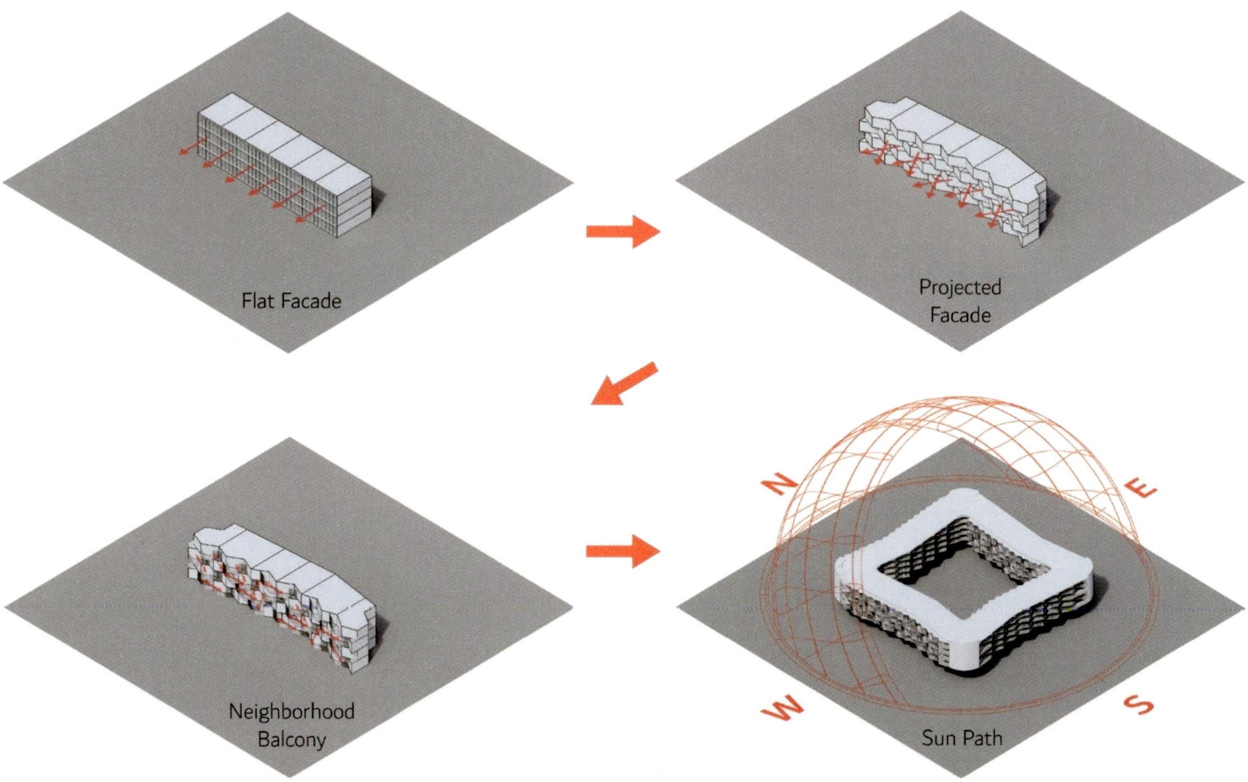

MRMU PROGRAM

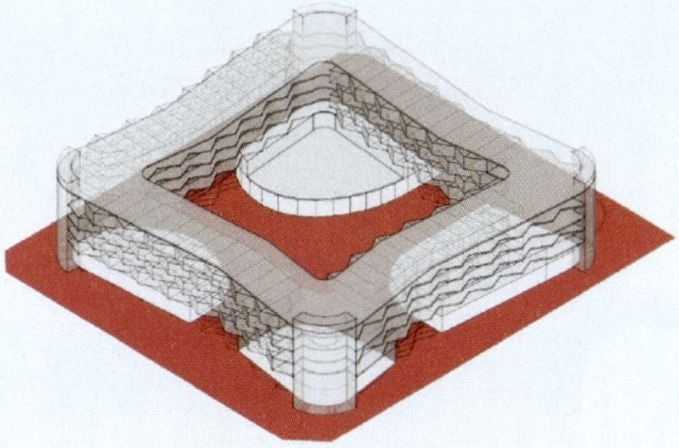

PUBLIC 25%
6,800sqm

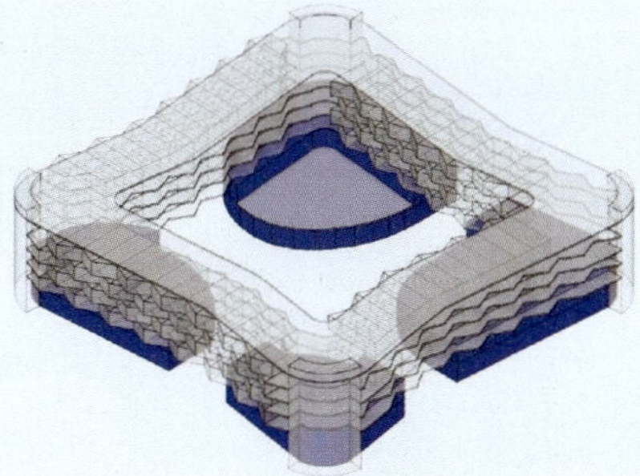

COMMERCIAL 20%
5,500sqm

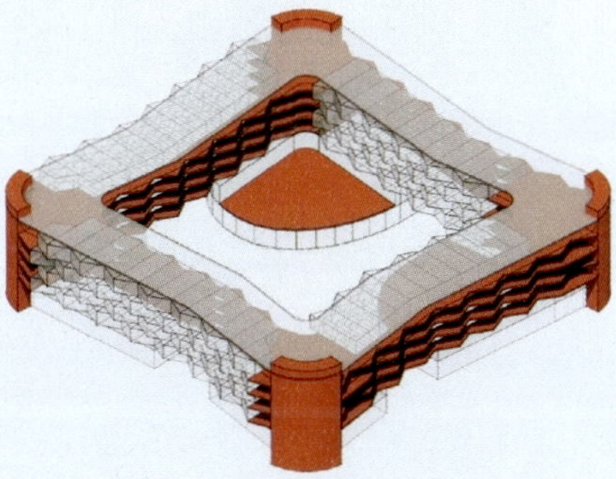

COMMUNAL 24%
7,500sqm

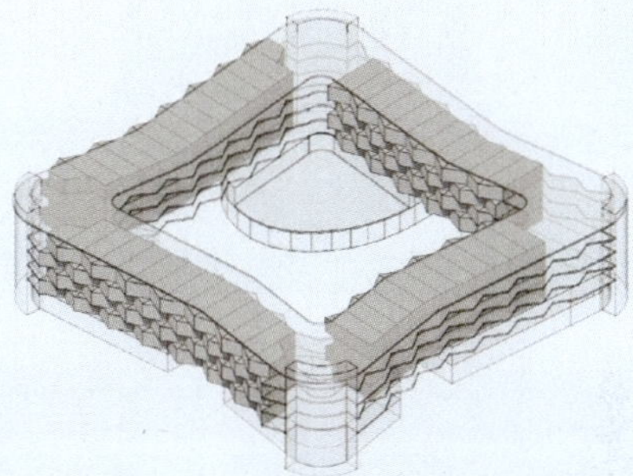

DOMESTIC 30%
8,000sqm

MODULE TYPE

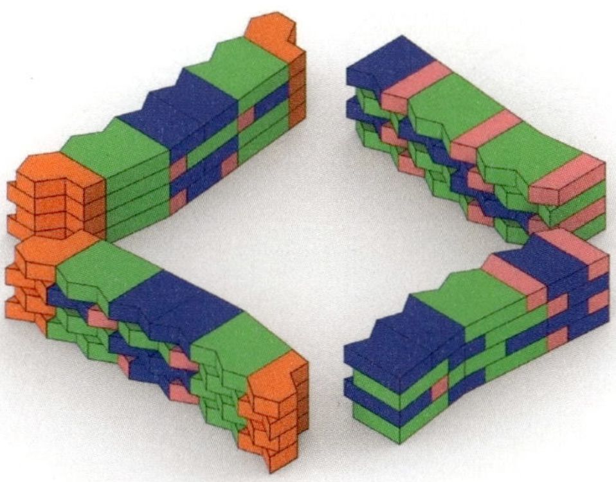

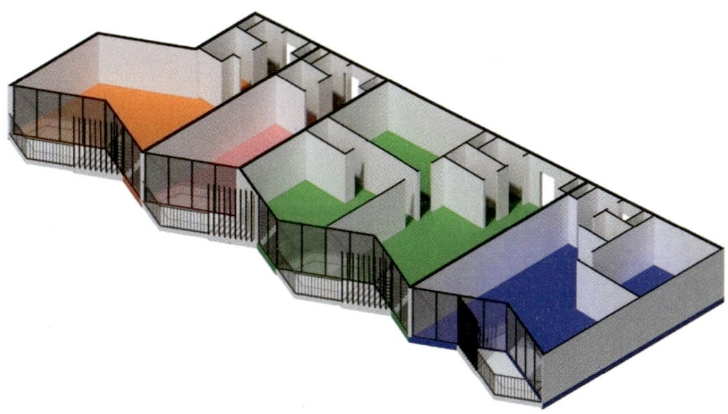

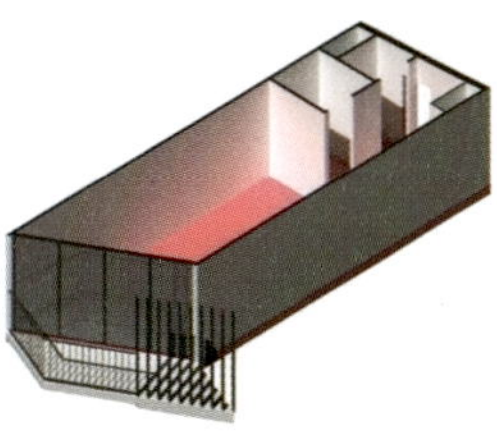

MY HOUSE

45 A TYPE
22households

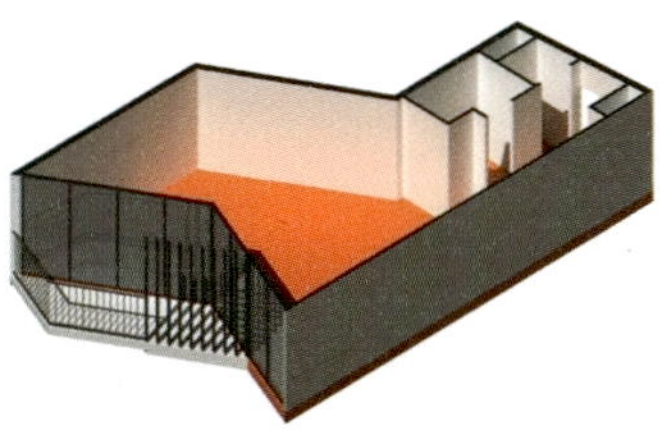

MULTI HOUSE

60 A TYPE
16households

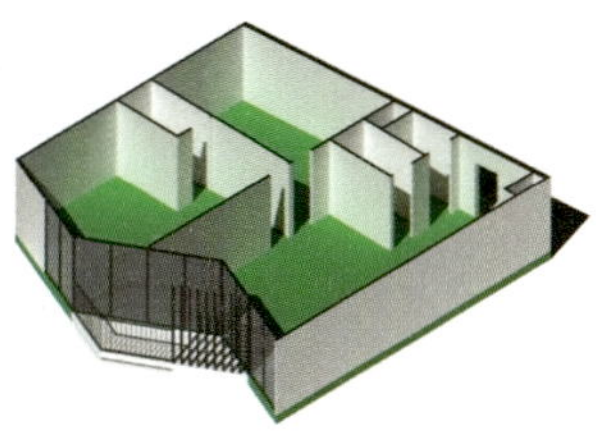

DUAL HOUSE

95 B TYPE
28households

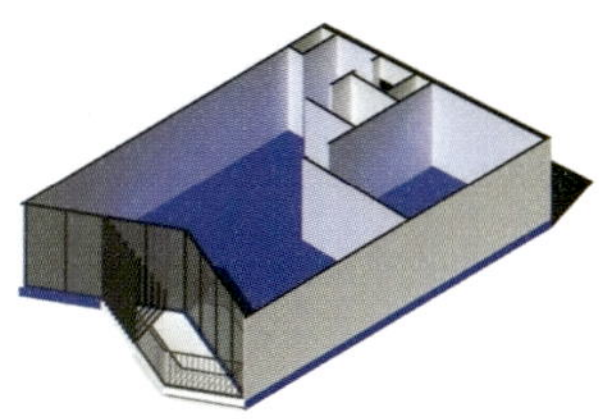

GROWING HOUSE

130 C TYPE
30households

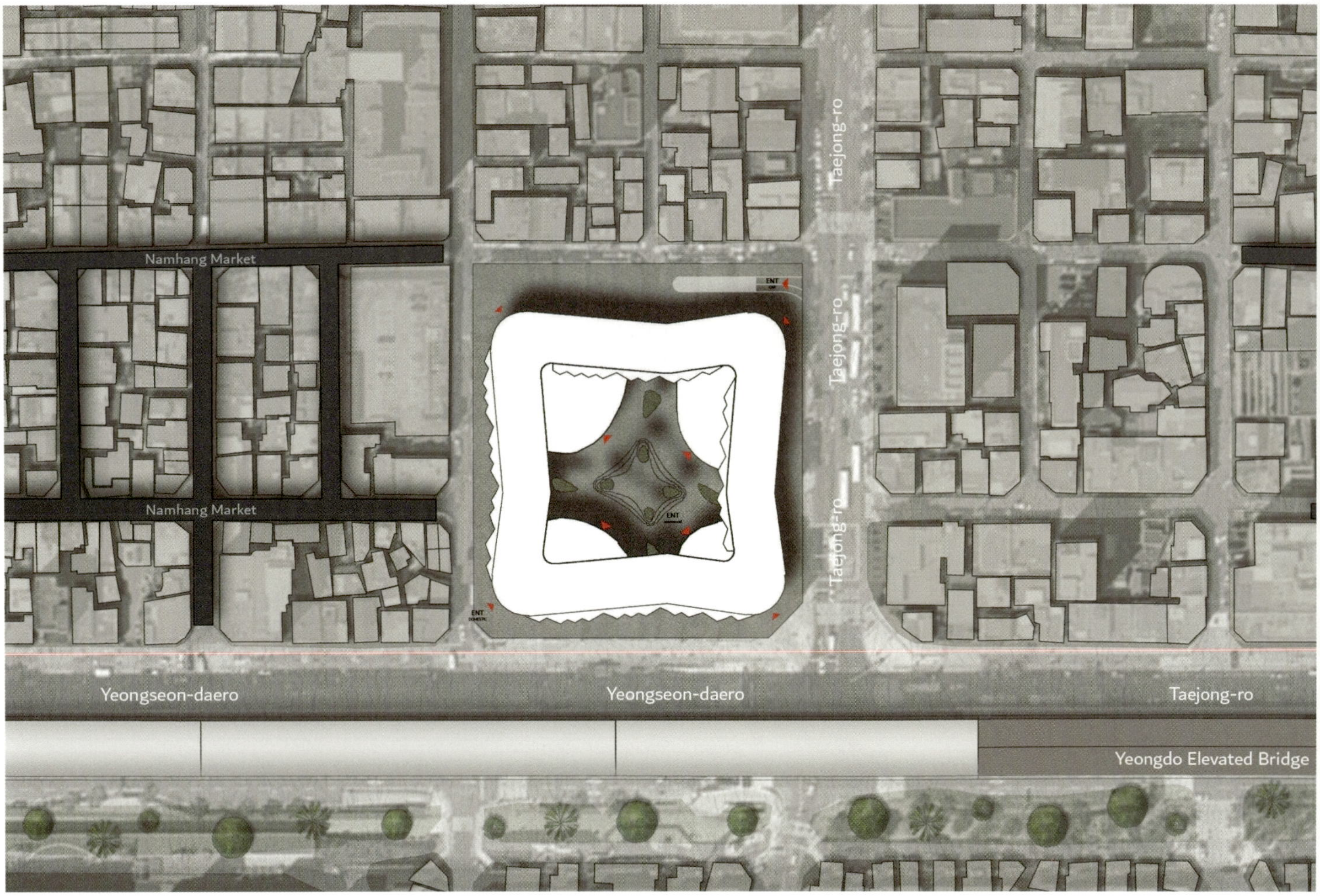
Namhang Market
Namhang Market
Taejong-ro
Taejong-ro
Taejong-ro
Yeongseon-daero
Yeongseon-daero
Taejong-ro
Yeongdo Elevated Bridge

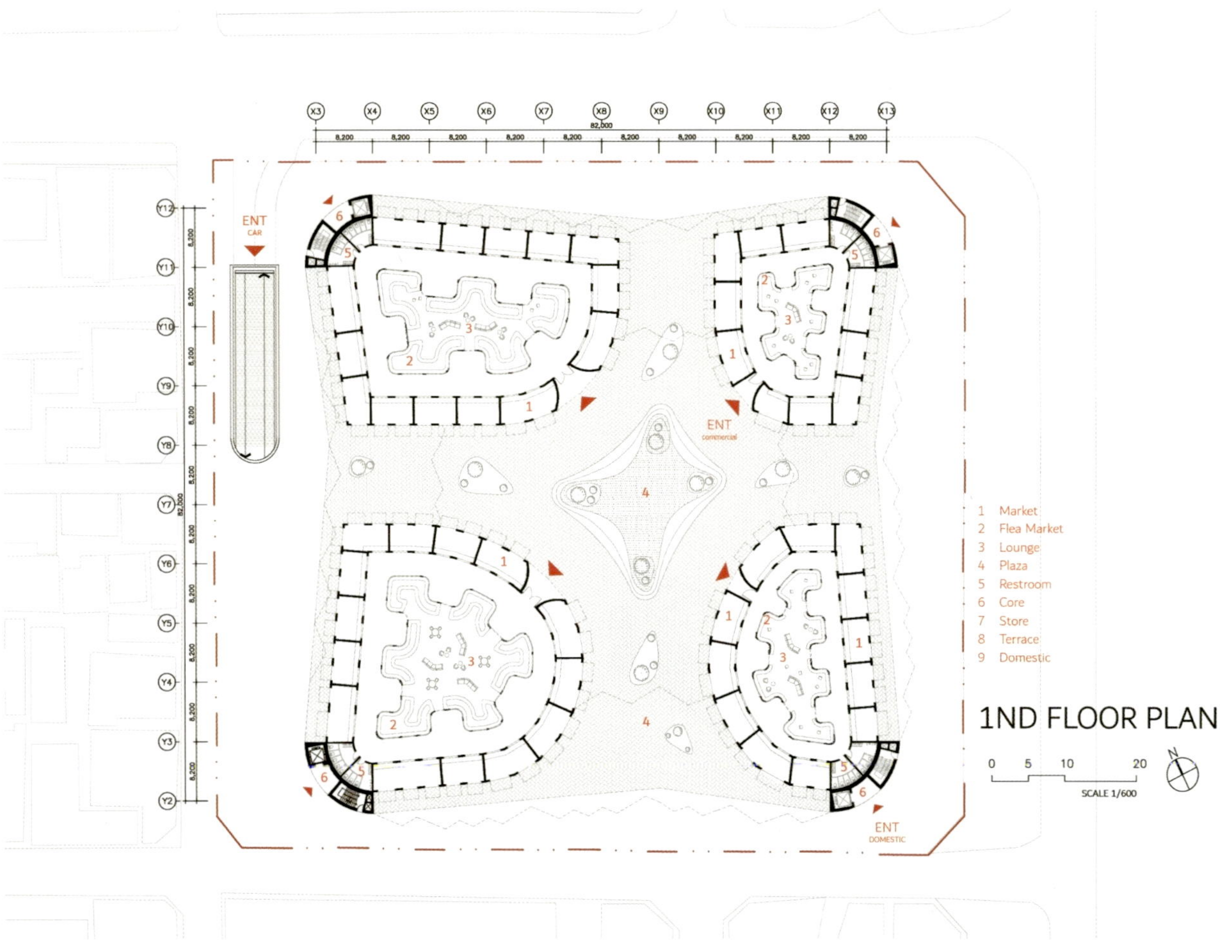
ENT CAR
ENT commercial
ENT DOMESTIC
1 Market
2 Flea Market
3 Lounge
4 Plaza
5 Restroom
6 Core
7 Store
8 Terrace
9 Domestic
1ND FLOOR PLAN
SCALE 1/600

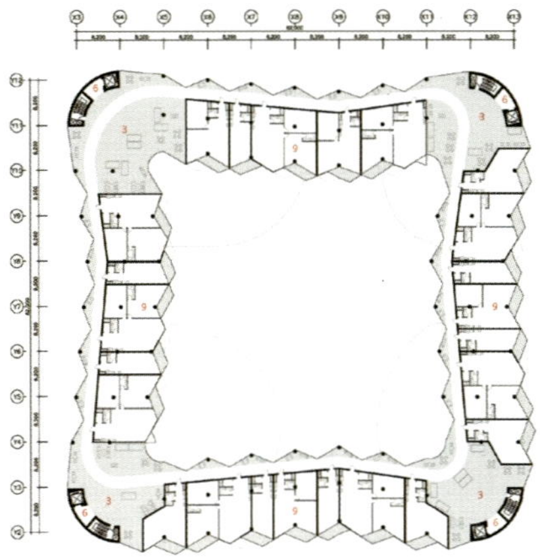

1 Market
2 Flea Market
3 Lounge
4 Plaza
5 Restroom
6 Core
7 Store
8 Terrace
9 Domestic

3ND FLOOR PLAN

0 5 10 20
SCALE 1/600

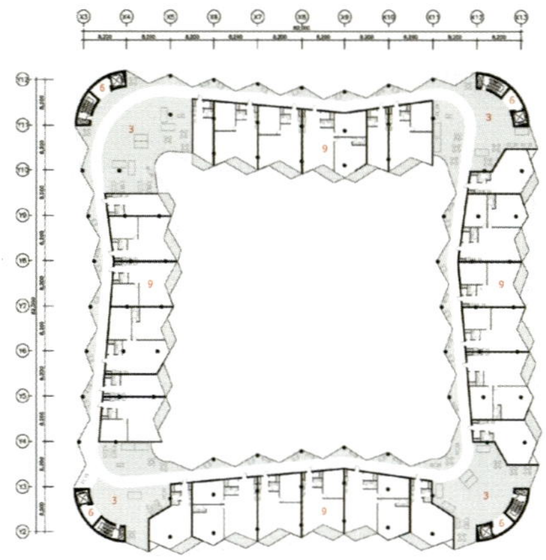

1 Market
2 Flea Market
3 Lounge
4 Plaza
5 Restroom
6 Core
7 Store
8 Terrace
9 Domestic

4ND FLOOR PLAN

0 5 10 20
SCALE 1/600

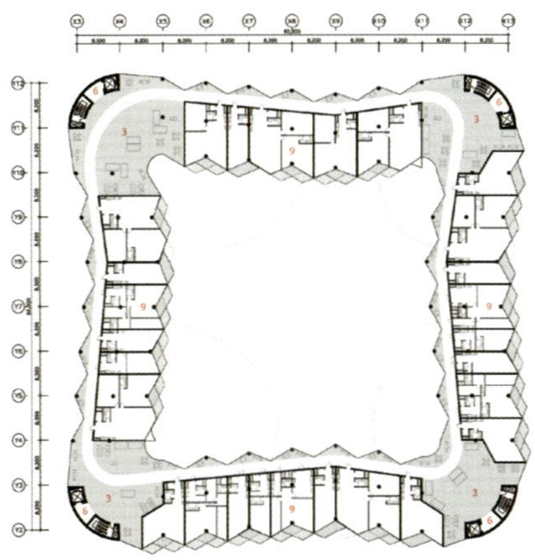

1 Market
2 Flea Market
3 Lounge
4 Plaza
5 Restroom
6 Core
7 Store
8 Terrace
9 Domestic

5ND FLOOR PLAN

0 5 10 20
SCALE 1/600

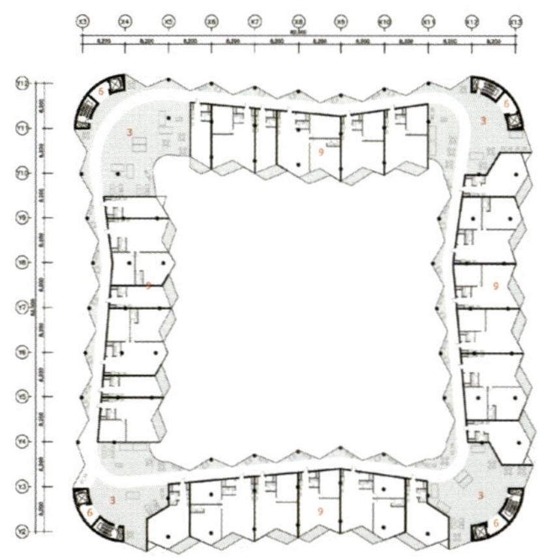

1 Market
2 Flea Market
3 Lounge
4 Plaza
5 Restroom
6 Core
7 Store
8 Terrace
9 Domestic

6ND FLOOR PLAN

0 5 10 20
SCALE 1/600

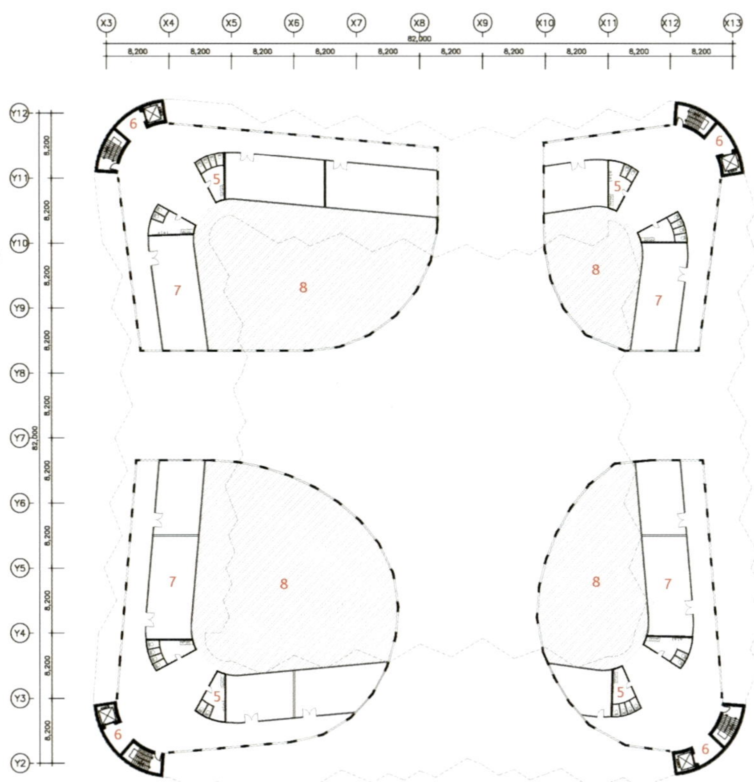

1 Market
2 Flea Market
3 Lounge
4 Plaza
5 Restroom
6 Core
7 Store
8 Terrace
9 Domestic

2ND FLOOR PLAN

0 5 10 20
SCALE 1/600

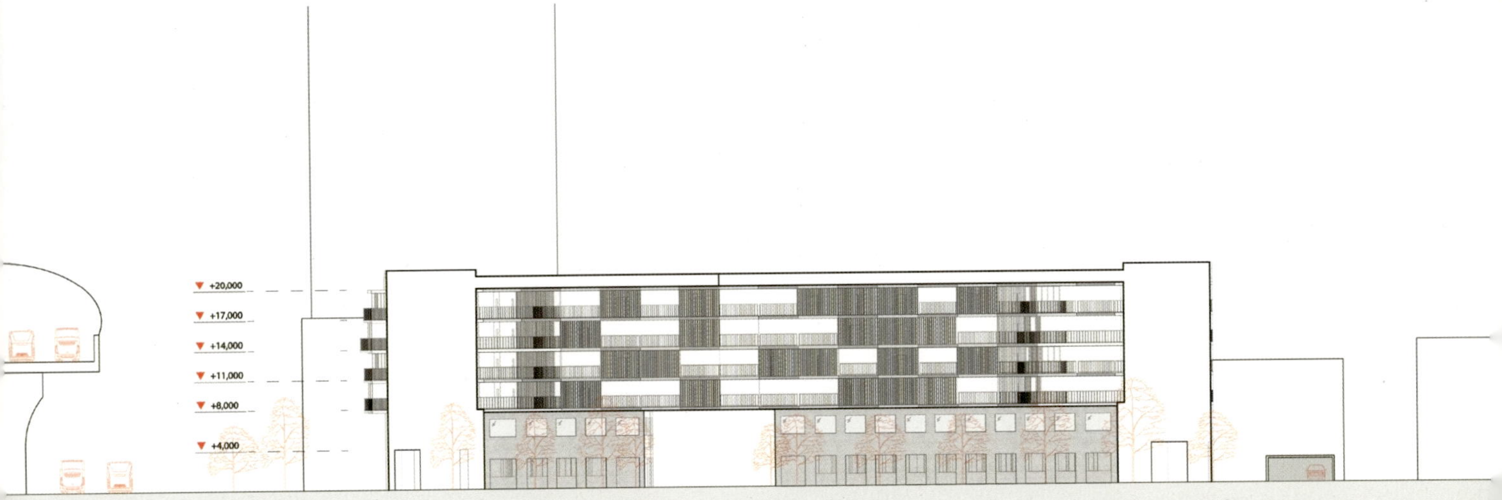

East Side Elevation

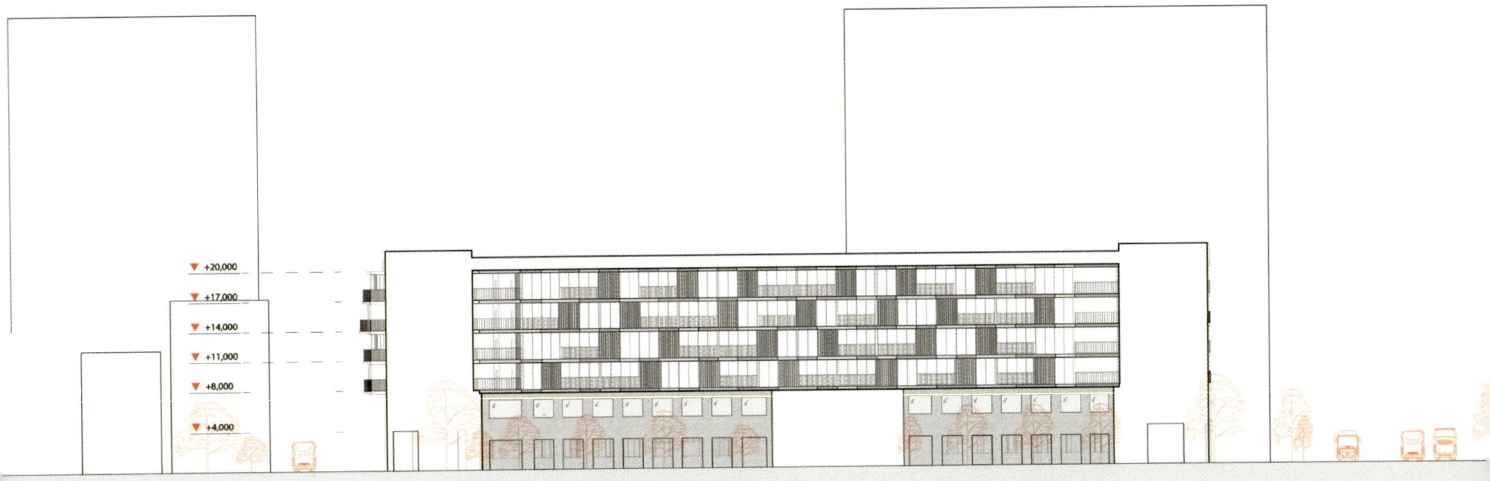

Front Elevation

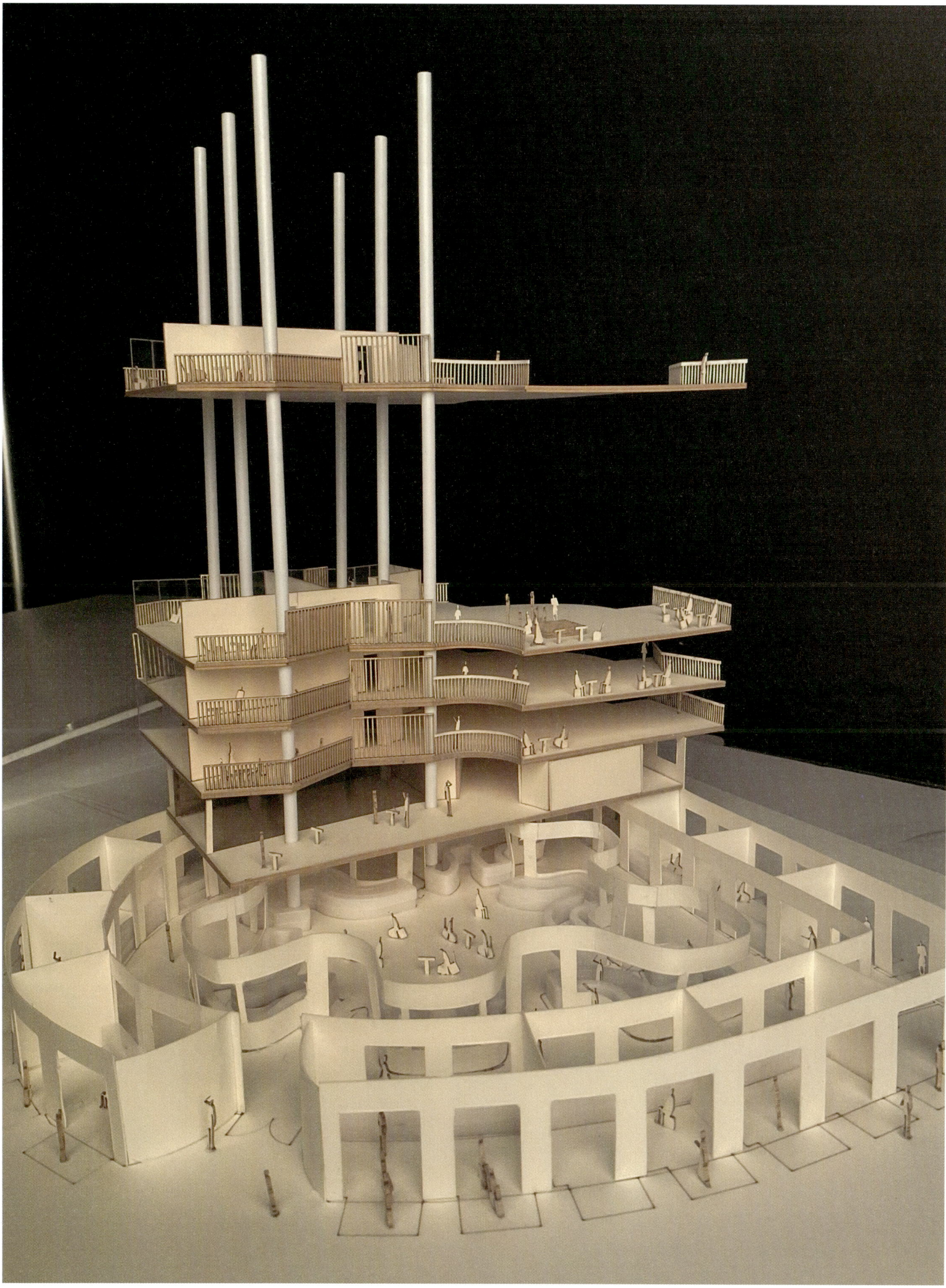

COMMUNITY CANYON PARK APARTMENTS

PROJECT BY
MYEUNGBIN KIM
UNGHUI JUNG

PROJECT INFORMATION
Location : 49, Jeoryeong-ro 49beon-gil, Yeongdo-gu, Busan, Republic of Korea
Use : Mid-Rise Mix-Use
Building area : 3,621sqm
Total Floor area : 19,690sqm
Building coverage ratio : 36%
Floor area ratio : 200%
Structure : RC
Floor : B2 - 6F
Height : 22m

Located in Yeongseon-dong, Yeongdo-gu, Busan, the project site is a neighborhood living district with Namhang Market, a traditional market with a history of 100 years. The market, which has long contained the daily lives of residents of Yeongdo-gu, has a familiar and friendly atmosphere due to a mixture of commerce and housing. We decided to continue this attractive atmosphere unique to the region.
In this project, we envisioned a village where a local community was created naturally by forming streets and squares with images of people's lives. People will pass through the roads and squares facing residential, commercial, and office spaces toward their respective destinations, and in the process, the daily appearance of local people is intuitively shown. It is expected that three rich external spaces, parks, roads, and squares, will be opened throughout the region, making it the center of the local community where residents' eyes are touched and various meetings take place.

부산광역시 영도구 영선동에 위치한 프로젝트 대지는 100년 역사의 전통시장인 남항시장을 주변에 둔 근린생활지구다. 오래 전부터 영도구 주민들의 일상을 담아 온 이 시장에는 상업과 주거가 혼재해 익숙하고 정겨운 분위기가 물씬 풍긴다. 우리는 이 같은 지역 고유의 매력적인 분위기를 고스란히 이어가기로 했다.
이 프로젝트에서 우리는 사람들의 삶의 모습이 담긴 거리와 광장을 형성함으로써 자연스럽게 지역 공동체가 만들어지는 마을을 구상했다. 사람들은 주거, 상업, 사무 공간이 면하는 길과 광장을 스치며 각자의 목적지를 향해 갈 것이며, 그 과정에서 지역 사람들의 일상적인 모습이 직관적으로 나타난다. 공원, 길, 광장이라는 세 가지의 풍부한 외부공간이 지역 전체에 열려 주민들의 시선이 닿고 다양한 만남이 이루어지는 지역 커뮤니티의 중심이 될 것으로 기대한다.

URBAN ISSUES

Around the site, there were complex areas where traditional market streets and small-scale residential and commercial buildings were concentrated. The sequence of moving from public space to private space was mainly found to be the process of moving through alleys to residential spaces, and these alleys were narrow and dark, so there were no elements that could cause encounters and activities. Even large open spaces such as apartments and schools were surrounded by walls. It was also characteristic that the two areas were cut off based on a wide road with an overpass. On the other hand, on some streets, human-scale buildings were gathered and people's daily lives were interacting in them, and we noted this. In this project, we would like to propose a space where the traditional market street and the residential and commercial complex are harmonized.

대지 주변에는 전통시장 거리와 소규모의 주상복합 건물들이 밀집된 복잡한 구역들이 있다. 공적 공간에서 사적 공간으로 이동하는 시퀀스는 주로 골목을 통해 주거 공간으로 이동하는 과정의 형태로 나타났는데, 이들 골목들은 좁고 어두워 쾌적하지 않았으므로 만남과 활동이 일어날 만한 요소들을 찾아볼 수 없었다. 아파트나 학교 등 그나마 대규모의 오픈스페이스들조차 담장으로 둘러싸여 있었다.
고가도로가 놓인 넓은 도로를 기준으로 두 지역이 단절시키는 점도 특징적이었다. 반면에 일부 가로에서는 휴먼스케일의 건물들이 모여 그 속에서 사람들의 일상이 상호작용하고 있었는데, 우리는 이 점에 주목했다. 이번 프로젝트에서는 전통시장 거리와 주상복합 건물군과의 조화가 이루어지는 공간을 제안하고자 한다.

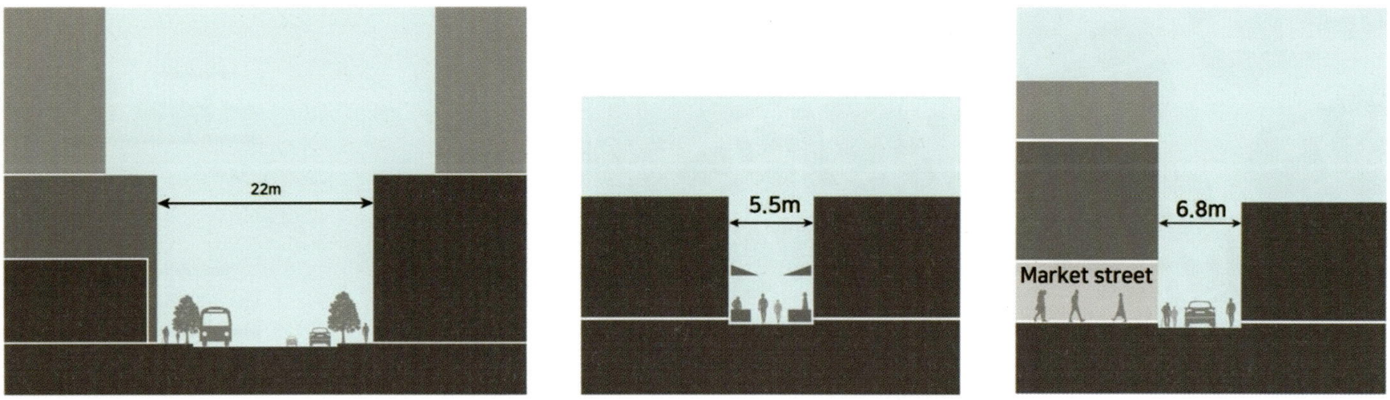

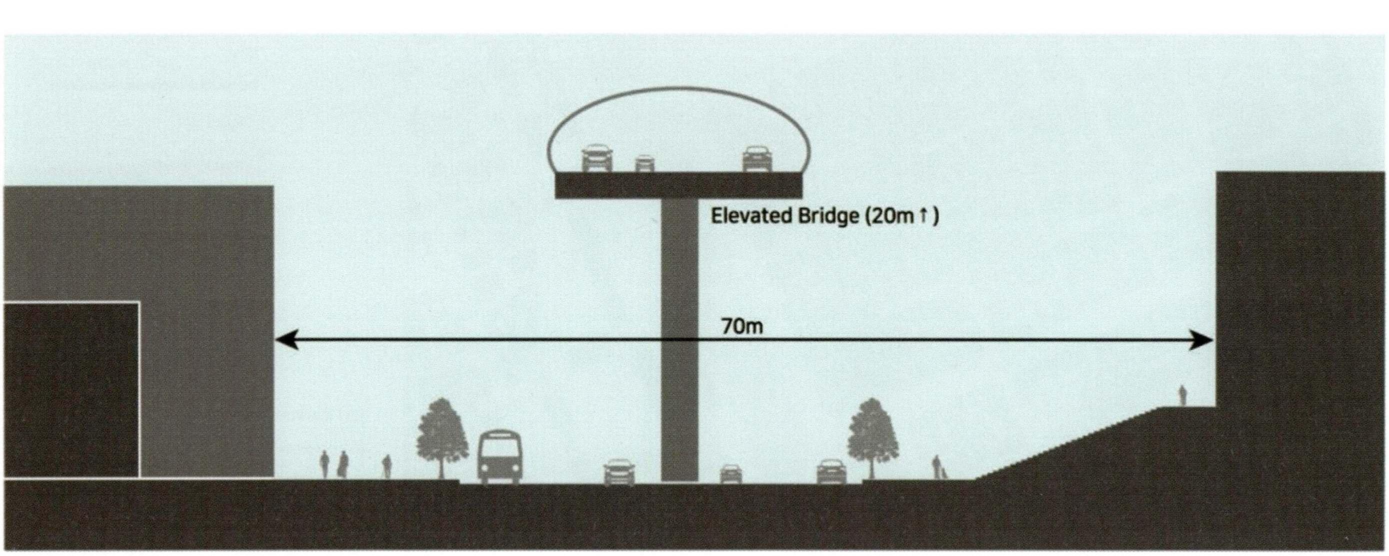

SITE ANALYSIS

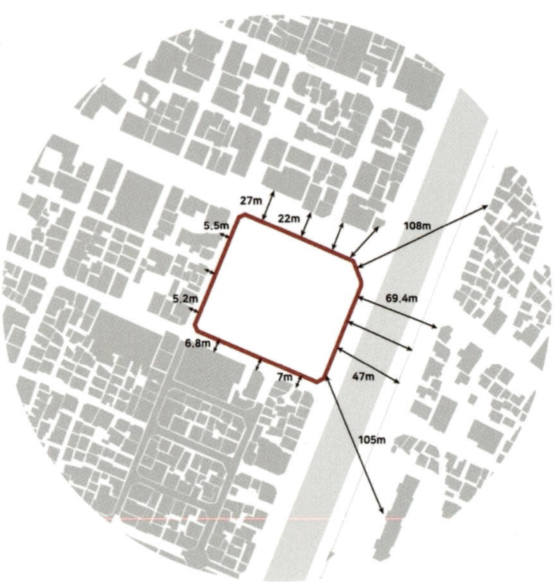

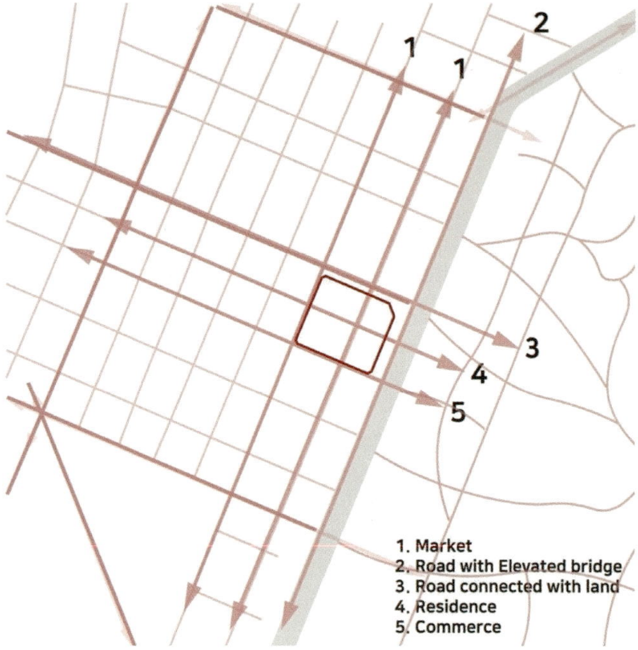

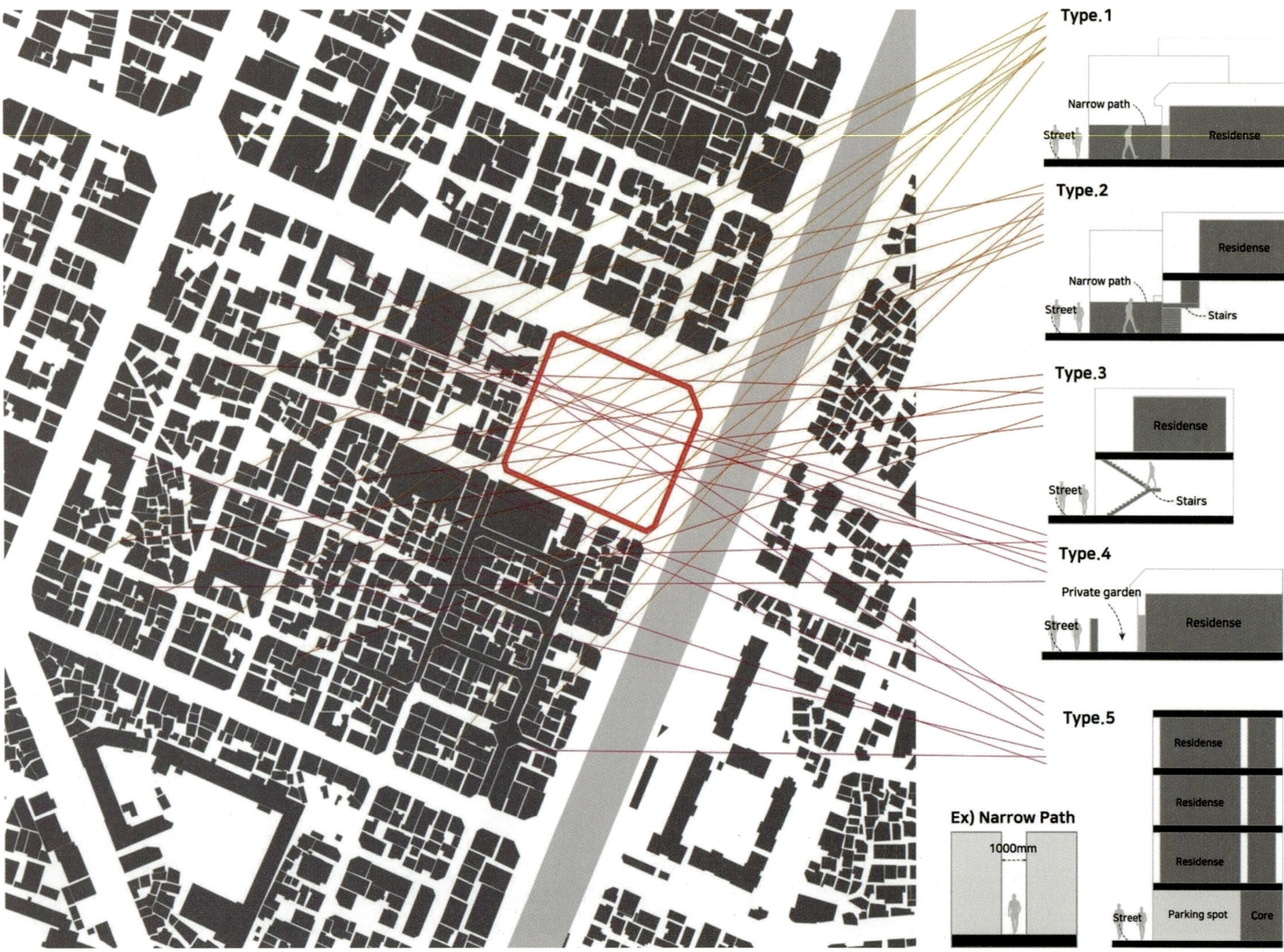

CONCEPT MODEL

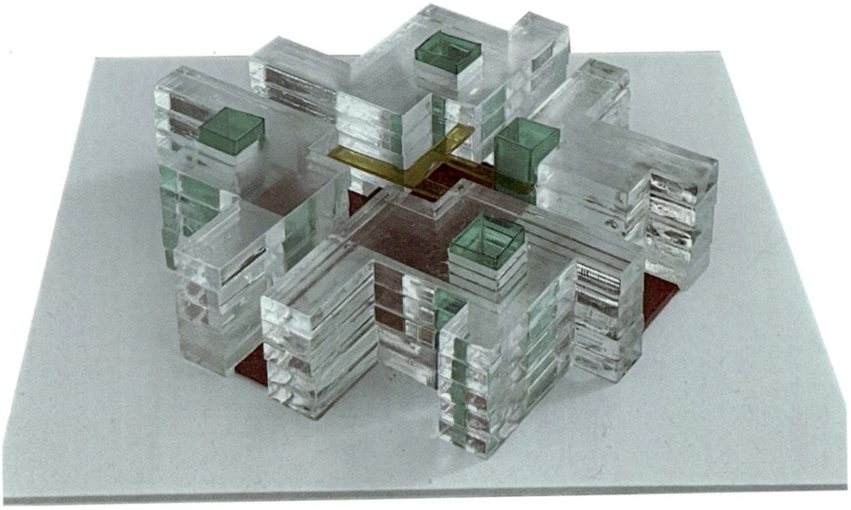

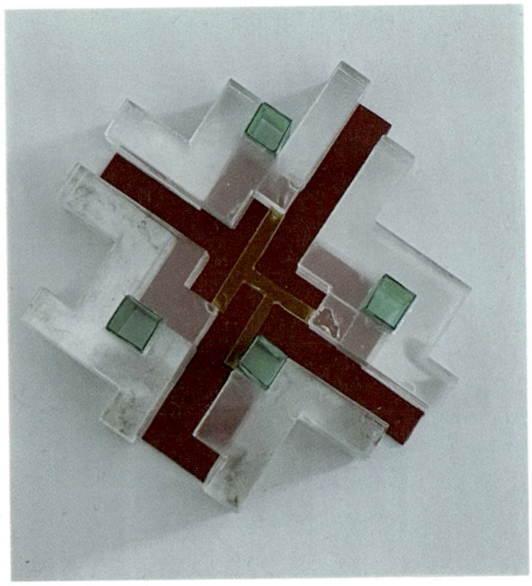

ARCHITECTURAL DEVICES

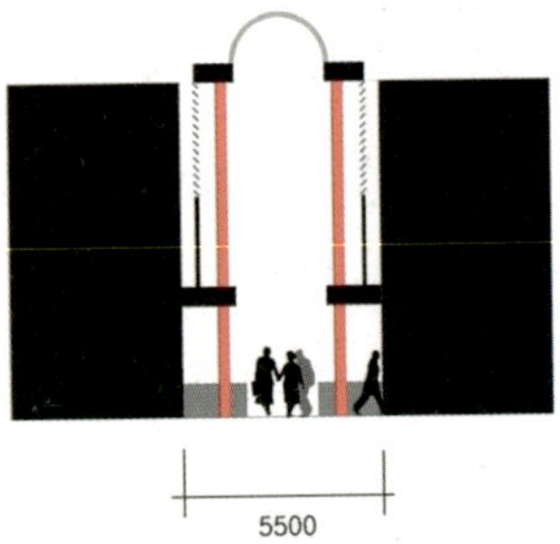

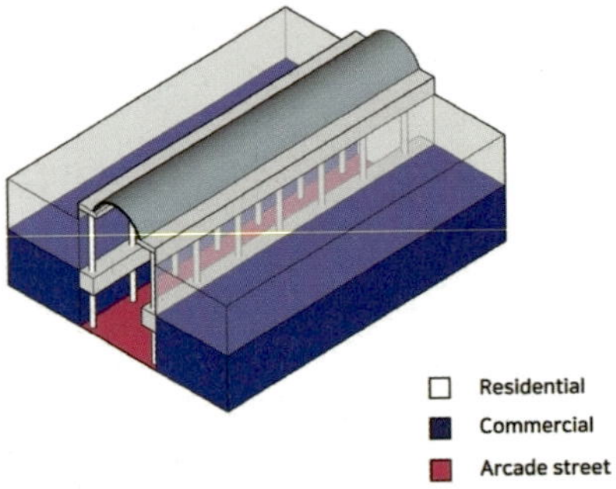

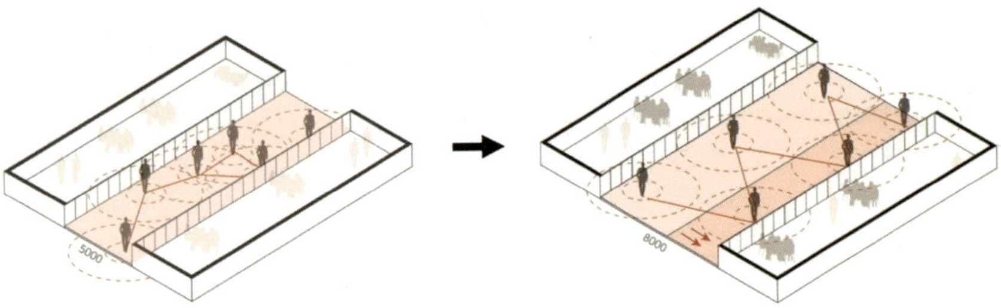

The arcade space of Namhang Market faces a street, so people's daily appearance can be seen intact. We want to use this to create a space where people can form a community at the local level without being isolated.
While creating an environment surrounded by buildings higher than the 5-meter-wide Namhang Market Street, it was planned to be 8 meters wide so that it would not become too complicated. The distance people can have social relationships is 3.6m in radius, so people will be able to communicate more leisurely at a distance of 8m.

남항시장의 아케이드 공간은 하나의 거리를 면하고 있어 사람들의 일상적인 모습들이 고스란히 보인다. 우리는 이를 활용하여 사람들이 고립되지 않고 지역 단위의 커뮤니티가 형성될 수 있는 공간을 만들고자 한다.
5m 폭의 남항시장 거리보다 높은 건물이 둘러싸고 있는 환경을 만들면서도 지나치게 복잡해지지 않도록 8m폭으로 계획했다. 사람들이 사회적 관계를 맺을 수 있는 거리는 반경 3.6m이므로, 8m 폭의 거리에서 사람들은 좀 더 여유롭게 소통할 수 있을 것이다.

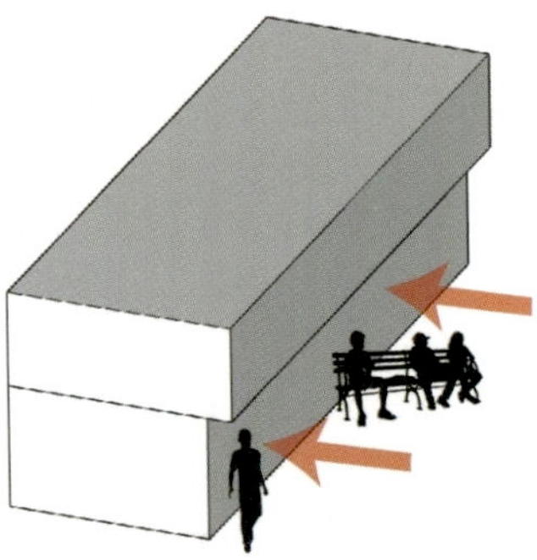

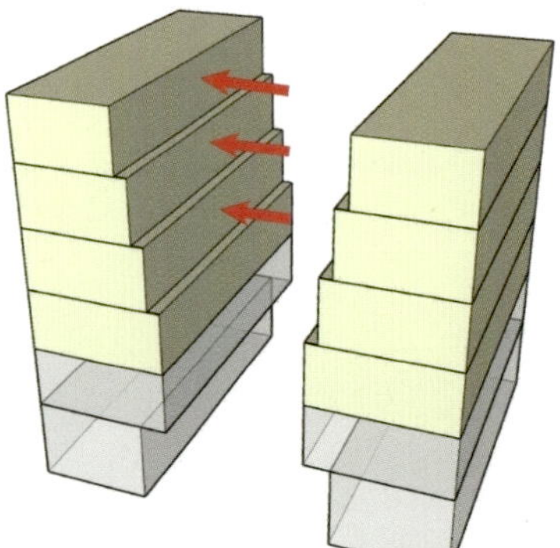

In order to create a daily and relaxed atmosphere, the walls of the ground floor are set-back to add depth to the space. Public spaces with shades have access roads, spaces to sit, and street planting to make people's daily lives the main factor, and upper floors, which are residential spaces, also use the same device to eliminate visual frustration and protect privacy.

일상적이고 여유로운 분위기의 거리를 만들기 위해 지상층의 벽체를 셋백(set-back)하여 공간에 깊이를 더한다. 차양이 있는 공공공간에는 출입로, 앉을 공간, 거리의 식재를 두어 사람들의 일상이 주된 요소가 되도록 하며, 주거 공간인 상층부 역시 같은 장치를 활용하여 시각적인 답답함을 없애고 프라이버시를 보호한다.

SECTION DETAILS

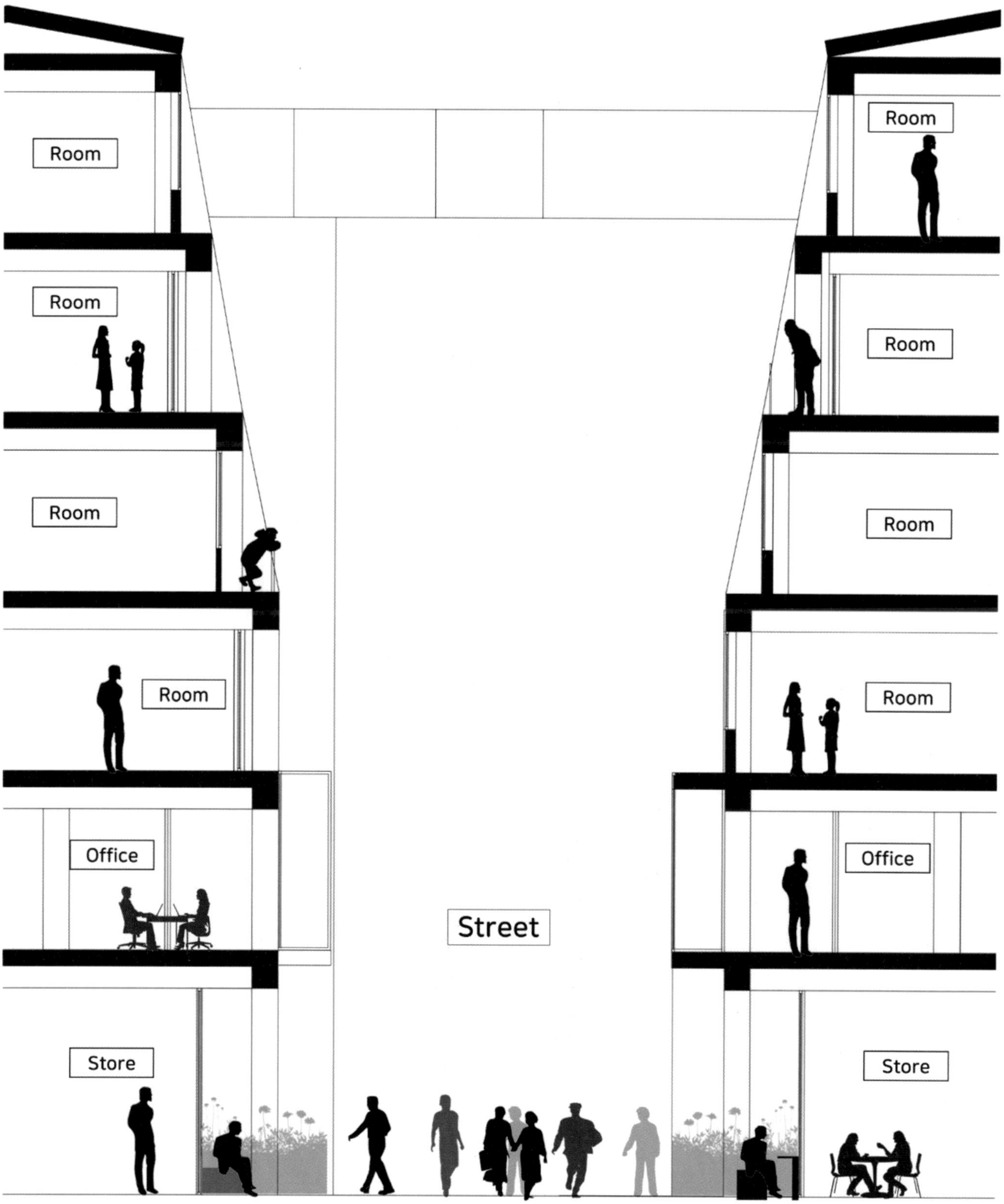

MASS PROCESS

Grid structure

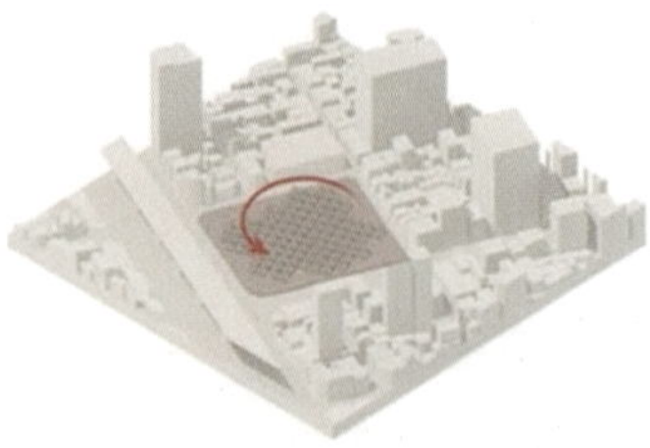

Turn the axis

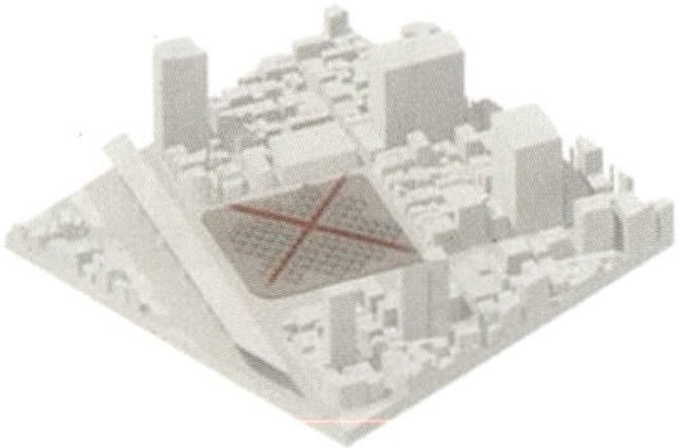

Make a path

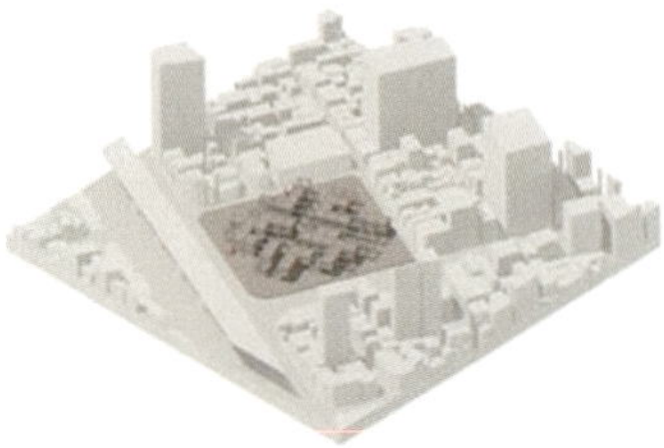

Fill the mass

Stack the mass

MRMU PROGRAM

Through the park and piloti facing the street, residents can naturally meet with local people and interact or stay for a while during the flow of entering their private residential space. The process of moving from the public domain to the private domain was to gradually create a private atmosphere by configuring the floor height to be lower. These spaces are visually open to each other, allowing for continuous experiences.

가로에 면한 공원과 필로티를 통해 거주민이 개인 주거 공간으로 진입하는 흐름 중 자연스럽게 지역 사람들과 만나 교류하거나 잠시 머물 수 있도록 했다. 공적 영역에서 사적 영역으로 이동하는 과정은 층고가 점점 낮아지게 구성해 점진적으로 사적인 분위기를 만들고자 했다. 이 공간들은 서로 시각적으로 개방되어 연속적인 경험을 할 수 있게 한다.

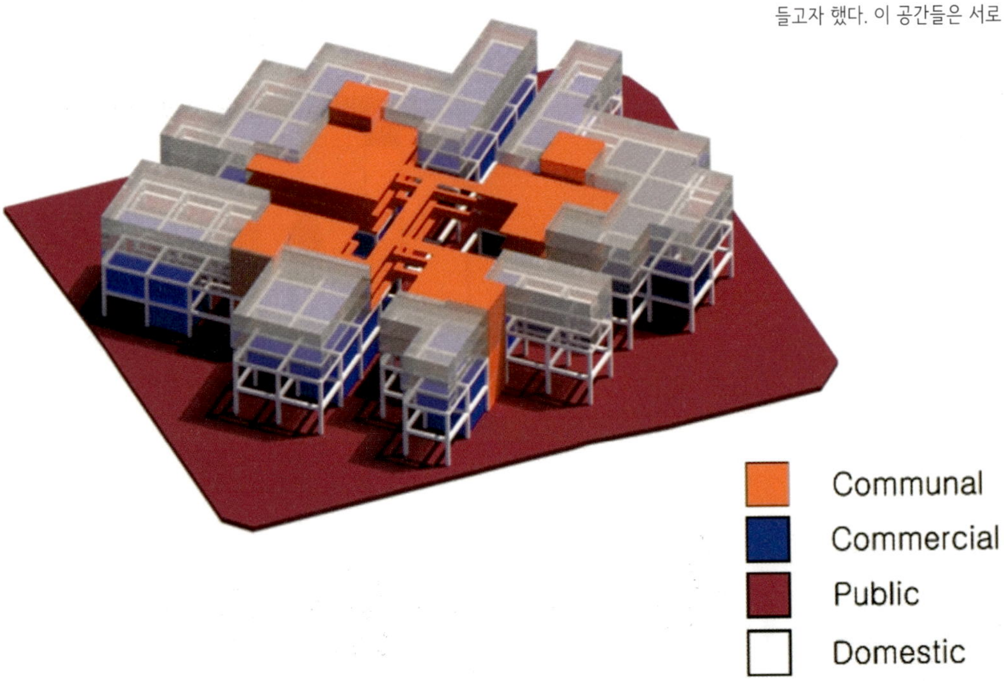

PUBLIC TO PRIVATE SEQUENCE

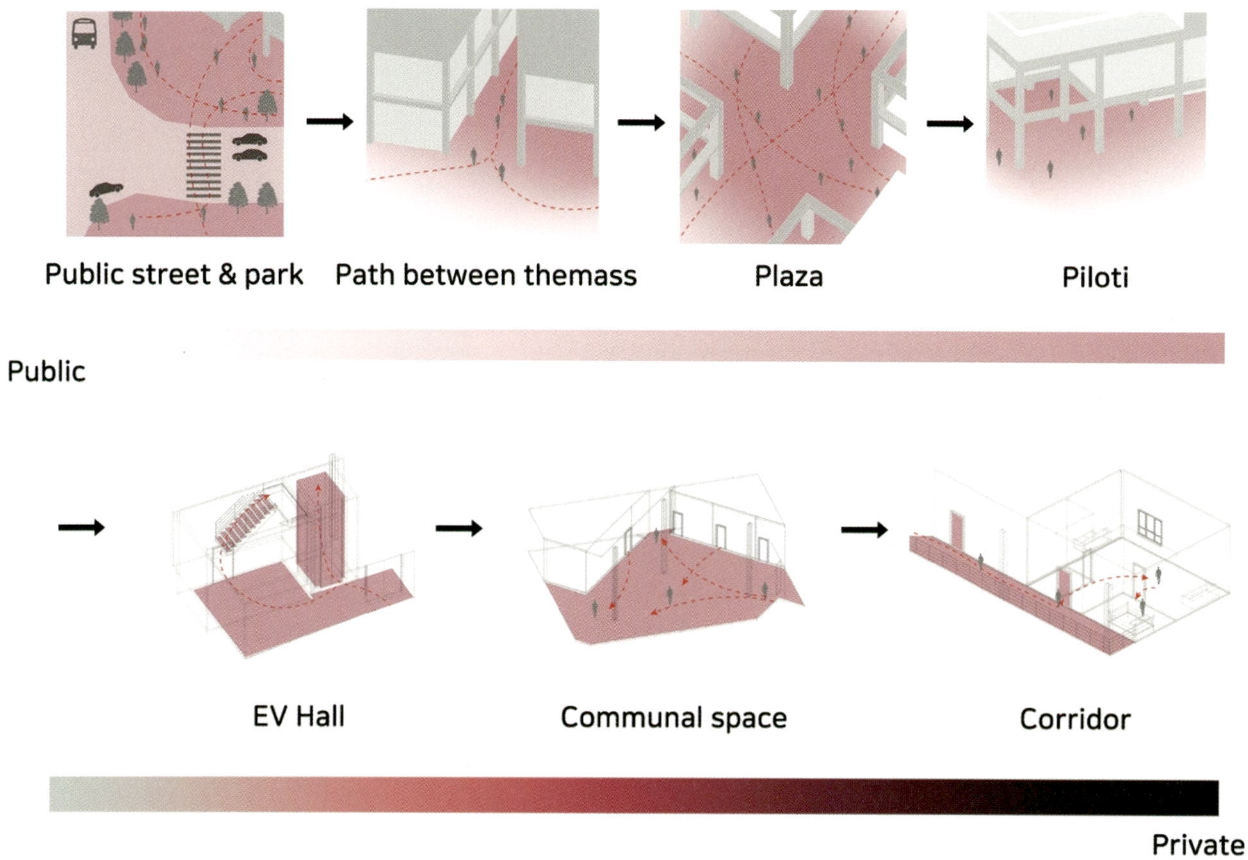

SECTION DETAILS

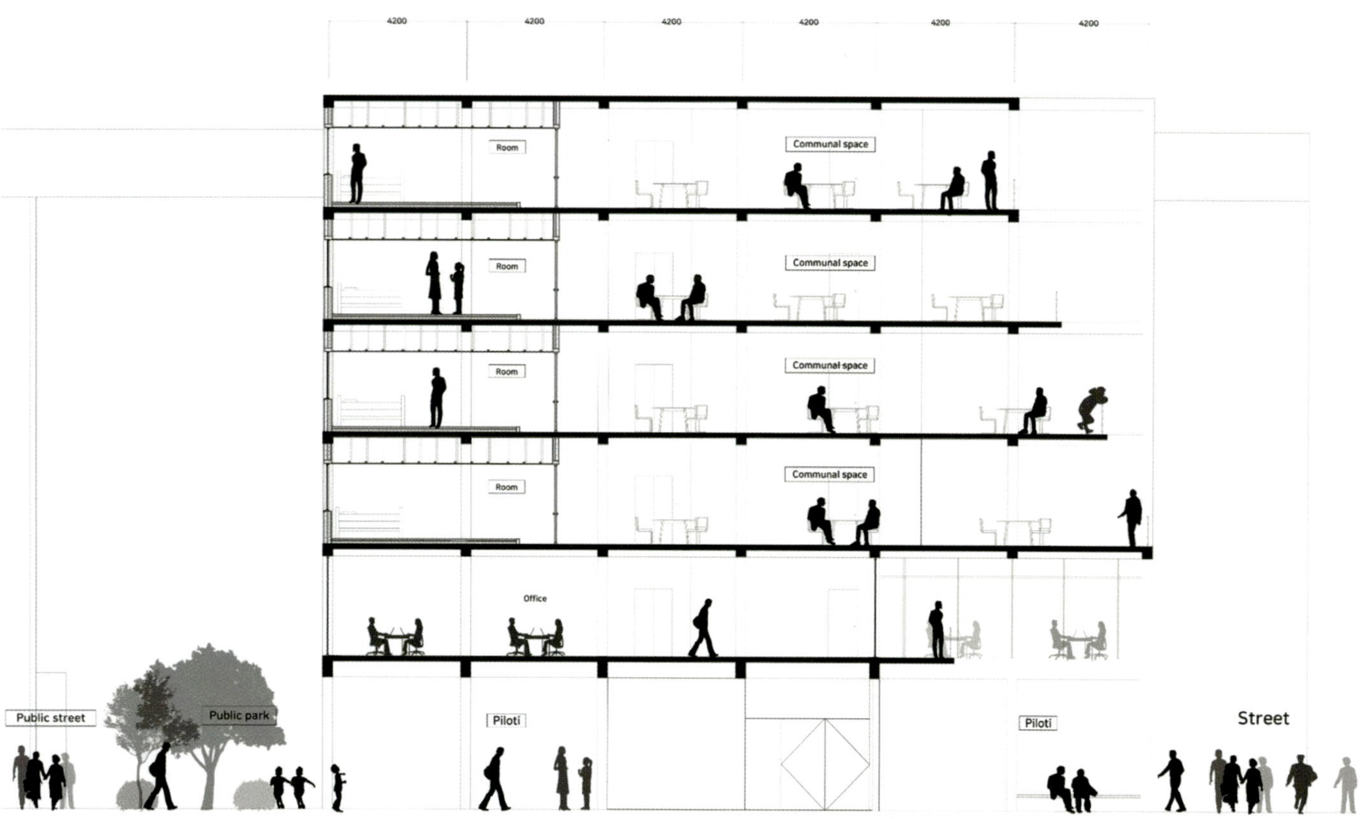

ENT
ENT
ENT

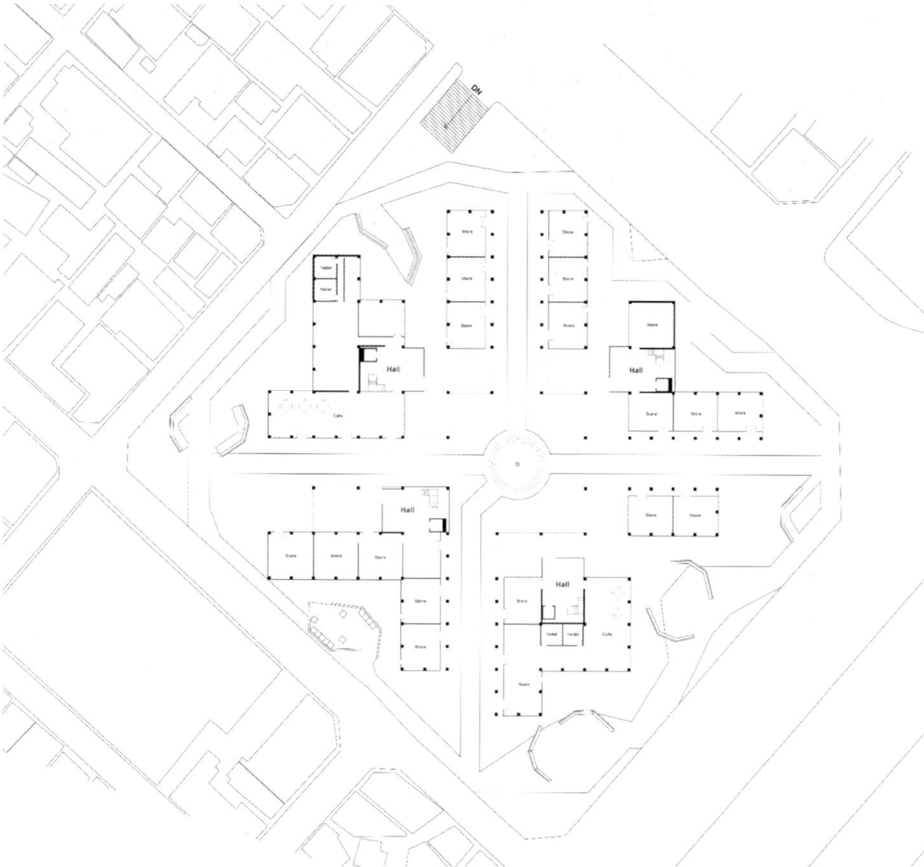
DN
Toilet
Hall
Cafe
Store

1F

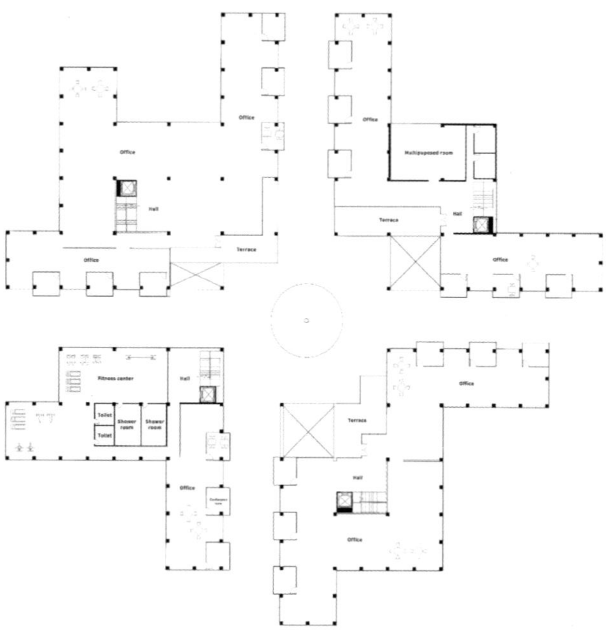
Office
Hall
Terrace
Multipurpose room
Fitness center
Toilet
Shower room

2F

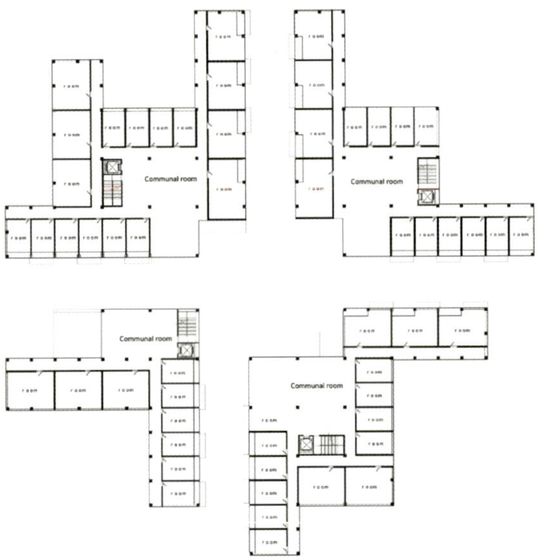
Communal room
Communal room
Communal room
Communal room

3 F

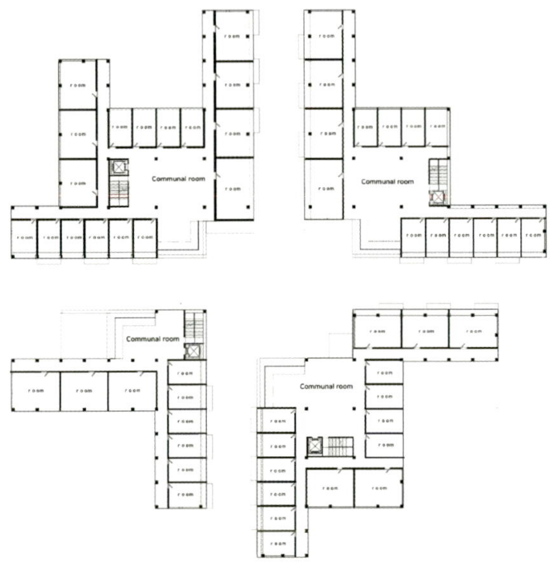
Communal room
Communal room
Communal room
Communal room

4 F

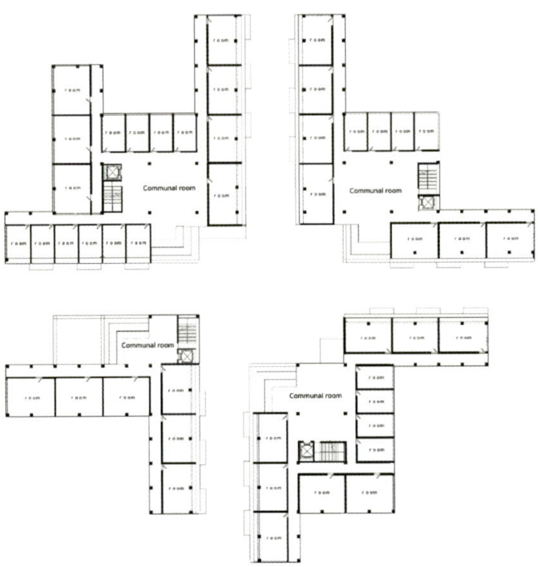
Communal room
Communal room
Communal room
Communal room

5 F

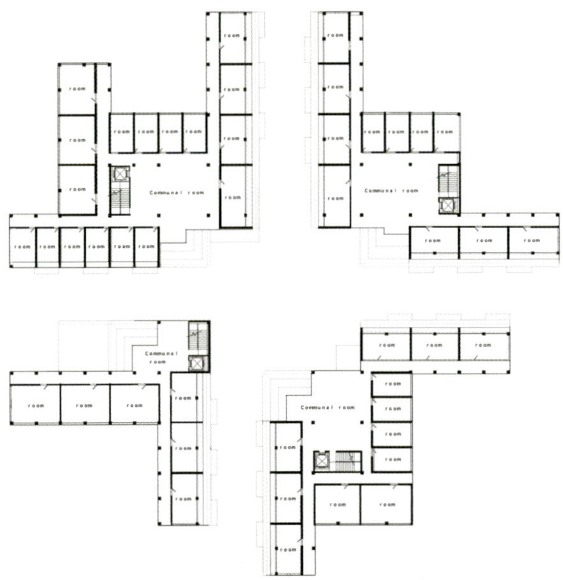

6 F

North east Elevation

South west Elevation

North west Elevation

SECTION & DETAILS

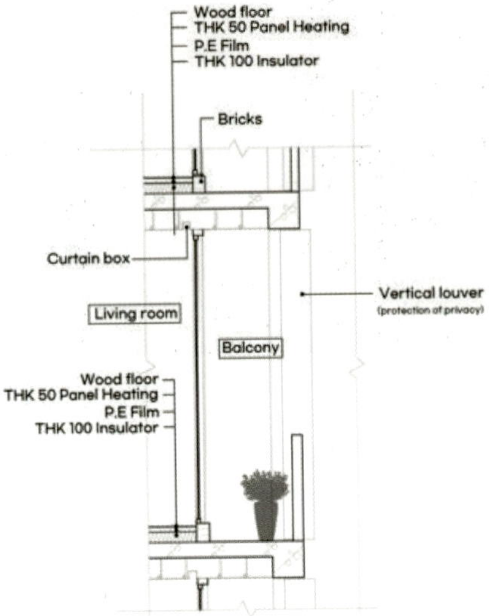

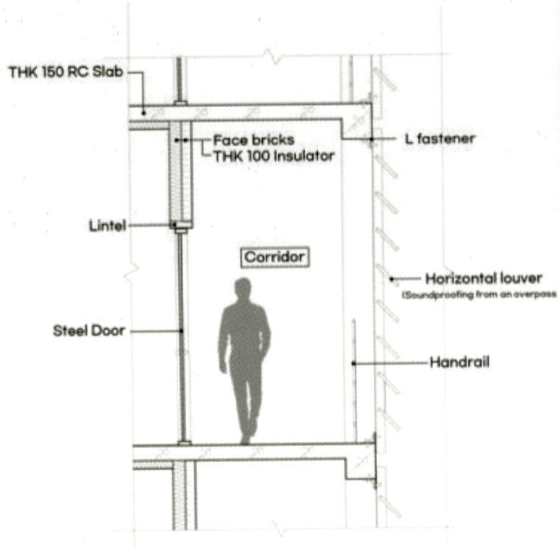

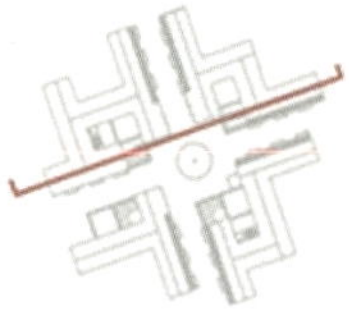

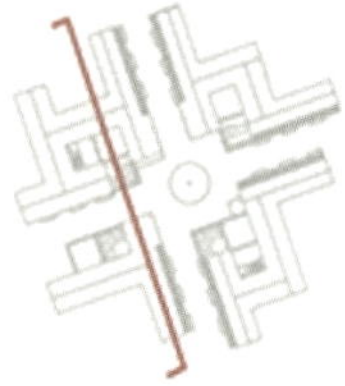

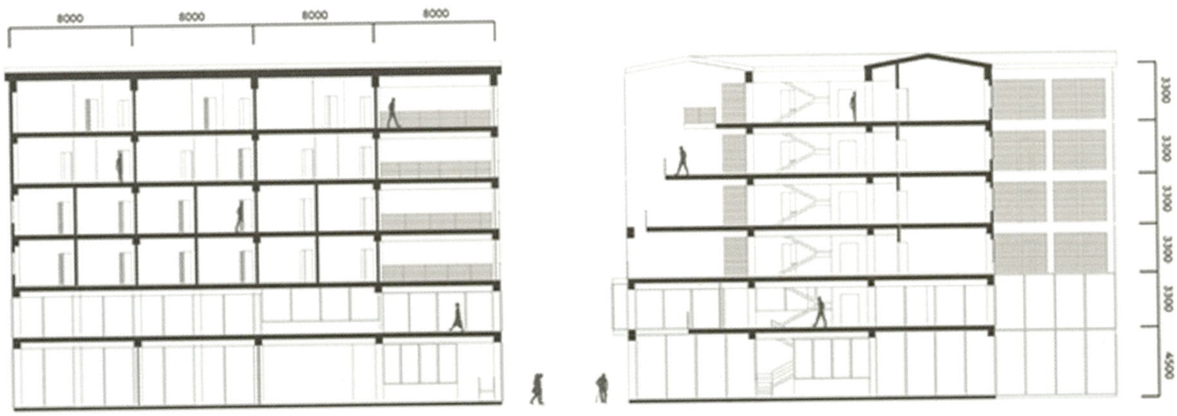

COMMUNITY CANYON: FINAL MODEL

The residential floor is gradually pushed back to alleviate the frustration of the outside distance, and in the inner space, balcony spaces and residential units of various sizes are created. The external distance facing such a residential space becomes a place to form a community.

주거층을 차츰 뒤로 밀어 외부거리의 답답함을 완화하고, 내부공간에서는 발코니공간과 다양한 크기의 주거유닛을 생성한다. 이러한 주거공간이 면한 외부 거리는 공동체를 형성하는 장이 된다.

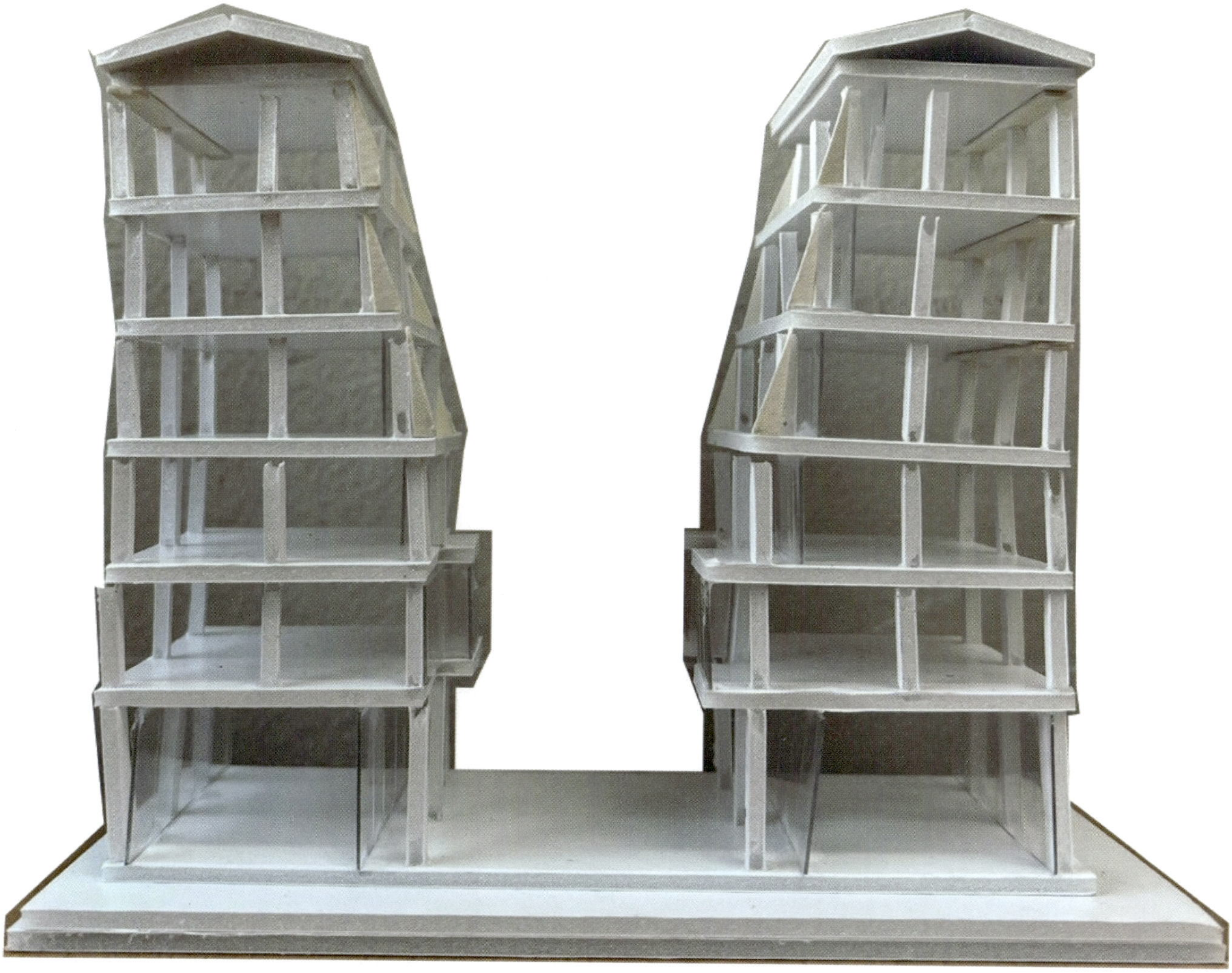

LOS ANGELES

THE FIRST AMERICAN CITY

As noted by Richard Weinstein, Los Angeles is the first American city to separate itself decisively from European models. With a population of 18.5 million, the region is famous for its sprawling highways and low-density residential neighborhoods.

리처드 웨인스타인의 언급에서처럼, 로스앤젤레스는 유럽식 모델들로부터 결정적으로 분리된 최초의 미국식 도시이다. 인구 1,850만 명의 이 지역은 고속도로와 저밀도 주거 지역으로 유명하다.

Photo by Dillon Shook on Unsplash

We choose the Greater Los Angeles region to locate our sites for two reasons, the uniqueness of the second-largest metropolitan region in the United States, and the universality of all the urban problems of American cities, from congested highways to lacking public spaces, only presented here in extreme ways. We also see recent popularity of MRMU in this region with numerous such projects being proposed or built in the past a few years. This proves a real need of higher density urban environment regardless of the decentralization caused by COVID-19 pandemic.

우리는 미국에서 두 번째로 큰 대도시라는 지역의 고유성과, 혼잡한 고속도로부터 공공 공간 부족에 이르기까지 극단적으로만 제시된 미국의 모든 도시 문제들의 보편성이라는 두 가지 이유에서 그레이터 로스앤젤레스를 선택했다. 지난 몇 년간 제안되거나 구축된 수많은 프로젝트를 통해 이 지역에서의 MRMU의 인기를 확인할 수 있다. 이는 COVID-19 팬데믹으로 분산화와 관계없이 고밀도 도시 환경의 진정한 필요성이 입증된 셈이다.

Lawrence Davis
Daekwon Park
Yo-Ichiro Hakomori
Edgar Rodriguez
Konrad Holtsmark
Paul Tang
Elizabeth Kamell
Fei Wang
Joel Kerner
Liang Wang
Julie Larsen
Jiong Abingo Wu
Flora Lee

RANCHO CUCAMONGA, SAN BERNADINO COUNTY, CALIFORNIA

RESEARCH BY

TIANCHONGHUI FANG
PETER WANG
YITING ZHONG

On the foot of Mount San Antonio, Rancho Cucamonga is a city of 180k+ residents in the outer Greater Los Angeles Region. With miles of commercial along Historic Route 66, industries and offices in the south, convenient transportation, diverse cultures, and booming development, Rancho Cucamonga, under rapid urbanization progress, attracts many people to find homes there.

샌 안토니오 산기슭에 위치한 란초 쿠카몽가는 18만 명 이상의 주민들이 거주하는 로스앤젤레스 외곽 도시이다. 역사적인 66번 국도를 따라서 수 마일 펼쳐진 상업 구역, 남부에 자리 잡은 산업과 사무실들, 편리한 교통, 다양한 문화, 그리고 활발한 개발로 빠른 도시화의 진전 아래에 있는 란초 쿠카몽가는 많은 사람들의 정착을 이끈다.

Cucamonga

RANCHO CUCAMONGA,

The city of Rancho Cucamonga will build a new train station adjacent to the existing Metrolink Station in Azusa Court and operate under San Be[rnardino] County Transportation Authority.

HART District Is Coming To Reality For Rancho Cucamonga

"The future of California is right here, and with Homecoming, we're offering a preview of what that future looks like".

By Sharjeel Sohaib

There is a great emphasis by emerging leaders and influencers of the Inland Empire on supply chain and logistics innovation. The fact of the matter is that the US still is behind other countries in using the latest technologies on supply chain networks like rail cars and trying to make it easy to move cargo-bound goods.

The City of Rancho Cucamonga has long been envisioned to be a place with a more urban feel – with walkability and easy access to transportation partners like Metrolink, Ontario International Airport, and Omnitrans.

The Hart District is a transit-oriented, mixed-use district that elevates the area into a walkable, culture-rich experience embracing economic opportunity. This area is within:

H – Haven Avenue, A – Arrow Route, and R[–] Rochester Avenue, forms our T– transit district.

Rancho Cucamonga Station will b[e] the key entry point from Las Vegas to Southern California for Brightline West – just two hours from Las Vegas, avoiding the hassle of a long drive on congested freeways. It will aim to be a "green station". The environmental benefits of high-speed rail, a rapid bus zero-emission fleet, and sustainable construction and operations are in line with the City of Rancho Cucamonga and San Bernardino County Transportation Authority's (SBCTA) vision and commitment to reducing carbon footprint.

In addition, part of the HART area will be residential. New housing projects are starting near the new Rancho Cucamonga train station. Named as 'Homecoming at the Resort,' it is an 867-unit next-generation housing community in Rancho Cucamonga. Also called a 'new urban village' in southeast Rancho Cucamonga, the housing project will be built for single-family residences, townhomes, apartment homes, and retail stores.

Rancho Cucamonga and SBCTA Approve Sale of Propert[y] of Brightline West Full-Service Transit Sta[tion]

Third-party analysis by Beacon Economics, a well-respected Inland Empire-based firm

By Inland Empire Business Journal

The City of Rancho Cucamonga City Council and the San Bernardino County Transportation Authority (SBCTA) Board of Directors have approved the first step of the sale of a five-acre portion of the jointly owned property at Cucamonga Station to Brightline West. The agreement clears the path for Brightline West to develop the property at the northwest corner of Milliken Avenue and Azusa Court, and build the first high-speed rail station in the Inland Empire connecting Rancho Cucamonga, the High Desert, and Las Vegas.

Brightline is the only private provider of modern, eco-friendly intercity rail service in America. The railroad has already announced that plans are now set to bring its trains into Rancho Cucamonga for a connection to Los Angeles Union Station. The proposed Brightline West schedule calls for 25 daily round trips, leaving 45 minutes apart. Travel time for the 235-mile journey would be two hours, making the average speed about 115 mph. The top speed will be 180 mph.

A 259-unit apartment complex with some commercial space was approved by the Planning Commission on April 13, 2022, for the northeast part of Rancho Cucamonga. (Image from city of Rancho Cucamonga staff report)

Cucamonga Station will go through a transformation in the coming years, making it the region's global connectivity point and destination location for everything California has to offer. The station will provide service through high-speed rail, the Metrolink San Bernardino Line, and Omnitrans Bus Rapid Transit and local bus service, as well as serve carpool commuters. Plans are also moving forward

California cows are leaving the state and that won't help glob[al]

California is nearly halfway to a 2030 goal for methane reduction. But that's in part thanks to cows, and their methane, 'leaking' into other states.

By Brooke Staggs

California has set the most ambitious goals in the nation for reducing methane emissions. These greenhouse gasses don't stick around in the atmosphere nearly as long as carbon dioxide but are much more potent, which means they contribute significantly to global warming. And since dairies are responsible for more than half of the state's methane emissions, legislators in 2016 set a goal to reduce emissions from farms by at least 40%, or some 9 million metric tons, by 2030.

Since passage of that law, the state's 1,200 dairies have made huge strides in reducing methane emissions. A report this spring by the California Air Resources Board estimated state farms are nearly halfway to the goal.

But while half of those reductions dairies changing the ways they process other half has come because California simp[ly] And when cows are moved, global methane actually drop. They just shift, or "leak," in where lighter regulations mean the green likely will get worse.

The way of sovling this aside from hav the most proven way to reduce methane at da add an anaerobic digester system to deal with are placed over manure lagoons to capture meth gasses as they're released. Those gasses are th into renewable biogas, which can be used to gas-equipped vehicles, injected into existing pipelines or used to produce electricity.

Of course, all of these promising program challenges and long way to go.

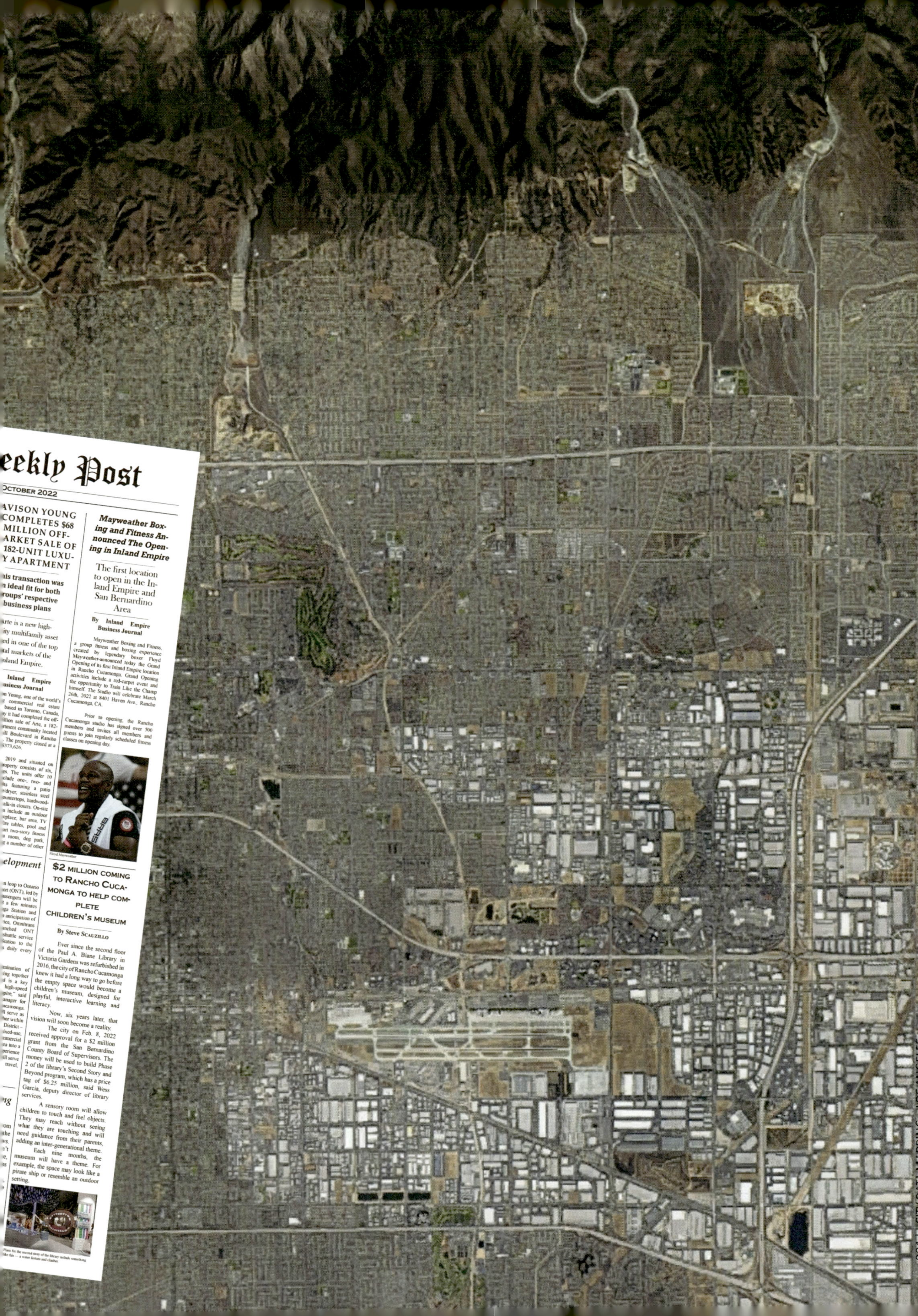
eekly Post
OCTOBER 2022
AVISON YOUNG COMPLETES $68 MILLION OFF-ARKET SALE OF 182-UNIT LUXU-Y APARTMENT
Mayweather Boxing and Fitness Announced The Opening in Inland Empire
The first location to open in the Inland Empire and San Bernardino Area
By Inland Empire Business Journal
Mayweather Boxing and Fitness, a group fitness and boxing experience created by legendary boxer Floyd Mayweather-announced today the Grand Opening of its first Inland Empire location in Rancho Cucamonga. Grand Opening activities include a red-carpet event and the opportunity to Train Like the Champ himself. The Studio will celebrate March 26th, 2022 at 8401 Haven Ave., Rancho Cucamonga, CA.
Prior to opening, the Rancho Cucamonga studio has signed over 500 members and invites all members and guests to join regularly scheduled fitness classes on opening day.
Floyd Mayweather
$2 MILLION COMING TO RANCHO CUCAMONGA TO HELP COMPLETE CHILDREN'S MUSEUM
By Steve Scauzillo
Ever since the second floor of the Paul A. Biane Library in Victoria Gardens was refurbished in 2016, the city of Rancho Cucamonga knew it had a long way to go before the empty space would become a children's museum, designed for playful, interactive learning and literacy.
Now, six years later, that vision will soon become a reality.
The city on Feb. 8, 2022 received approval for a $2 million grant from the San Bernardino County Board of Supervisors. The money will be used to build Phase 2 of the library's Second Story and Beyond program, which has a price tag of $6.25 million, said Wess Garcia, deputy director of library services.
A sensory room will allow children to touch and feel objects. They may reach without seeing what they are touching and will need guidance from their parents, adding an inter-generational theme.
Each nine months, the museum will have a theme. For example, the space may look like a pirate ship or resemble an outdoor setting.

ETHNIC

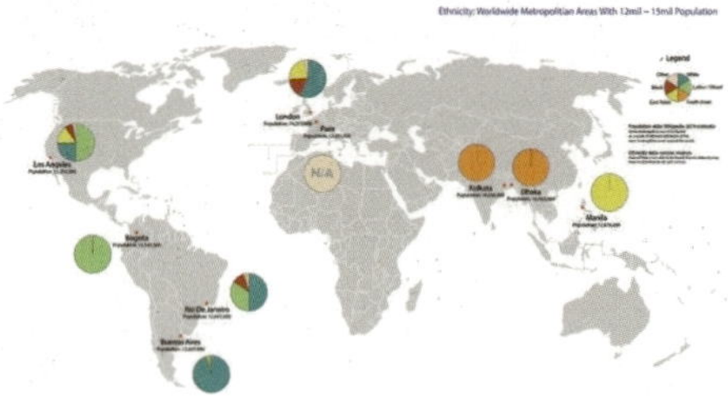

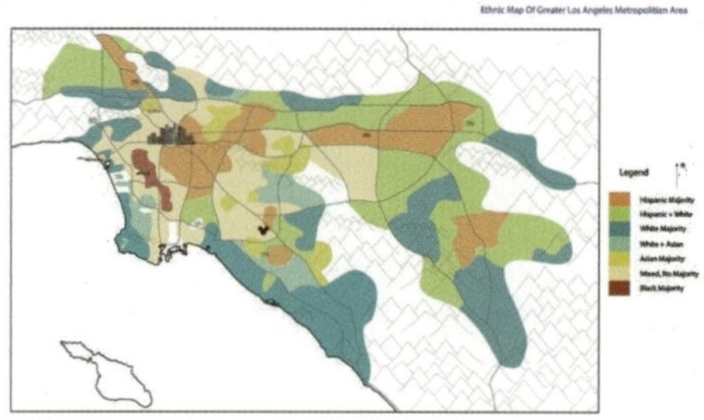

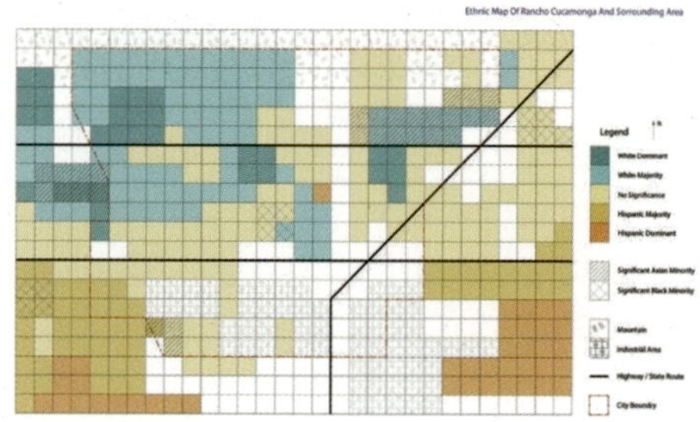

INCOME

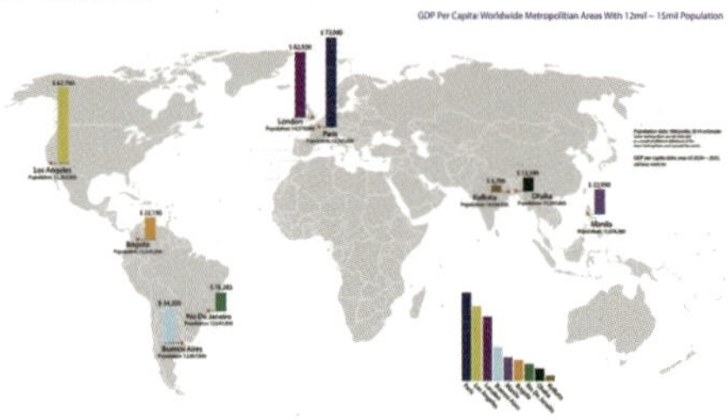

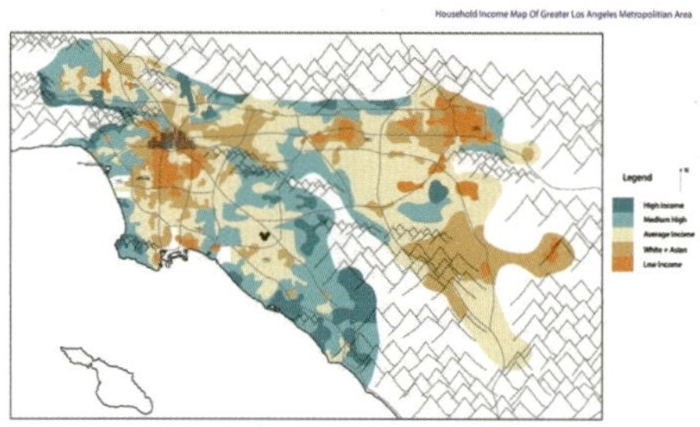

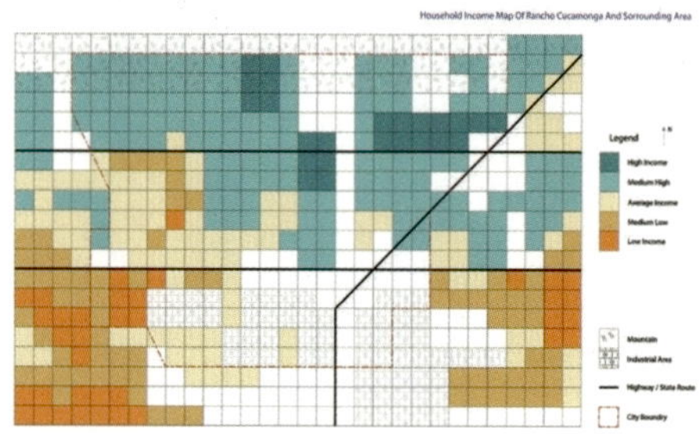

CLIMATE CHANGE

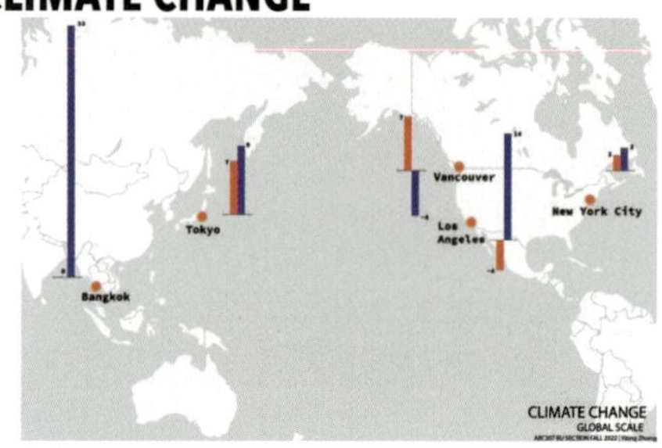

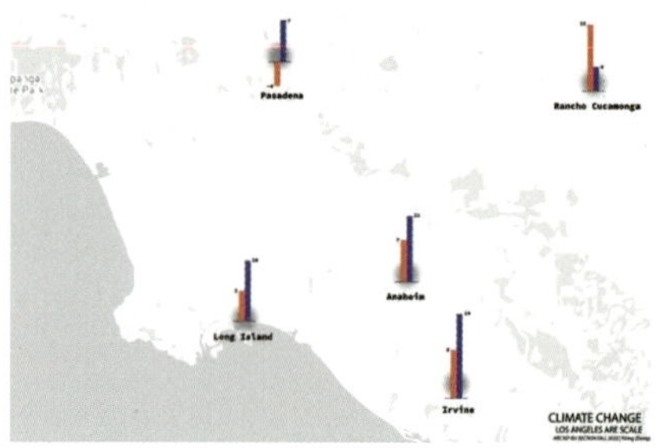

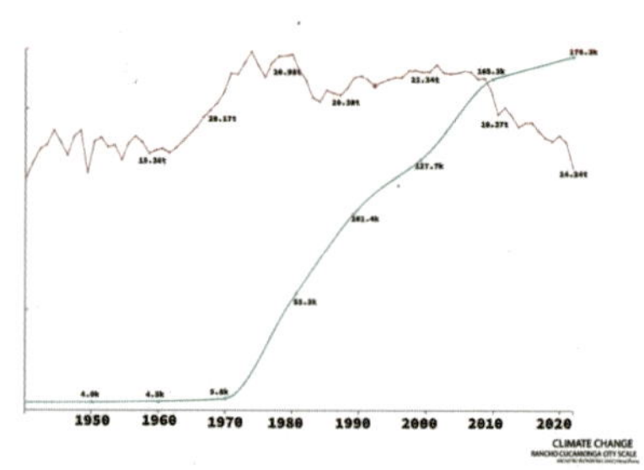

TRAIL SYSTEM

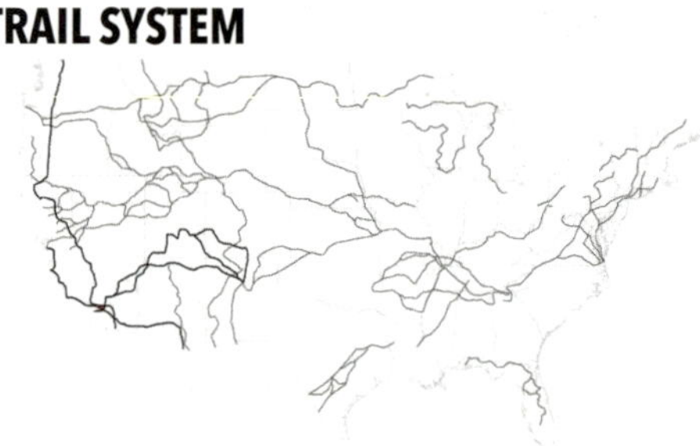

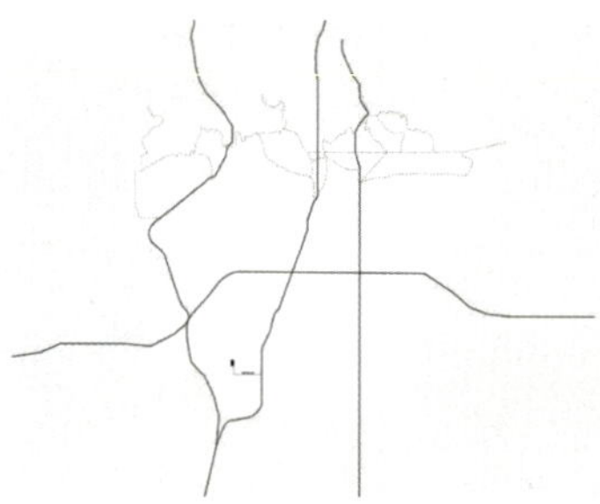

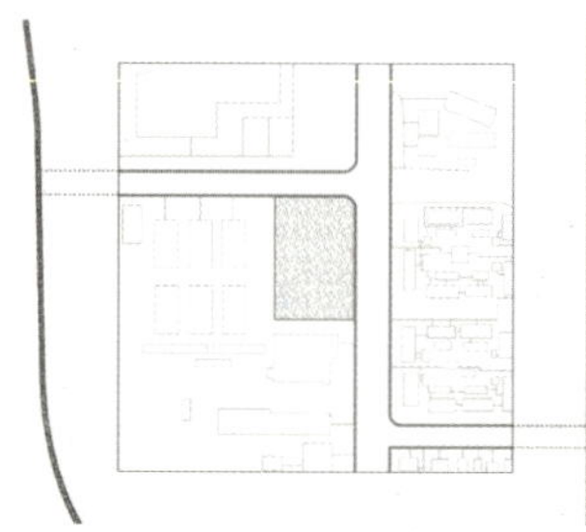

AIR QUALITY

BUILDING MORPHOLOGY

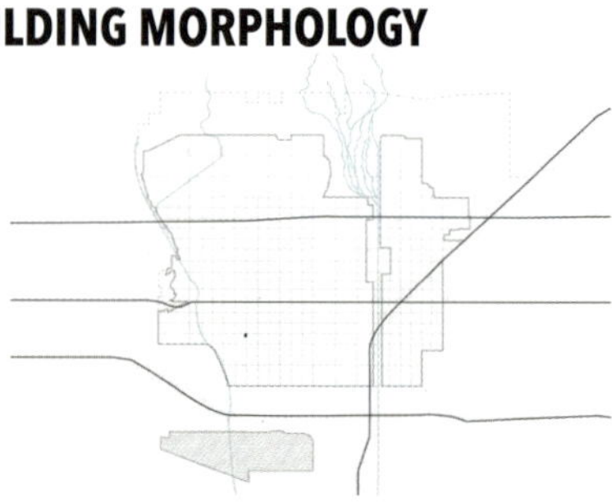

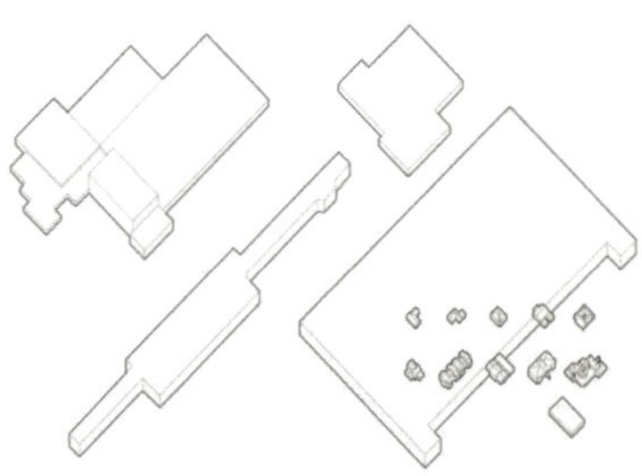

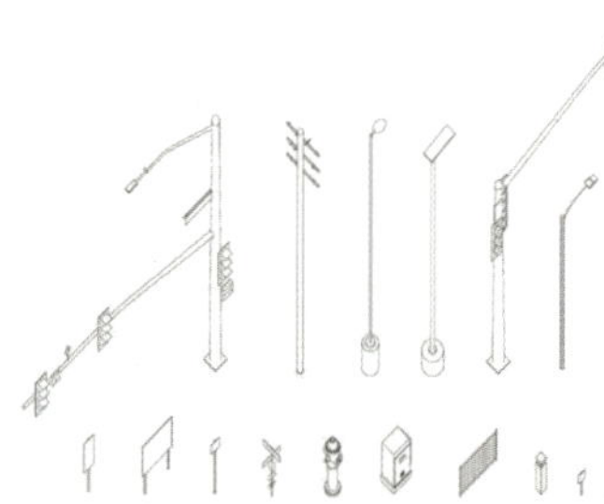

VEGETATION

VEGETATION
LA AREA SCALE

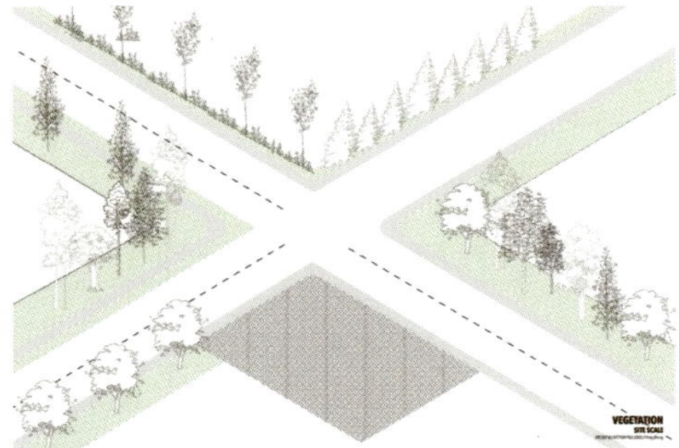
VEGETATION
SITE SCALE

VEGETATION
RANCHO CUCAMONGA SCALE
ARC307 BU SECTION FALL 2022 | Yiting Zhong

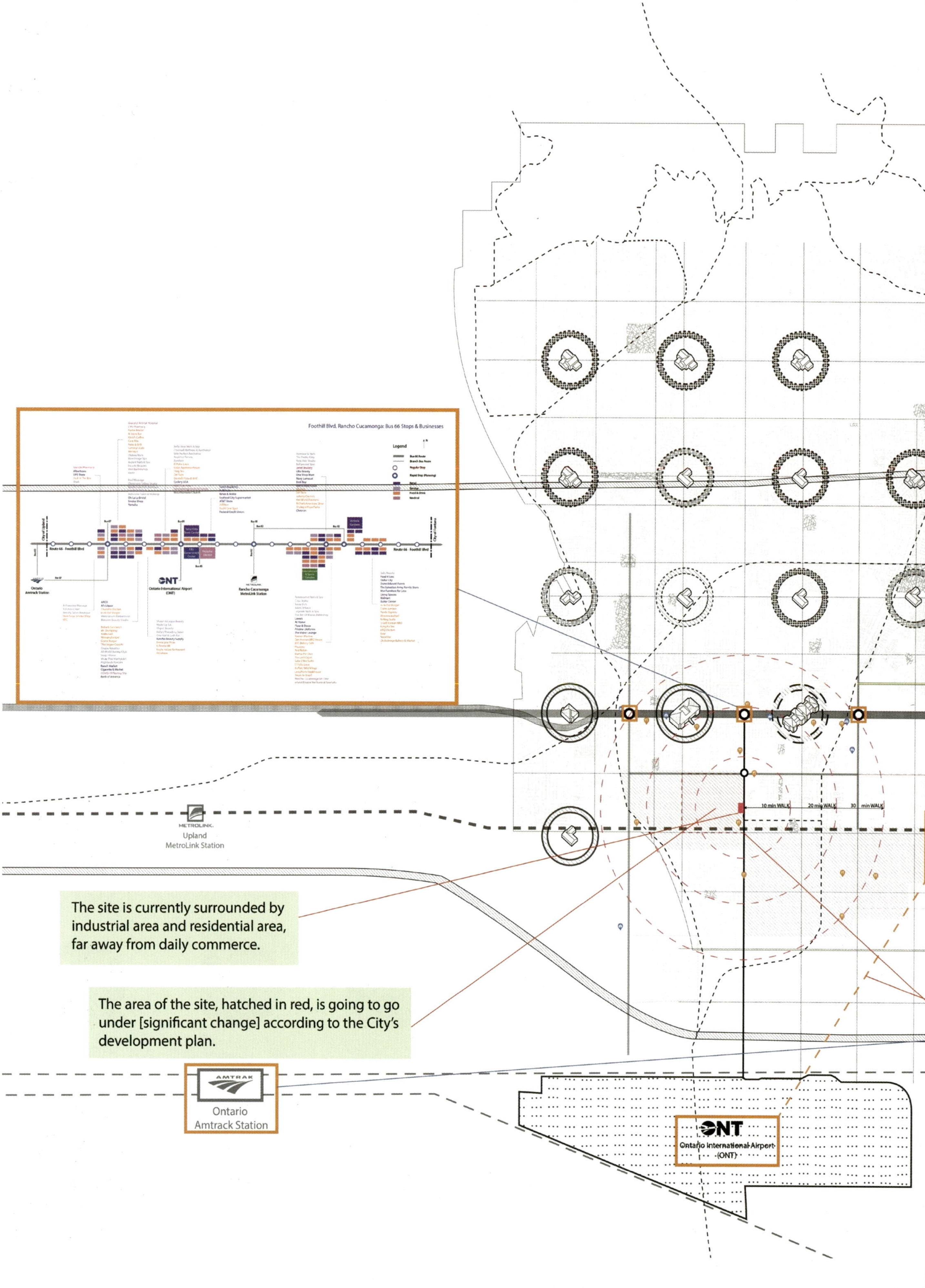
Foothill Blvd, Rancho Cucamonga: Bus 66 Stops & Businesses
Legend
Route 66 - Foothill Blvd
ONT
Ontario International Airport (ONT)
Ontario Amtrack Station
Rancho Cucamonga MetroLink Station
METROLINK
Upland
MetroLink Station
10 min WALK
20 min WALK
30 min WALK
The site is currently surrounded by industrial area and residential area, far away from daily commerce.
The area of the site, hatched in red, is going to go under [significant change] according to the City's development plan.
AMTRAK
Ontario
Amtrack Station
ONT
Ontario International Airport
(ONT)

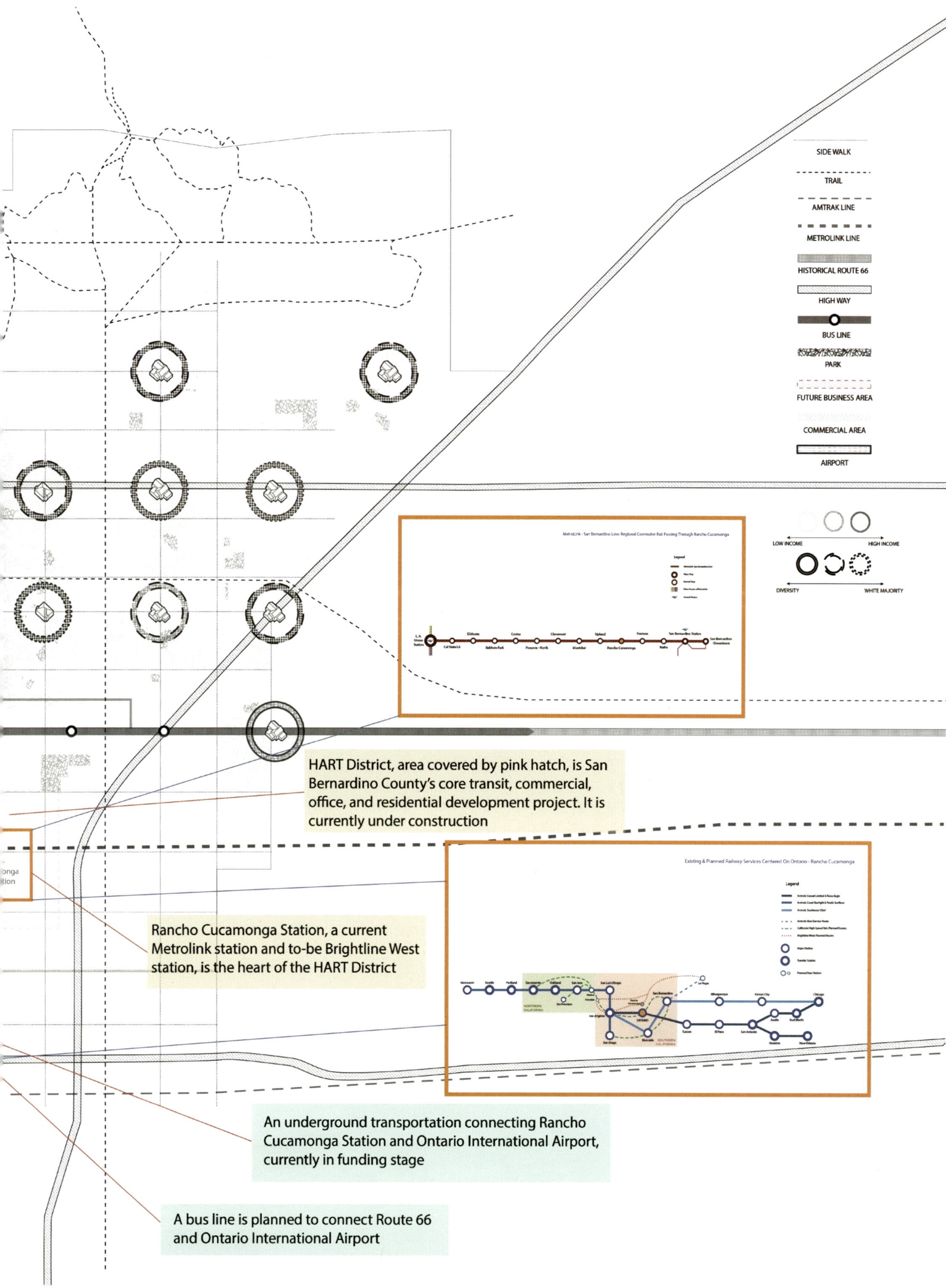
SIDE WALK
TRAIL
AMTRAK LINE
METROLINK LINE
HISTORICAL ROUTE 66
HIGH WAY
BUS LINE
PARK
FUTURE BUSINESS AREA
COMMERCIAL AREA
AIRPORT
LOW INCOME
HIGH INCOME
DIVERSITY
WHITE MAJORITY
HART District, area covered by pink hatch, is San Bernardino County's core transit, commercial, office, and residential development project. It is currently under construction
Rancho Cucamonga Station, a current Metrolink station and to-be Brightline West station, is the heart of the HART District
An underground transportation connecting Rancho Cucamonga Station and Ontario International Airport, currently in funding stage
A bus line is planned to connect Route 66 and Ontario International Airport

THE ASSEMBLY

PROJECT BY
TIANCHONGHUI FANG
PETER WANG
YITING ZHONG

PROJECT INFORMATION
Site area : 79,750sqf
Floor area ratio : 60%

To echo the urbanization theme the city is undergoing, our project, an "Assembly" of different programs and masses, was designed to be a mini-scaled city based on a road-grid system on the ground floor. The center of the "city" is a public swimming pool with commerce surrounding it to attract people with enjoyment and convenience. Different housing types are offered above the commercial spaces to meet clients' different preferences. A platform across the second floor is set to serve as a park with plantations and open spaces. These elements can bring us to achieve our goal of making the location a popular, connective, and convenient local center for people.

지역에서 진행 중인 도시화 주제를 반영하기 위하여, 다양한 프로그램과 대중의 "집합"인 우리 프로젝트는 지상 1층의 기존 도로망 시스템을 기반으로 하는 작은 규모의 도시로 설계되었다. 우리의 '도시'의 중심은 상업이 주변을 둘러싸고 즐거움과 편리함으로 사람들을 끌어들이는 공공 수영장이다. 이용자들의 다양한 선호를 충족시키기 위해 상업 공간 위에 여러 주거 유형이 제공된다. 2층 전체를 가로지르는 플랫폼은 플랜테이션과 개방된 공간이 있는 공원의 역할을 하도록 설정되어 있다. 이러한 요소들은 대상 장소를 사람들에게 인기 있고, 서로 연결되며 편리한 지역 중심지로 만들려는 우리의 목표를 달성하게 할 수 있다.

Transportation is a focus of Rancho Cucamonga's development plan. The city currently has a MetroLink (Greater LA light rail) station and a bus system mostly around the major roads. Ontario International Airport and two Amtrak stations are located right outside the city's boundary. In the near future, the railway station (currently only for MetroLink) will become a key station for Brightline West, an under-construction light rail line connecting Los Angeles and Las Vegas. The area around the station, "HART District", is the core transportation, commercial, office, and residential development project in San Bernardino County.

The current industrial area in the southern part of Rancho Cucamonga is undergoing a shift to a new neo-industrial zone, creating thousands of new office and factory jobs. The area between Route 66 and Ontario Airport is marked by the government's planning book to have "significant change" in the recent future. Our project site locates in the coverage of this area.

The city lacks common spaces for residents to gather and interact with each other in their communities. Although there are commercial spaces for people to relax with their families and friends, most public spaces, such as the local parks, have failed to attract residents by having almost nothing else than a green space often for sports. An area that can be used for streetside performances, exhibitions, communication, outdoor market, and more is needed in the neighborhood to bring people together and make the location a connective local center.

Following the urban development, Rancho Cucamonga has become a growing job center, creating the need for housing for employees with different household sizes and preferences. The site is located between neo-industrial and residential areas without commerce, making the lives of people working or living nearby inconvenient. A place with housing units, gathering points, and commerce is therefore needed to settle the arrivals and make the neighborhood more convenient and connected.

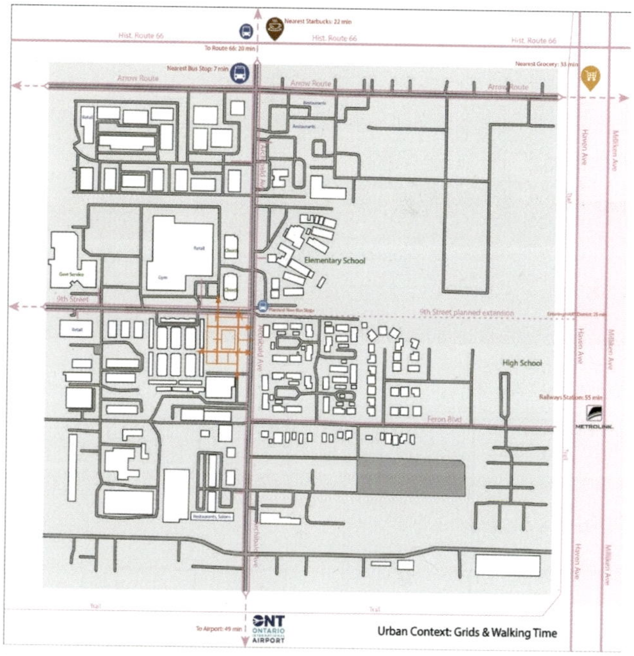

Urban Context: Grids & Walking Time

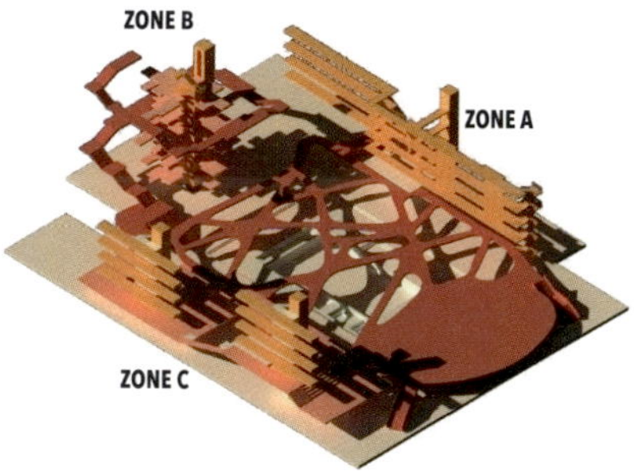

Zone A

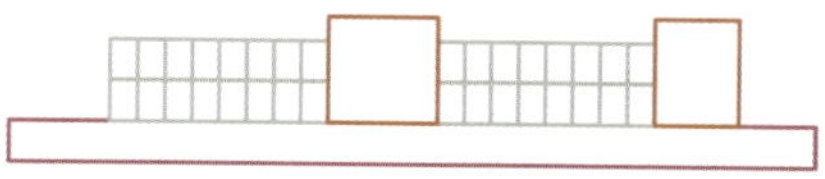

High density, compress living area, required more communal sapce

Zone B

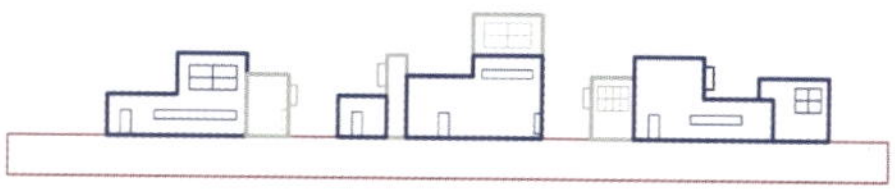

Complex and diverse area, various shape and unexpected counters

Zone C

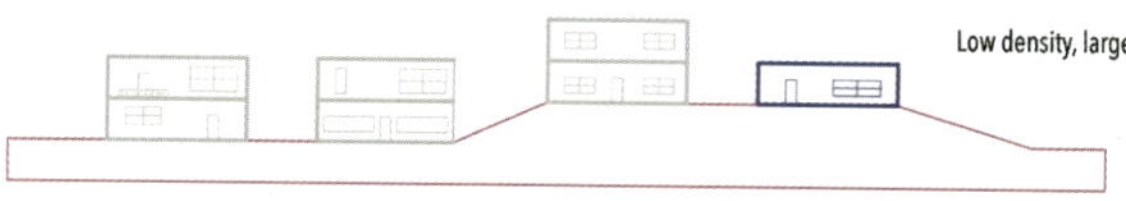

Low density, larger living area

ZONE A

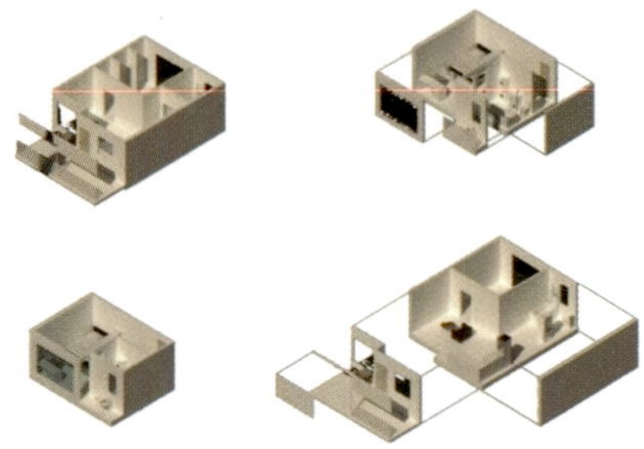

Communal Space Detial Diagram

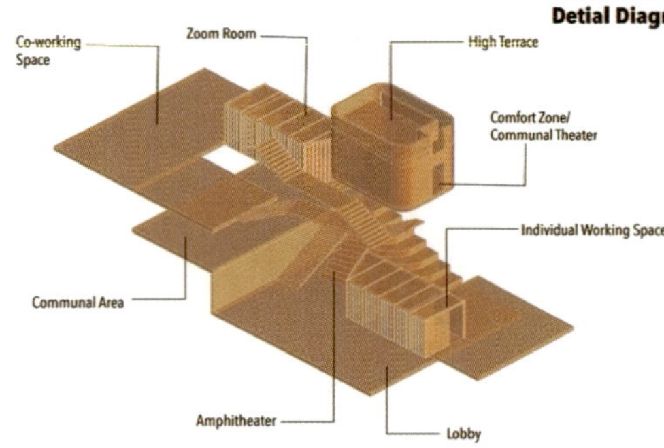

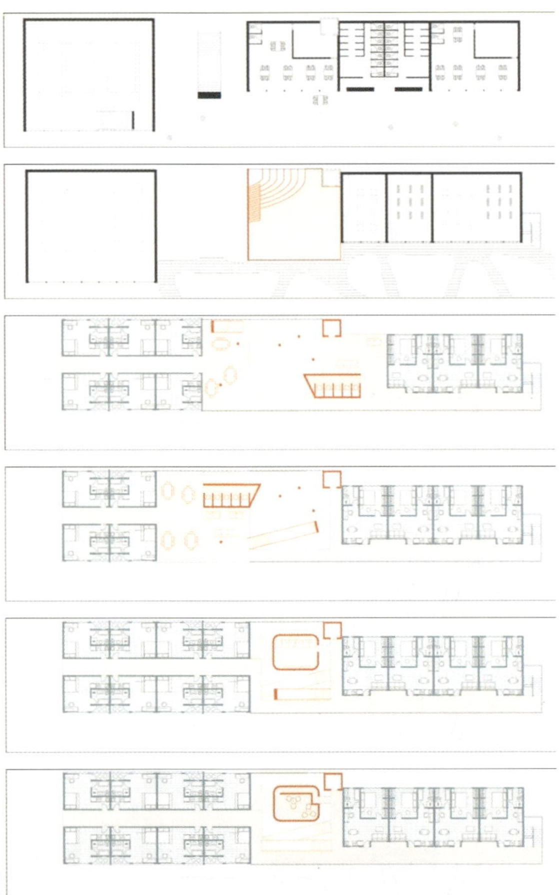

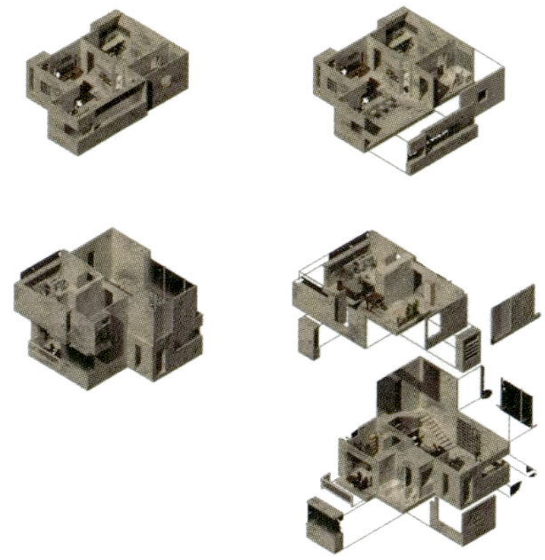

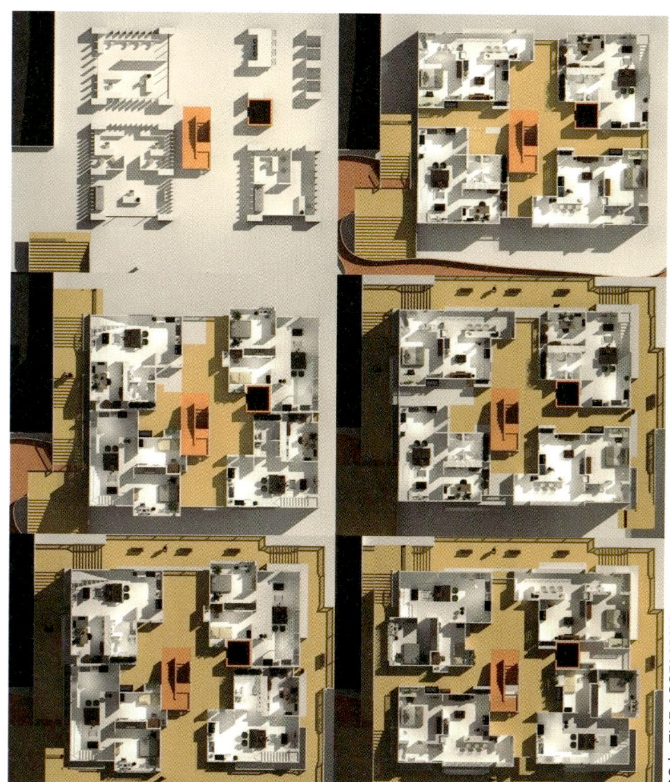

ZONE C

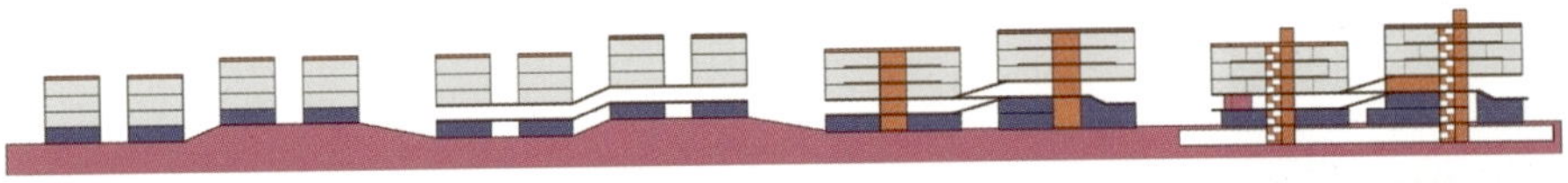

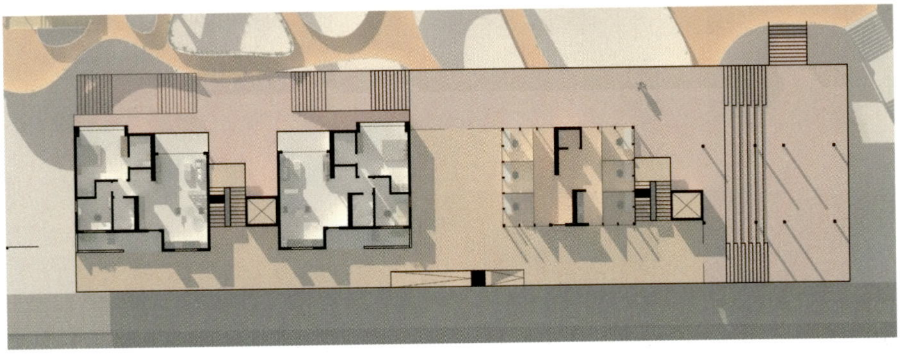

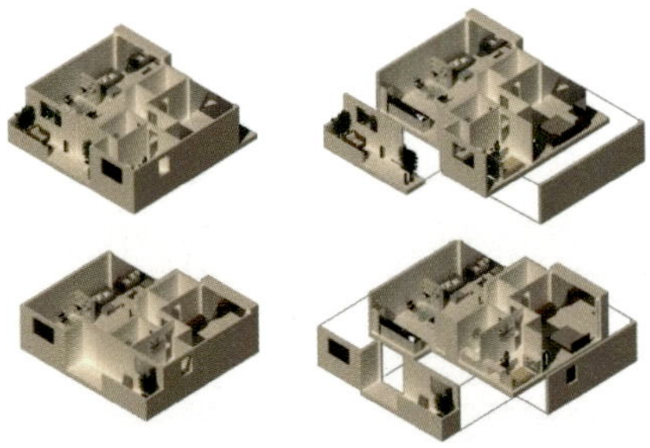

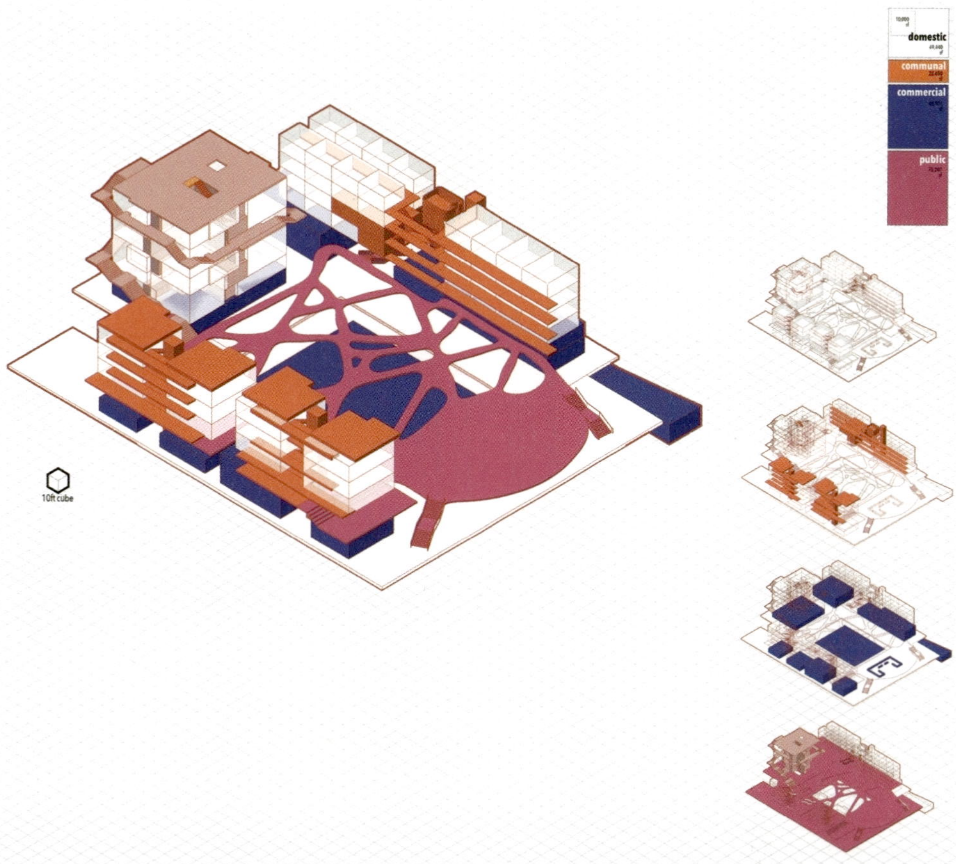
domestic
communal
commercial
public
10ft cube

MONTEREY PARK, LOS ANGELES COUNTY, CALIFORNIA

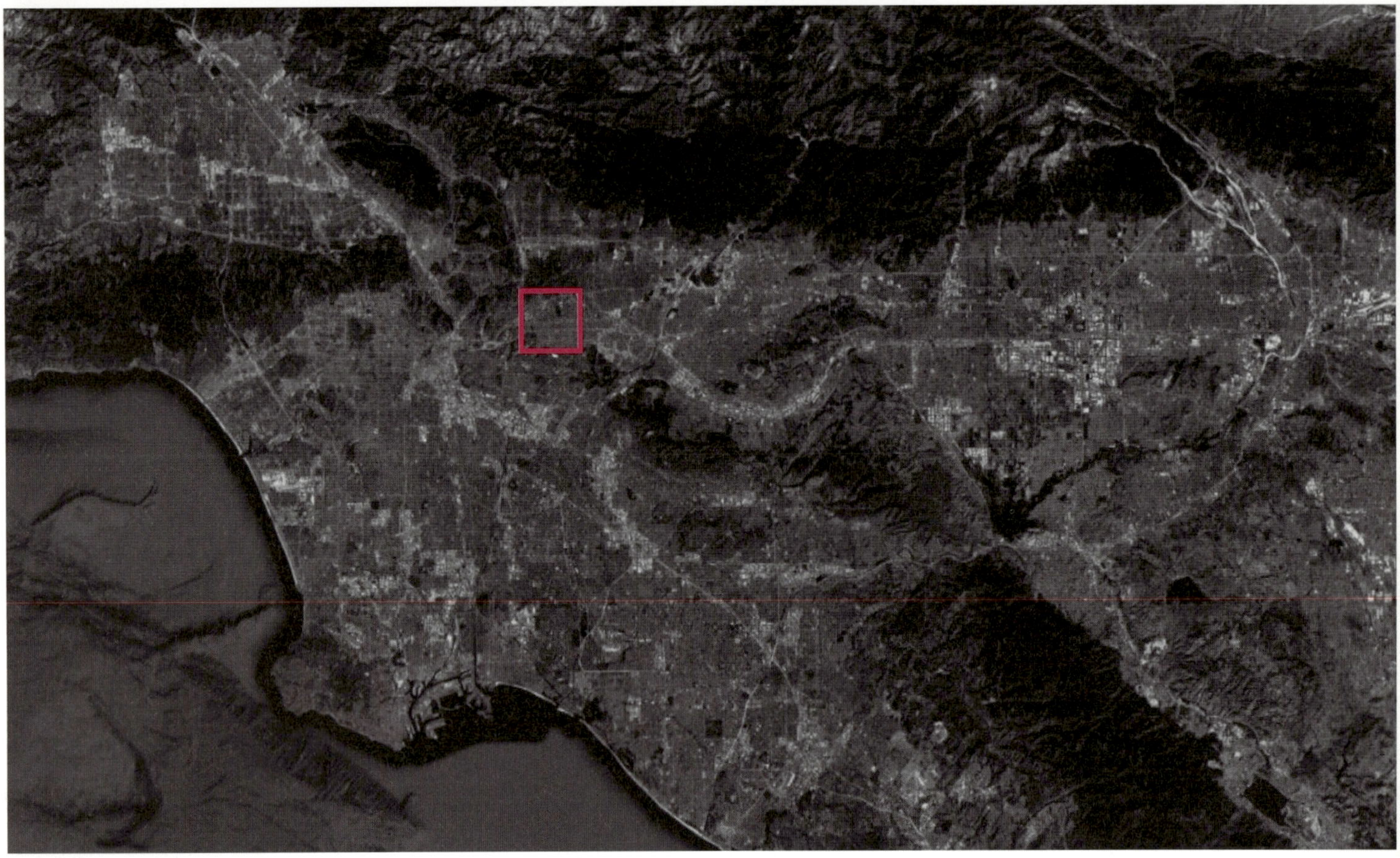

RESEARCH BY

MATTHEW WILLIAMS
ARYAN AMBANI
MICHELA BRITTIS-TANNENBAUM

Monterey Park is only 7–8 miles from Downtown Los Angeles and is known for being a cultural center, reflected in their diverse population with a majority of Asian Americans, primarily Chinese-Americans. They have a wealth of different asian cuisines and host large festivals from Lunar New Year to the Cinco de Mayo Celebration. Other popular venues of entertainment include parks and museums dotted around the city. Despite its high living cost, it's seen as a relatively desirable place to live.

로스앤젤레스와 7-8마일 떨어진 몬테레이 공원은 중국계 위주의 아시아계 미국인을 비롯해 다양한 인구가 모인 문화 중심지로 알려져 있다. 다양한 아시아 요리가 있고 음력 새해부터 신코 데 마요 축제까지 여러 축제가 개최된다. 다른 인기 있는 오락거리로는 도시 곳곳의 공원과 박물관이 있다. 비싼 물가에도 불구하고 비교적 살기 좋은 곳으로 여겨진다.

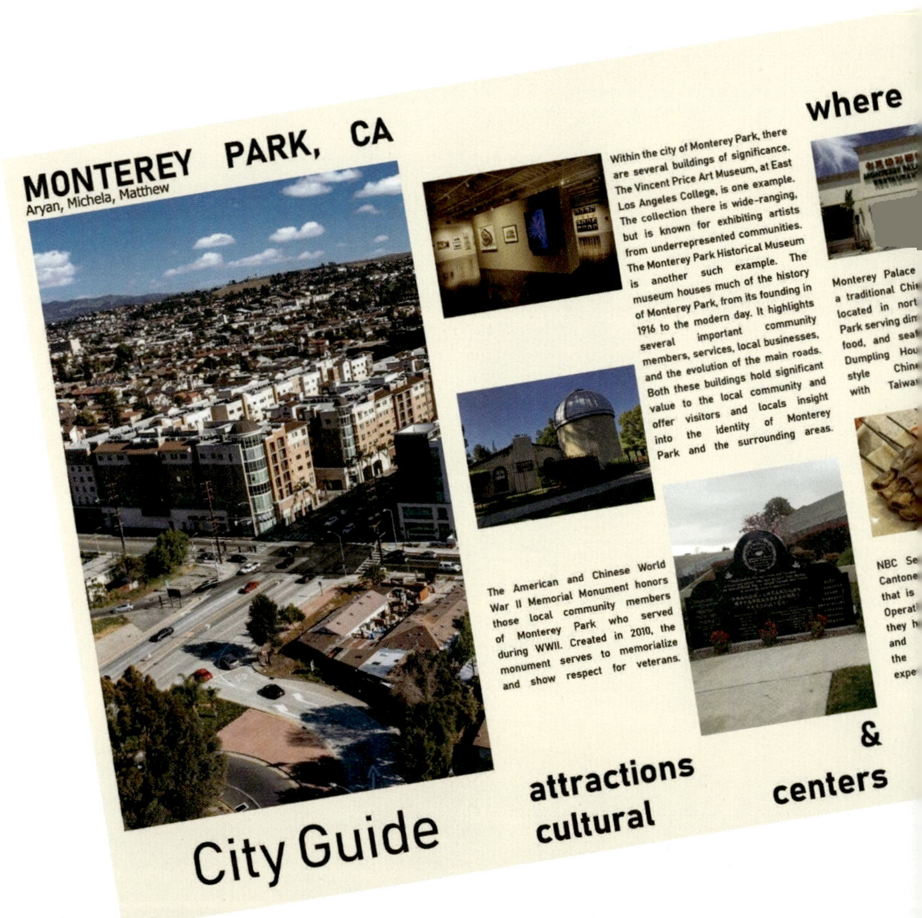

to eat

The Monterey Park area is known for its impressive food, specifically traditional Chinese food. With new restaurants popping up regularly, there is no shortage of good food to try. When in the city, make sure to stop at a few of these places.

T-Kebob is located right by Monterey Palace Restaurant and serves several different meats served on skewers. While they have several other food options, the Kebab is their specialty and a must have.

Monterey Park offers several services for residents and visitors and will soon be expanding thanks to increasing economic investment in the area. The Bruggemeyer Library is a local library and community staple. Also known as El Encanto, the library offers a wide-range of books and resources. The SPIRIT bus is the bus line that runs through Monterey Park and another important part of the city. It runs M-F from morning to mid afternoon and costs 25 cents to ride. It is currently being reviewed by the city to evaluate its performance and see how it can be improved.

s e r v i c e s
e d u c a t i o n

East Los Angeles College is a community college in the southwest corner of Monterey Park. It has a current student enrollment of 35,403 students, the largest of any California community college. The student body is primarily Hispanic students, but also has a significant number of white, Asian, and black students. It offers associates degrees in 27 different fields of study. It is one of the best colleges to get an associates degree from and is the best in California at awarding degrees to Hispanic students.

There are several parks within the city which host many events like birthday parties, family gatherings, barbecues, baseball games, concerts, and festivals. Barnes park is one of the largest parks in the city and is 17 acres. On the park grounds there is a community center, basketball courts, an amphitheater, and a swimming pool among other things. The city and parks are also host to several festivals including the Lunar New Year Festival. During the festival several main streets are shut down to accommodate for vendors setting up along whole city blocks. There is amusement rides and live entertainment as well as dragon dancers who kick-off the festival on the Saturday morning. Previous festival years have seen visitor numbers ranging from 50,000 to 100,000.

parks + events

the site

The city of Monterey Park is currently undergoing serious economic development as a result of its high quality of life ratings. With this massive influx of new residents there has been an increased demand in medium and high density housing options. Our project is designed to help alleviate this housing problem while also offering new jobs to help support the high number of new residents. Located on the corner of E Garvey Ave and S Garfield Ave, our site would take over the location of one parking lot and two abandoned lots. Being in the heart of Monterey Park means close proximity to many of the attractions listed in this pamphlet. For example, Mama Lu's Dumpling House and T-Kebob are both located less than a block from our site. There are also several bus stops located at the intersection giving quick and easy access to public transit for residents. Barnes park is also a 9 minute walk away.

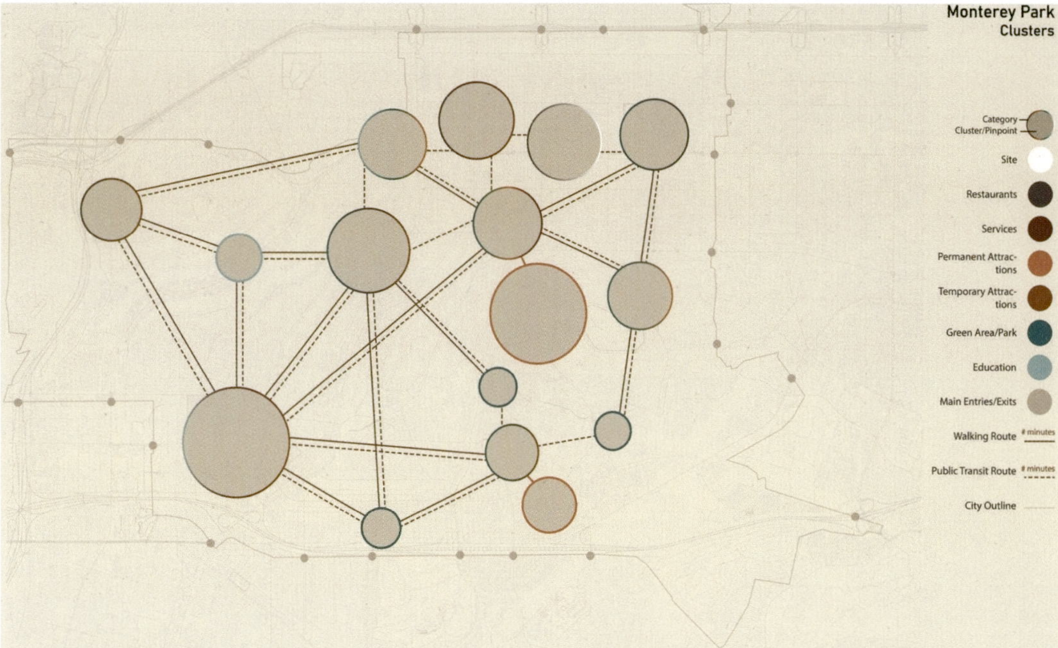

Monterey Park Clusters
Category Cluster/Pinpoint
Site
Restaurants
Services
Permanent Attractions
Temporary Attractions
Green Area/Park
Education
Main Entries/Exits
Walking Route
Public Transit Route
City Outline

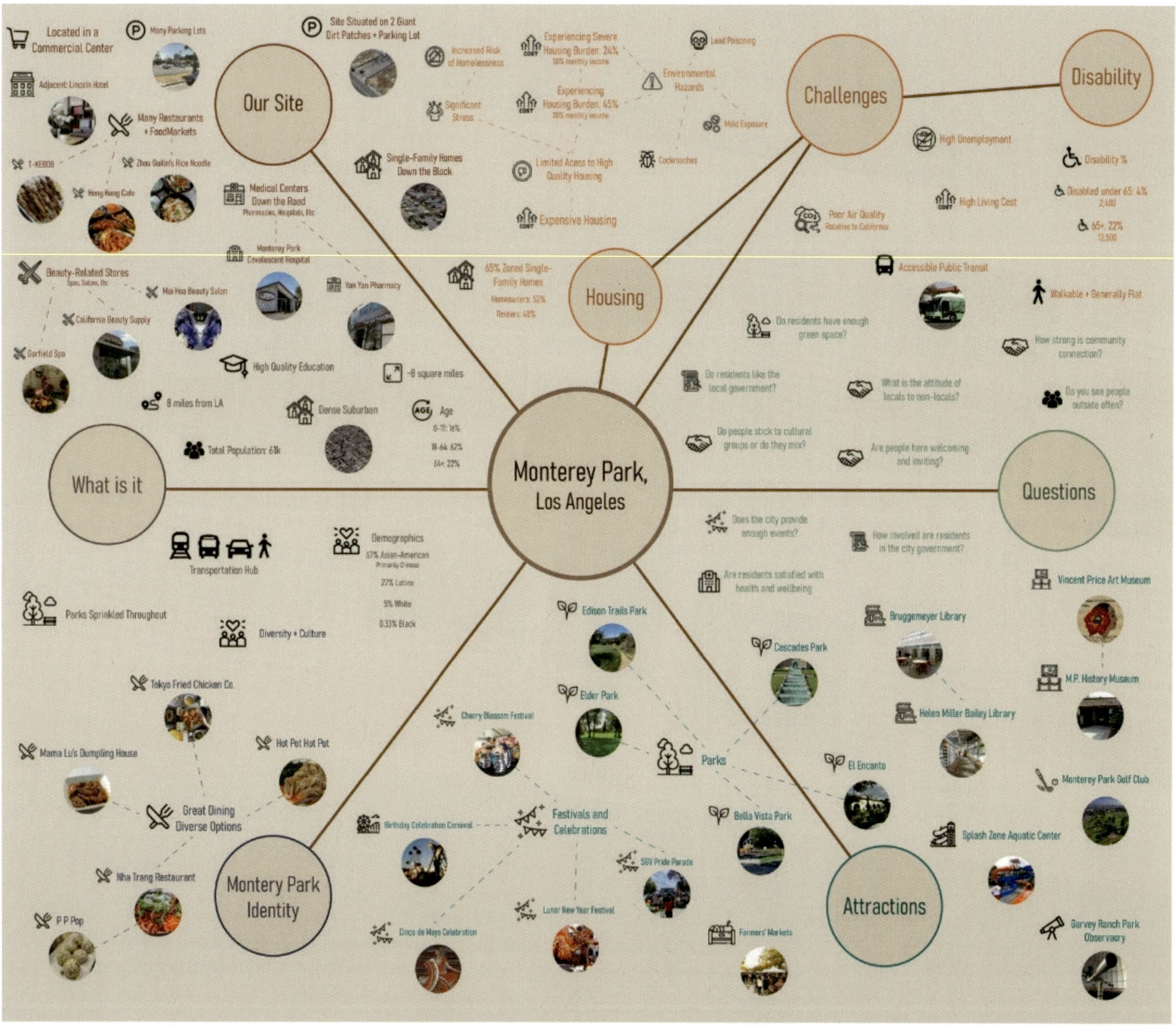

Monterey Park, Los Angeles
Our Site
Located in a Commercial Center
Many Parking Lots
Site Situated on 2 Giant Dirt Patches + Parking Lot
Adjacent: Lincoln Hotel
Many Restaurants + FoodMarkets
T-KEBOB
Zhou Guilin's Rice Noodle
Hong Kong Cafe
Medical Centers Down the Road
Pharmacies, Hospitals, Etc
Monterey Park Covalescent Hospital
Yan Yan Pharmacy
Single-Family Homes Down the Block
Beauty-Related Stores
Spas, Salons, Etc
Mai Hoa Beauty Salon
California Beauty Supply
Garfield Spa
High Quality Education
~8 square miles
8 miles from LA
Dense Suburban
Age
0-17: 16%
18-64: 62%
64+: 22%
Total Population: 61k
What is it
Transportation Hub
Demographics
67% Asian-American
Primarily Chinese
27% Latino
5% White
0.33% Black
Parks Sprinkled Throughout
Diversity + Culture
Increased Risk of Homelessness
Experiencing Severe Housing Burden: 24%
50% monthly income
Significant Stress
Experiencing Housing Burden: 45%
30% monthly income
Limited Acess to High Quality Housing
Expensive Housing
Lead Poisoning
Environmental Hazards
Mold Exposure
Cockroaches
Challenges
Disability
High Unemployment
Disability %
Disabled under 65: 4%
2,400
65+: 22%
13,500
High Living Cost
Poor Air Quality
Relative to California
Housing
65% Zoned Single-Family Homes
Homeowners: 52%
Renters: 48%
Accessible Public Transit
Walkable + Generally Flat
Do residents have enough green space?
How strong is community connection?
Do residents like the local government?
What is the attitude of locals to non-locals?
Do you see people outside often?
Do people stick to cultural groups or do they mix?
Are people here welcoming and inviting?
Questions
Does the city provide enough events?
How involved are residents in the city government?
Are residents satisfied with health and wellbeing
Vincent Price Art Museum
Bruggemeyer Library
M.P. History Museum
Helen Miller Bailey Library
Monterey Park Golf Club
Splash Zone Aquatic Center
Garvey Ranch Park Observaory
Attractions
Edison Trails Park
Cascades Park
Elder Park
Parks
El Encanto
Bella Vista Park
Farmers' Markets
Cherry Blossom Festival
Festivals and Celebrations
Birthday Celebration Carnival
SGV Pride Parade
Lunar New Year Festival
Cinco de Mayo Celebration
Tokyo Fried Chicken Co.
Hot Pot Hot Pot
Mama Lu's Dumpling House
Great Dining Diverse Options
Nha Trang Restaurant
P P Pop
Montery Park Identity

RACIAL DEMOGRAPHICS

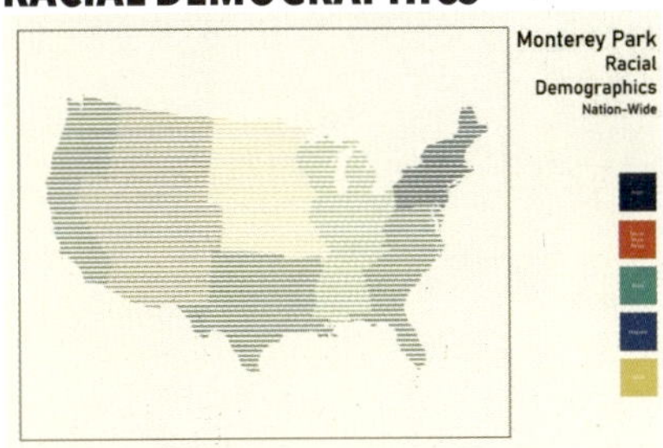

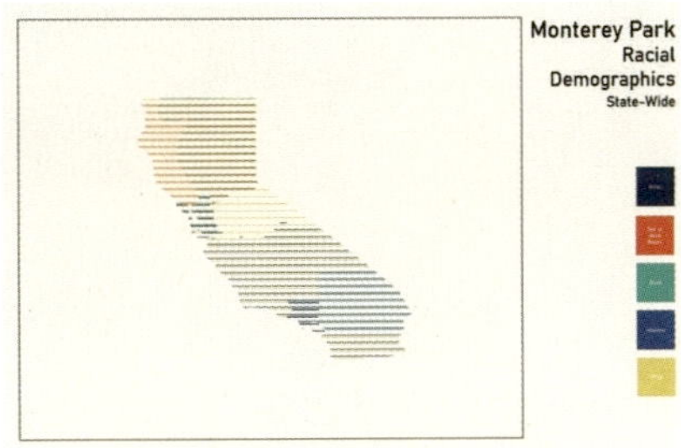

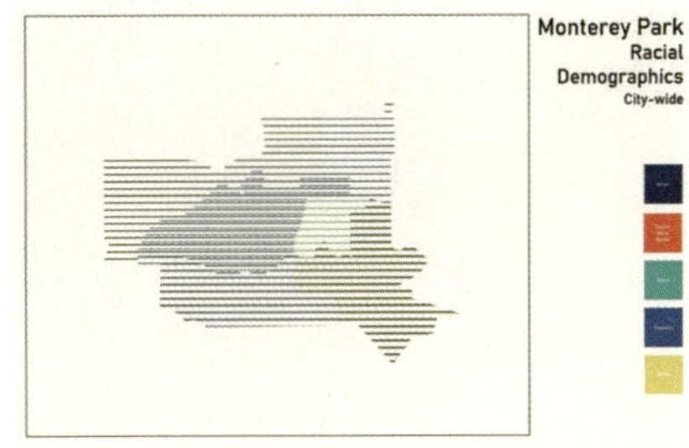

TOURSIT DESTINATIONS

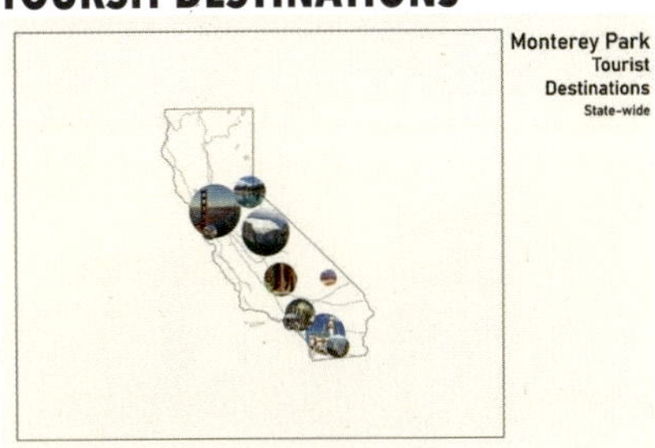

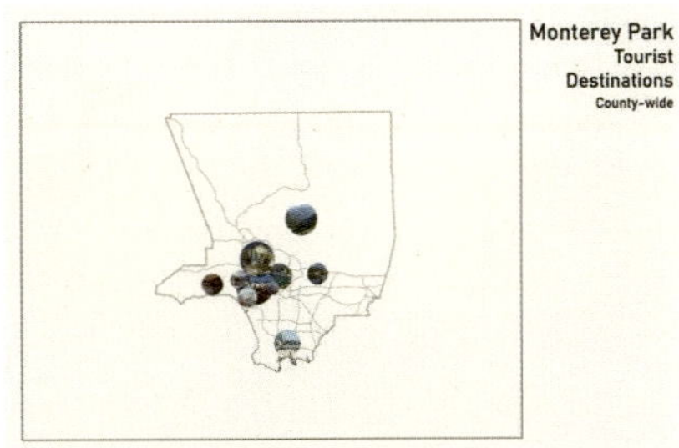

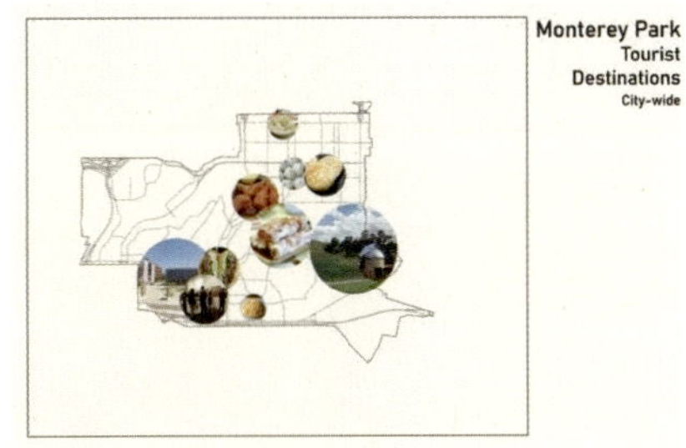

HOUSEHOLD INCOME

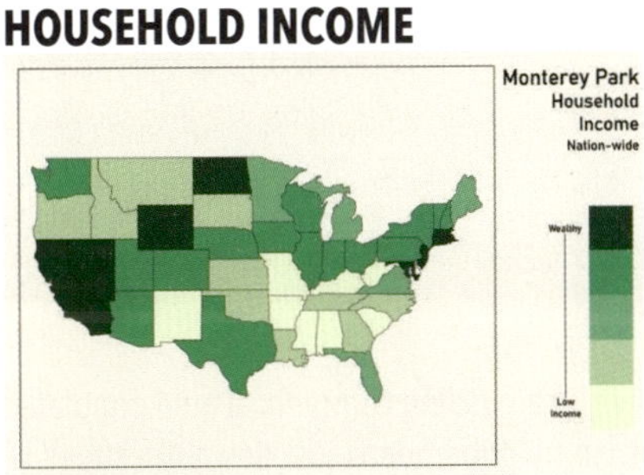

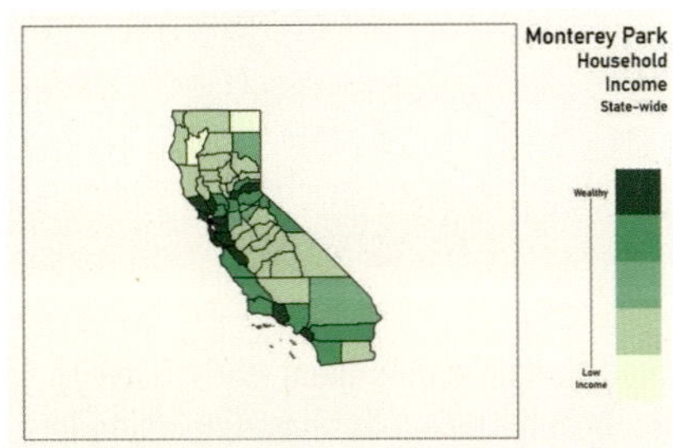

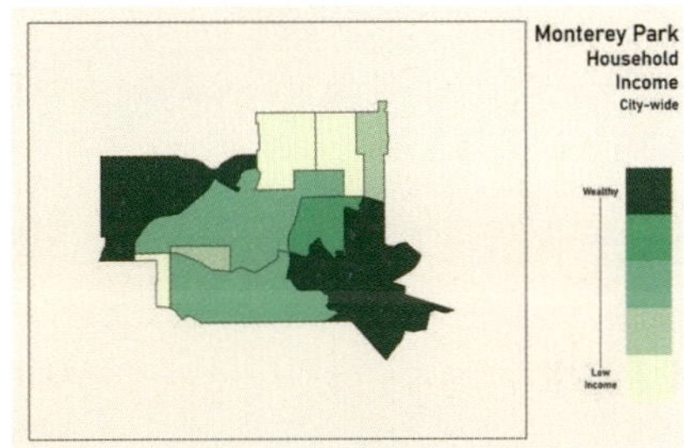

DISABILITY

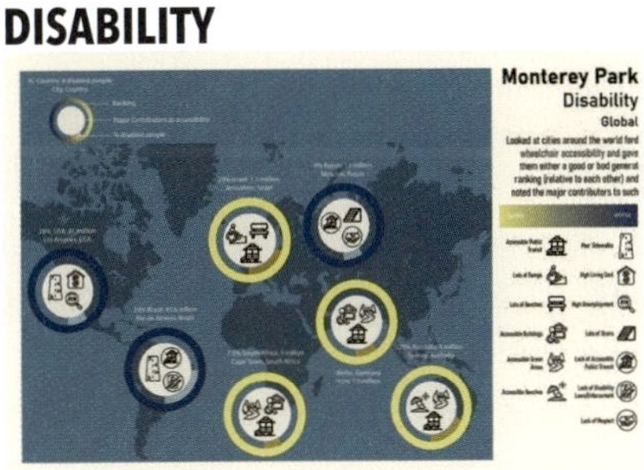

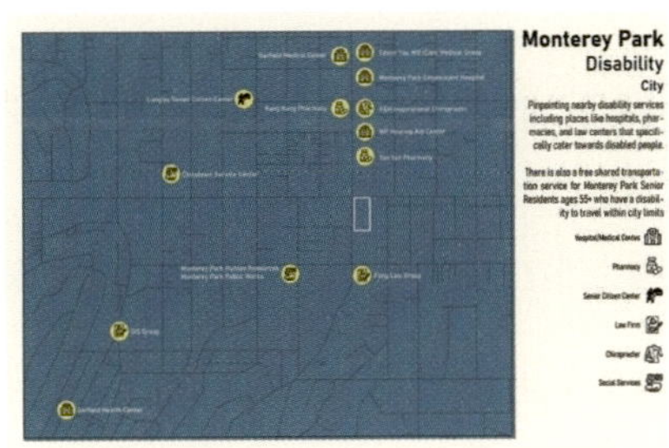

AIR QUALITY

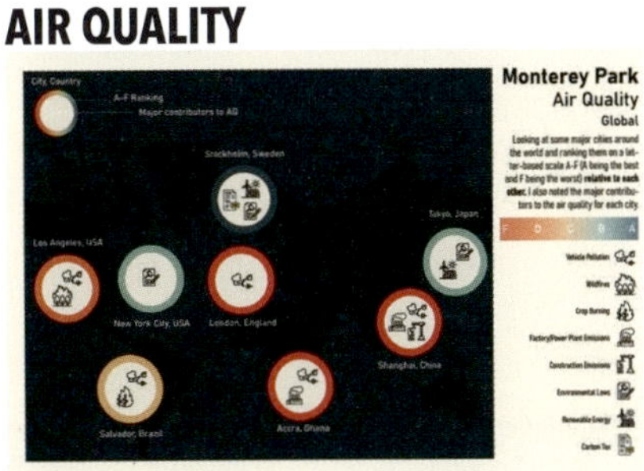

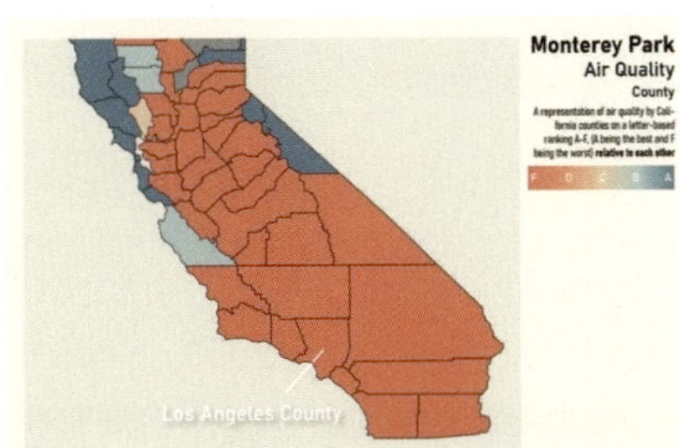

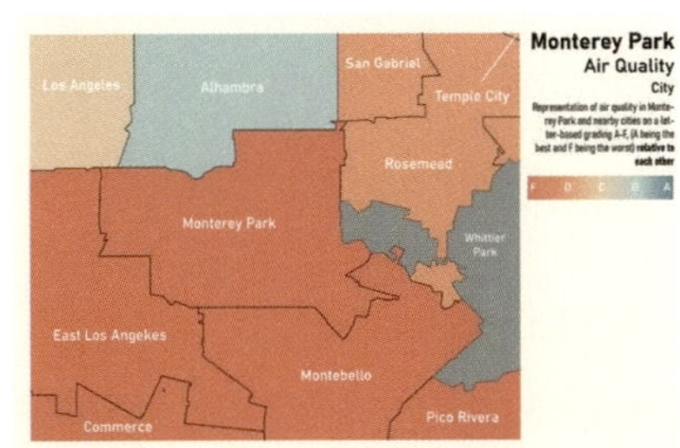

VEGETATION

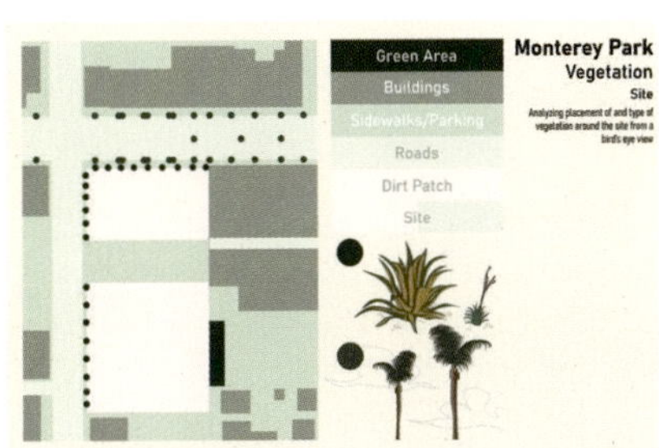

THE PARK

PROJECT BY

MATTHEW WILLIAMS
ARYAN AMBANI
MICHELA BRITTIS-TANNENBAUM

PROJECT INFORMATION

Location : Monterey Park, Ca
Use : Mid-Rise Mix-Use
Building area : 40,100sf
Total Floor area : 180,400sf
Building coverage ratio : 58%
Floor area ratio : 249%
Floor : 5F
Height : 65ft
The number of parking spaces : 45
The number of households : 80

Our site is located along two prominent roads, Garvey and Garfield Ave, in a bustling commercial area populated by various eateries, beauty-related businesses, parking lots, and suburban neighborhoods just down the street. Adjacent, the buildings mainly hold businesses that are only one to two stories, with the exception being a hotel climbing to about nine stories high.

From our research, we determined that our main focuses would be on allowing access to lively green space, disabled accessibility throughout our project, and fostering interactive relationships between residents.

우리 대지는 가비와 가필드라는 두 개의 눈에 띄는 도로를 곁에 두었으며, 바로 아래에는 식당과 미용 관련 사업체, 주차장, 교외 지역이 밀집한 번화한 상업 지역이 위치한다. 대지 인근의 건물들은 9층 규모의 호텔을 제외하고 대부분 1-2층 규모의 사업장이 운영되고 있다.

연구를 통해 우리는 활기찬 녹지 공간에 대한 접근을 허용하고 프로젝트 전체에 걸쳐 배리어프리를 적용함으로써 주민 간의 상호적인 관계를 이끄는 것을 최종 목표로 결정했다.

MASSING STUDY

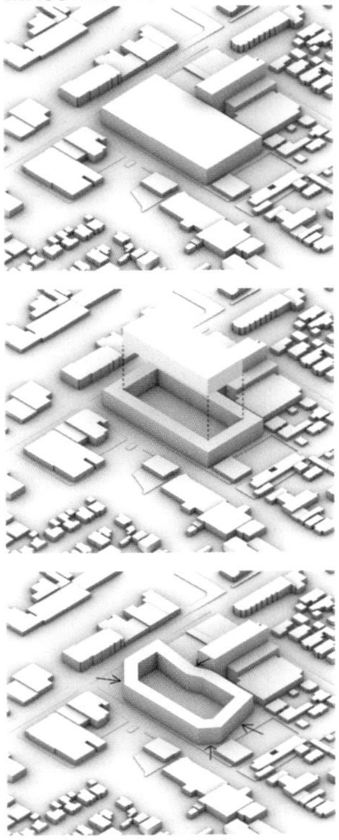

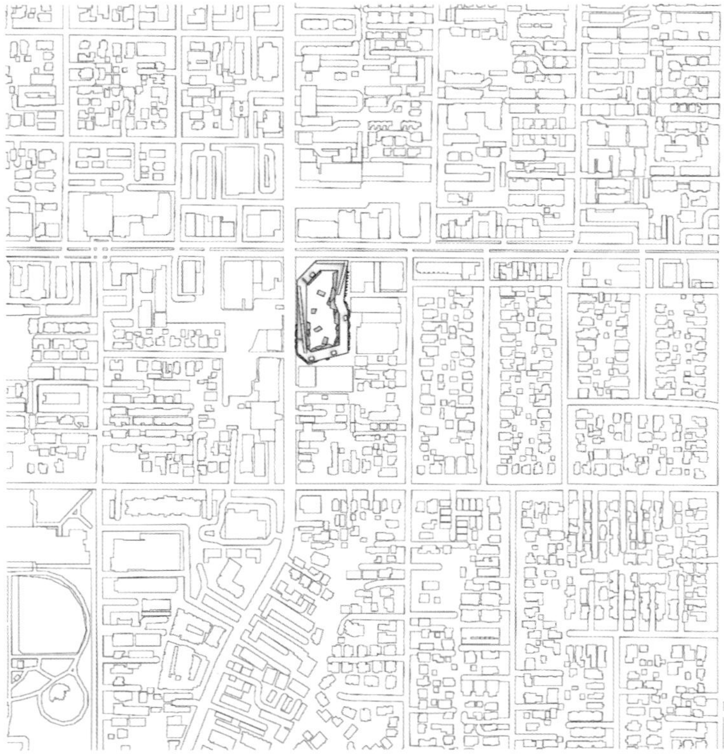

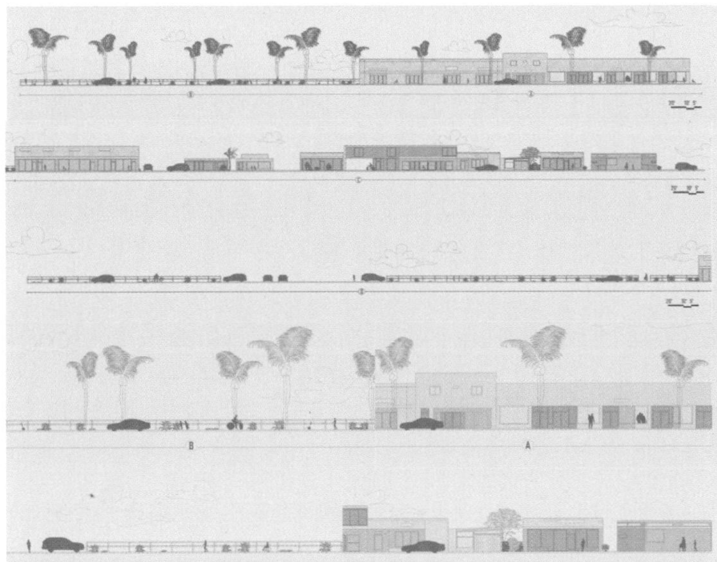

Street Elevations

While there are parks littered throughout the city, they are often static in that they serve few purposes such as an open, grass lands or sports field. Other areas of green space are reserved for the single-family homes which are private and regulated to specific plots of land. Because of this, it's much more difficult for those living in apartments to have easy access to meaningful green space that is fulfilling both visually and in terms of activity. To address this, we centered our form primarily on the idea of a central, public courtyard filled with native and drought-resistant vegetation due to the increasingly hot climate. To add to its appeal, we planted several small shops as well as intersecting paths which create small sections of spaces that could be used for a variety of activities.

We found that while the state of California ranked as some of the top performers in terms of income and percentage of employed disabled people, the overall expensive cost of living makes it one of the worst states for disabled people. Because they are often isolated from society for their disability and especially considering the lack of accommodations, we created a community in which disabled people are able to participate and enjoy life just as any other abled person. This is seen through the vertical transportation always being nearby, passageways and doorways being wide enough at all points, apartments allowing for space under sinks and tables for wheelchairs, shelving being within reaching distance, and so forth.

Lastly, we took the hallway and made it into something more animated and dynamic by introducing a pathway within a pathway. Where a hallway is usually used solely for a path to get to a destination, we turned it around by allowing residents to expand their homes and expression outside and into this space, blending the boundaries between the home and the communal. In this way, we hope to make this space something completed by human interaction and opportunity for people to take this space, and make it into what they want.

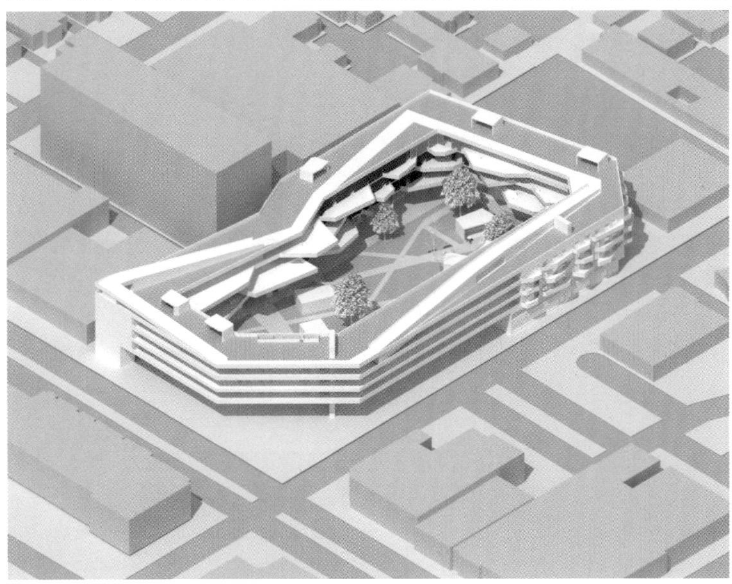

Ground Floor

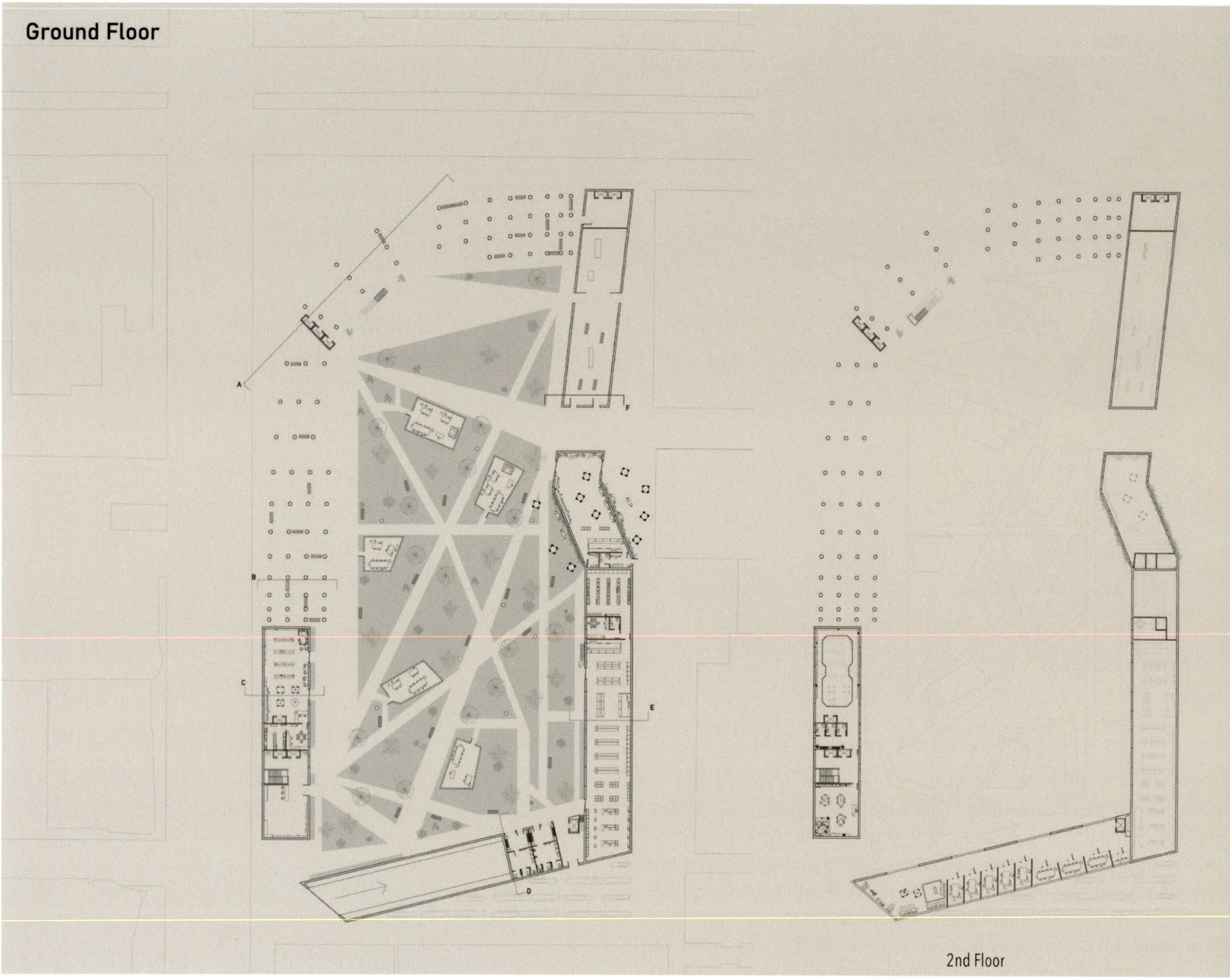

2nd Floor

Section C-C

Section E-E

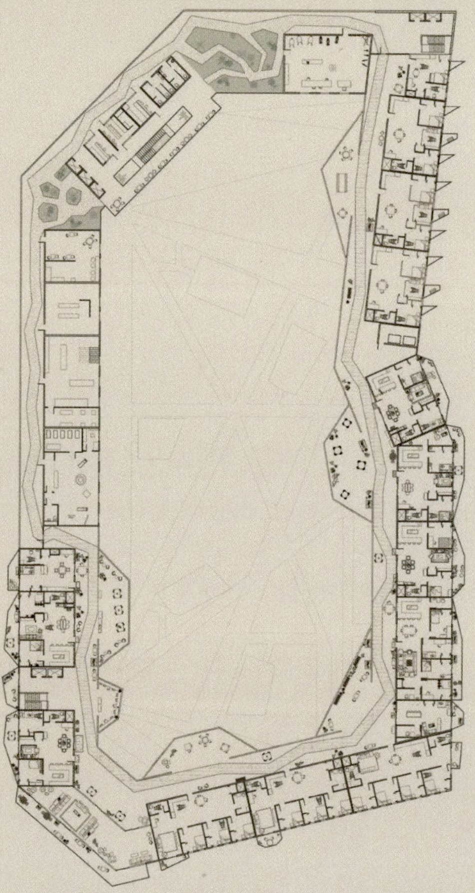

3rd Floor

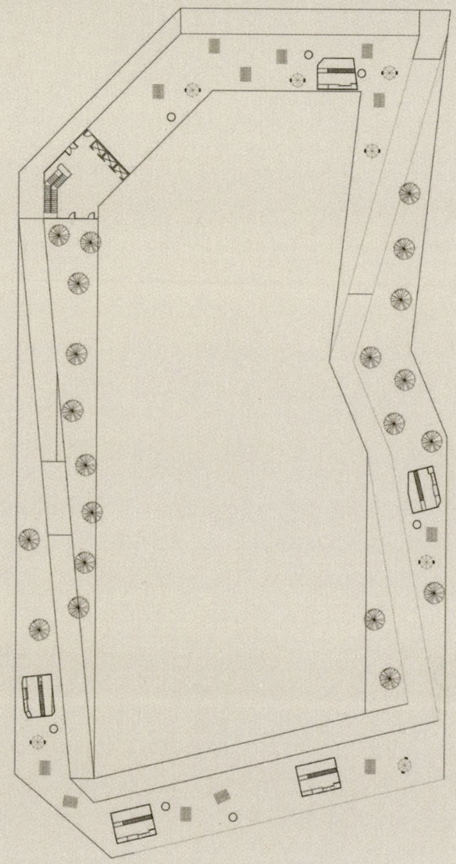

Roof

Unfolded Roof Elevation

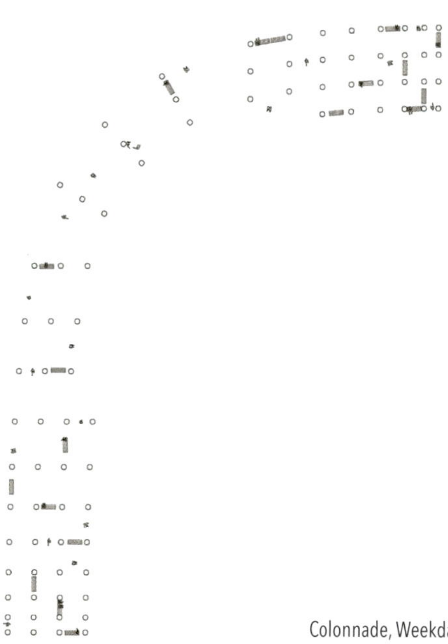

Colonnade, Weekday

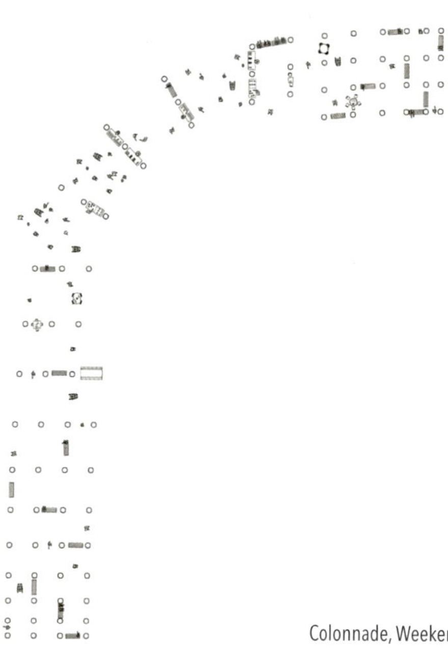

Colonnade, Weekend

Colonnade, Festival

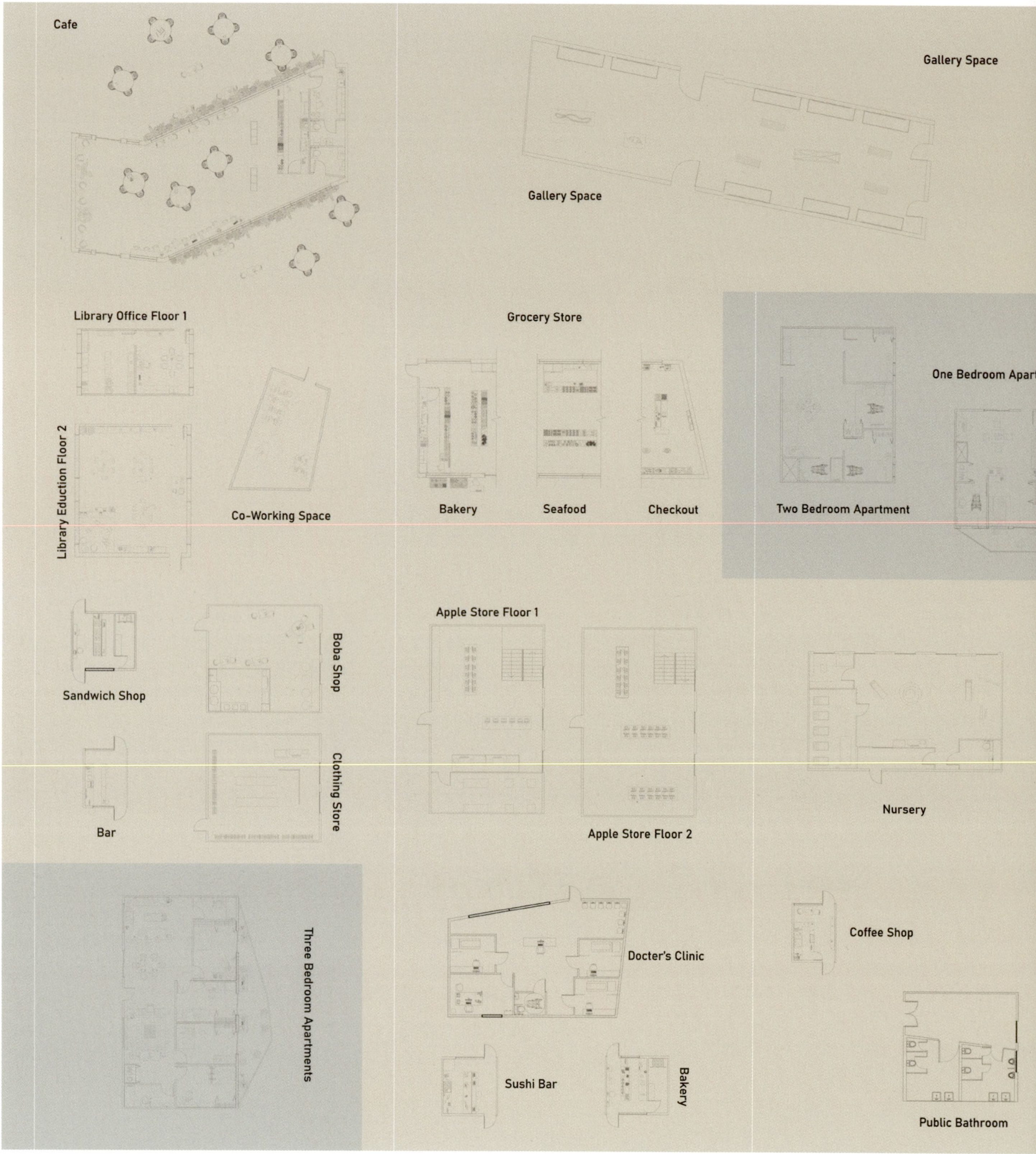
Cafe
Gallery Space
Gallery Space
Library Office Floor 1
Grocery Store
One Bedroom Apart
Library Eduction Floor 2
Co-Working Space
Bakery
Seafood
Checkout
Two Bedroom Apartment
Apple Store Floor 1
Sandwich Shop
Boba Shop
Bar
Clothing Store
Apple Store Floor 2
Nursery
Three Bedroom Apartments
Docter's Clinic
Coffee Shop
Sushi Bar
Bakery
Public Bathroom

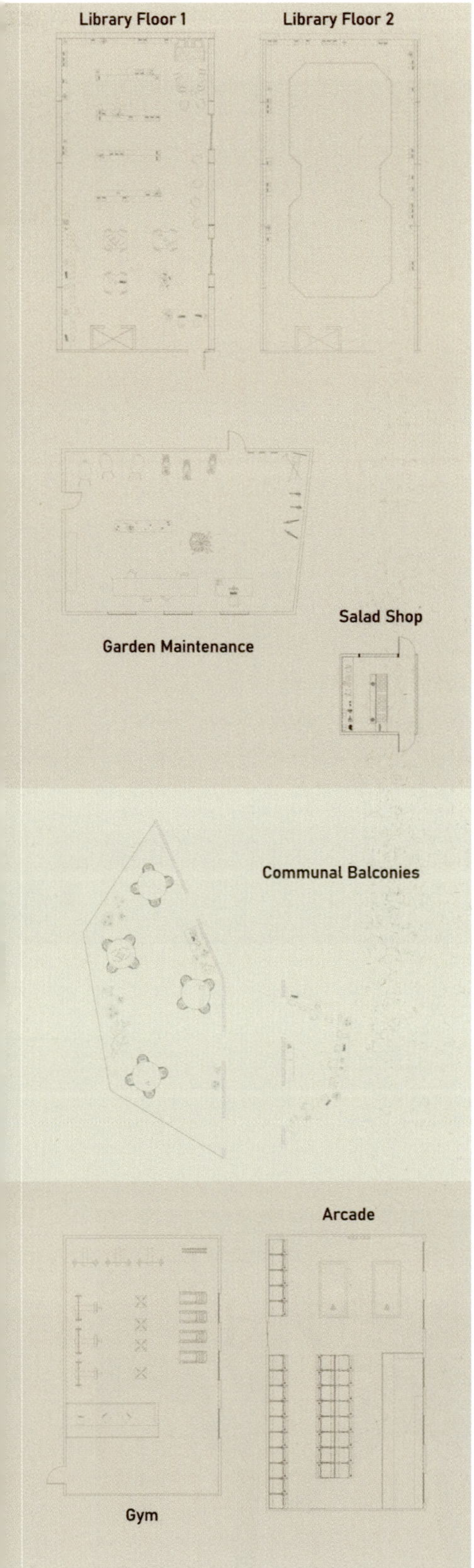
Library Floor 1
Library Floor 2
Salad Shop
Garden Maintenance
Communal Balconies
Arcade
Gym

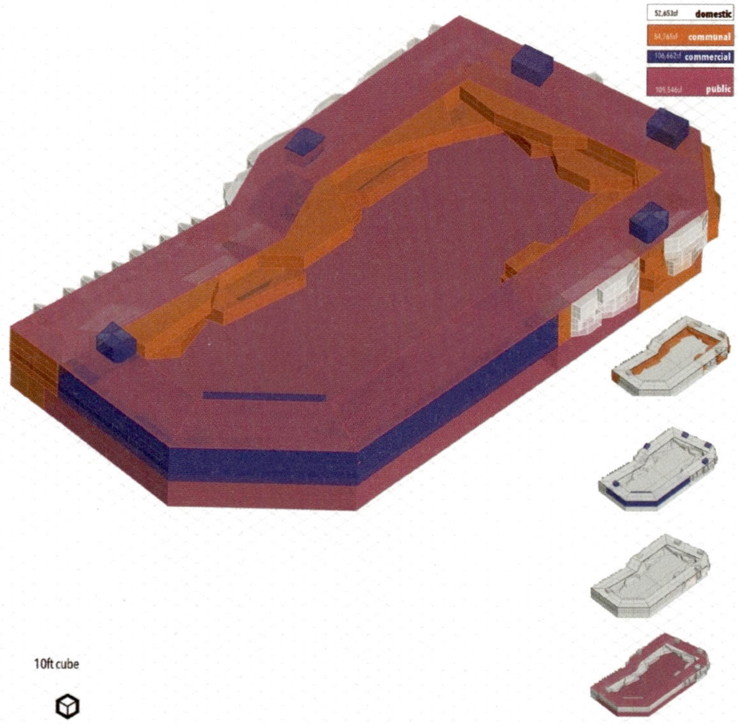
domestic
communal
commercial
public
10ft cube

WESTLAKE, LOS ANGELES CITY, CALIFORNIA

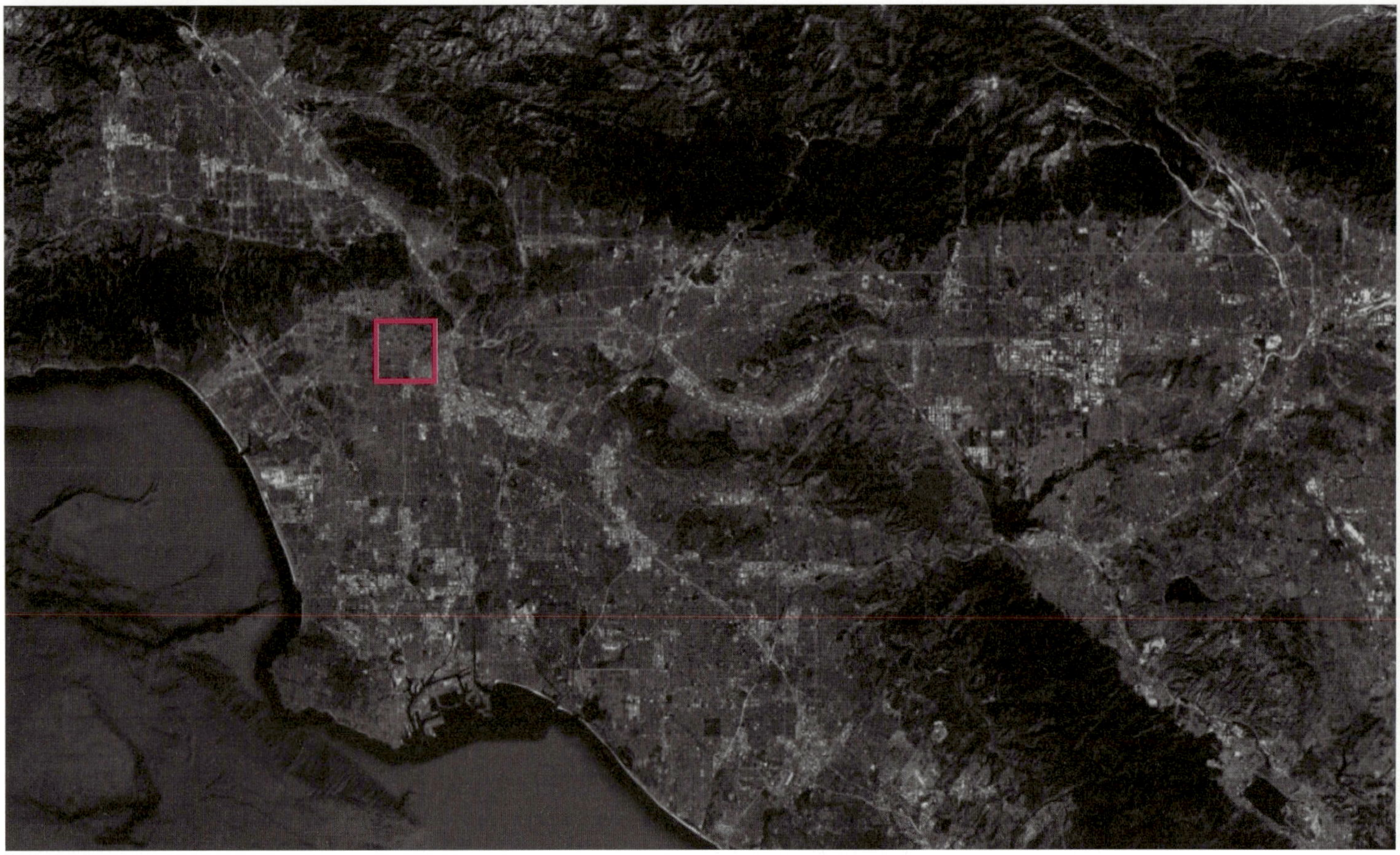

RESEARCH BY
SARAH BANNAN
DONOVAN HERNANDEZ
RUOHAN JIANG
ADRIENN VIRAG

Westlake is a neighborhood situated right next to downtown, on the brink of gentrification. As of 2022, 21 high profile projects are planned for the area surrounding the beautiful MacArthur Park, including 40 storey high-rises. But, Westlake is currently not very liveable. It's troubled by drug use, crime, gang violence, and a lack of sanitation, especially around MacArthur Park.

시내 바로 옆 웨스트레이크는 젠트리피케이션*의 위기에 직면해 있다. 2022년 현재 아름다운 맥아더 공원 주위로 40층 고층 건물을 비롯해 21개의 주요 프로젝트들이 계획되어 있다. 그러나 웨스트레이크는 살기 좋은 곳이 아니다. 맥아더 공원 주변에서 발생하는 마약, 범죄, 갱단 폭력, 위생 시설 부족으로 어려움을 겪고 있기 때문이다.

젠트리피케이션(En: gentrification)* : 도심 인근 낙후 지역의 활성화로 인하여 외부인과 돈이 유입되고, 임대료 상승 등으로 원주민이 밀려나는 현상.

DENSITY

LOCAL IDENTITY

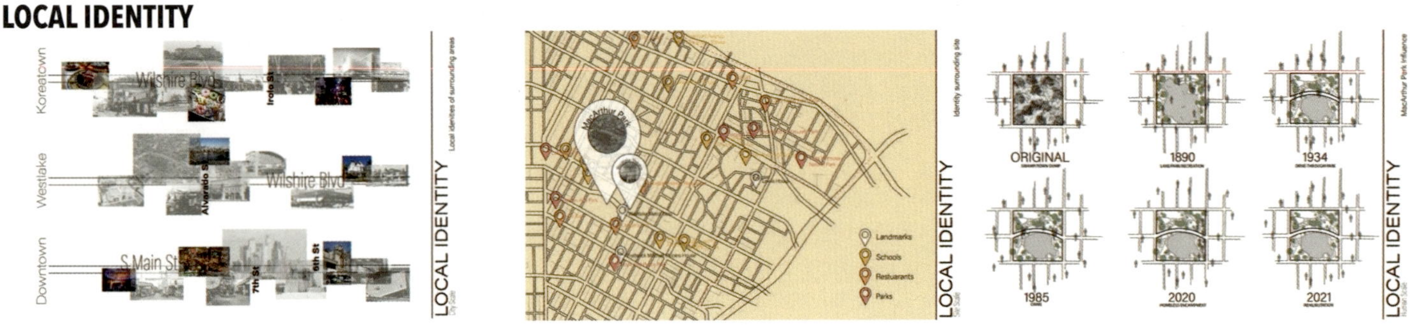

PUBLIC TRANSIT

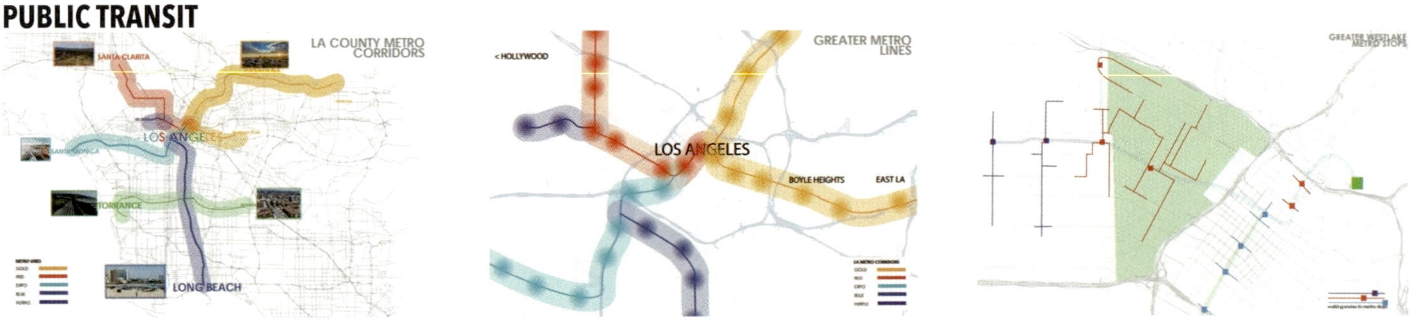

ETHNO RACIAL DISTRIBUTION

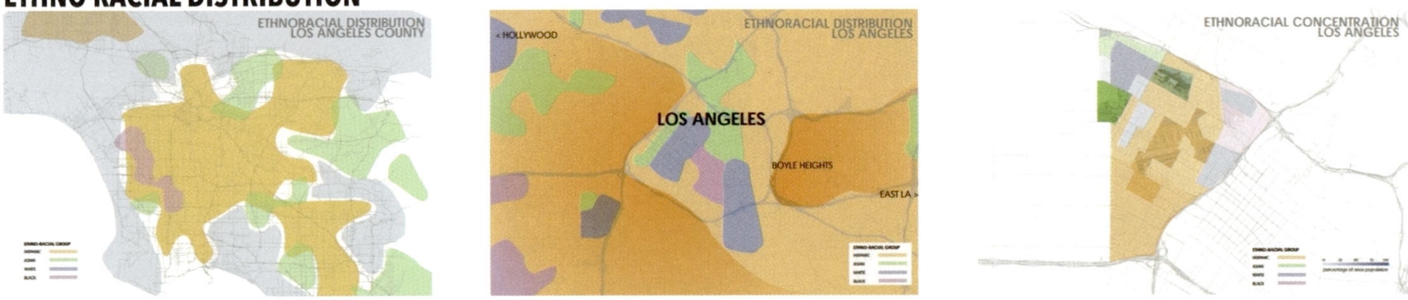

FOOD

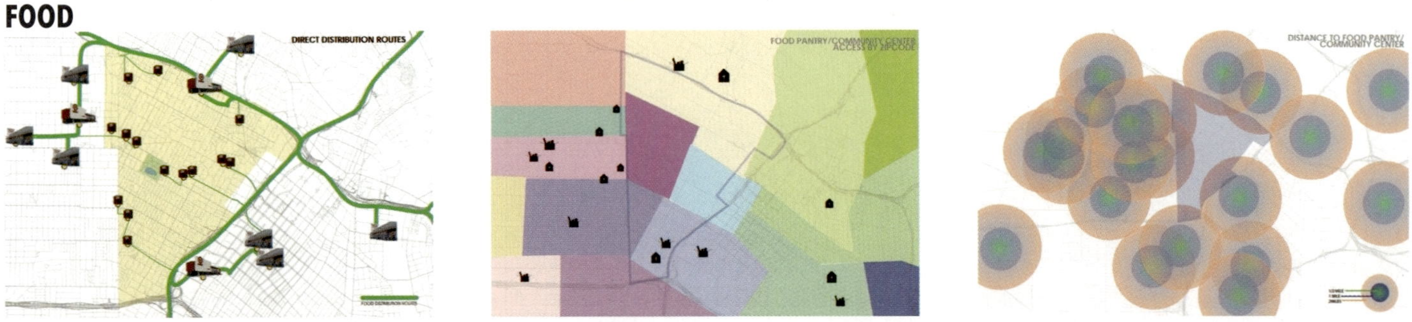

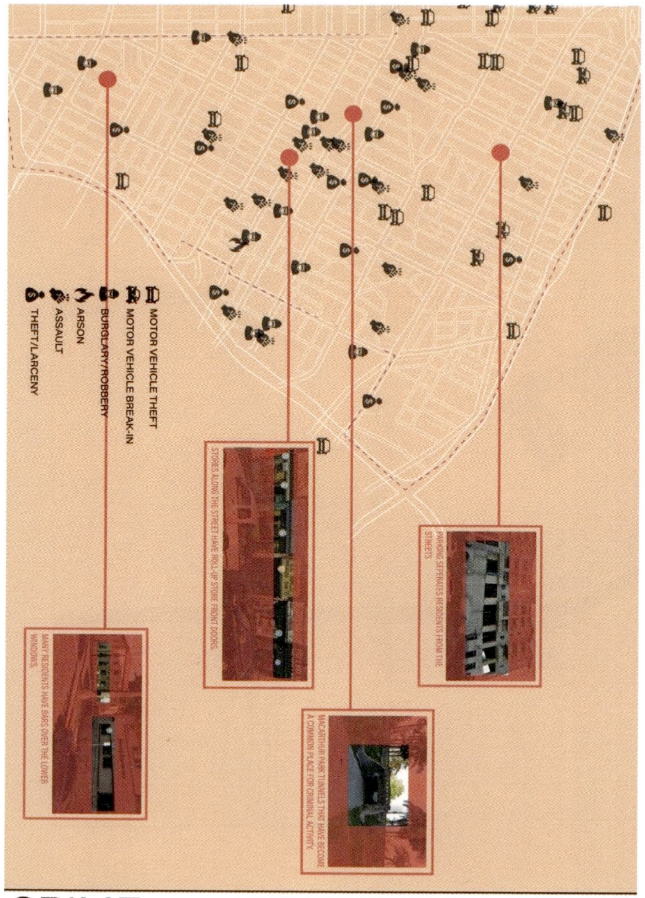

CRIME
Neighborhood Scale

Mapped Crimes in Westlake

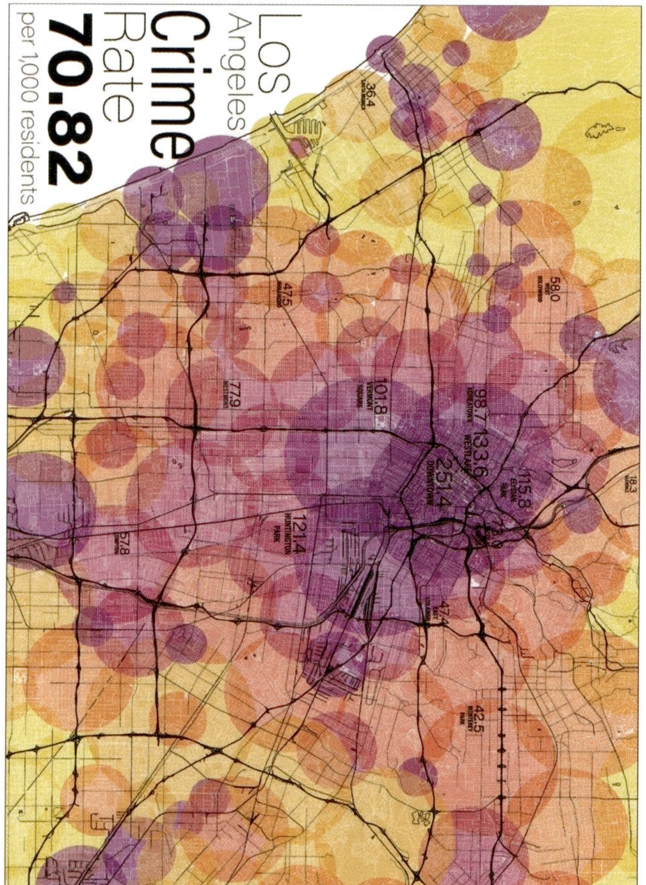

CRIME
City Scale

Crime per capita in Los Angeles

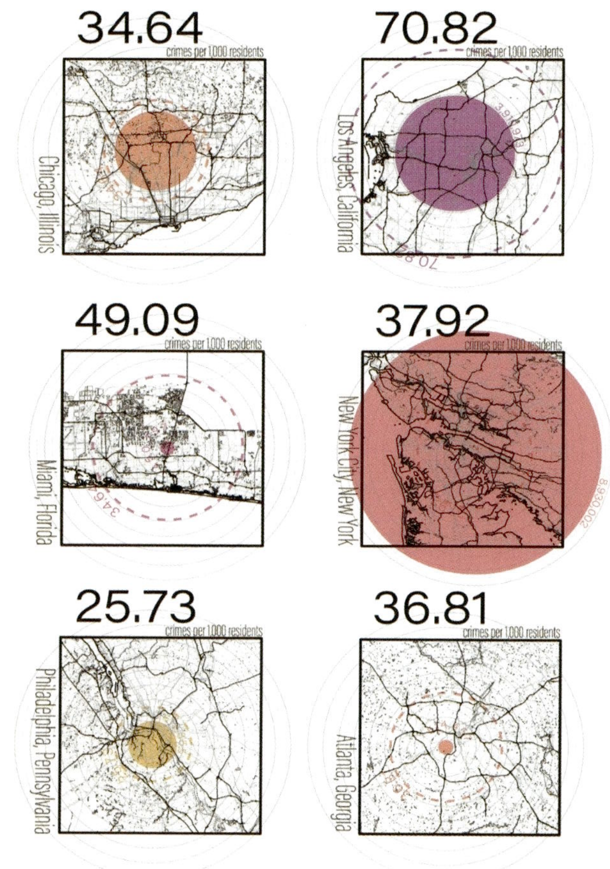

CRIME
National Scale

Crime per capita in cities in the U.S.

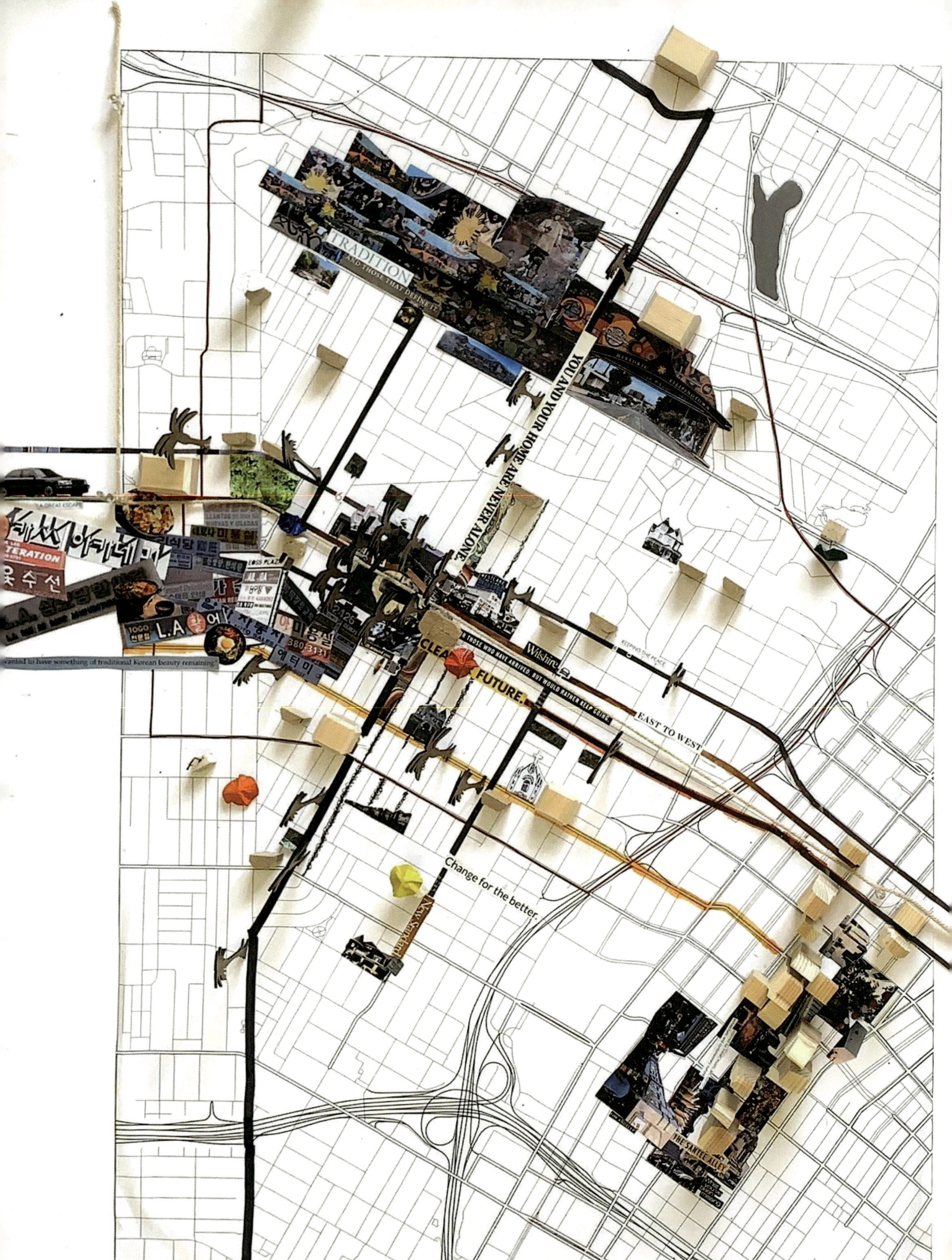
TRADITION
AND THOSE THAT DEFINE IT
YOU AND YOUR HOME ARE NEVER ALONE
Wilshire
EAST TO WEST
Change for the better.
THE SANTEE ALLEY
L.A.

Westlake Landmark

Westlake Landmark

Westlake Landmark

Gen. Douglas MacArthur Monument
MacArthur Park

Park Plaza Hotel
S Park View Street

Westlake Theatre
S Alvarado St

Westlake Landmark

Westlake Landmark

Westlake Landmark

Grier Musser Museum
S Bonnie Brae St

La Fonda
Wilshire Blvd

Westlake/MacArthur Park Station
S Alvarado St

Westlake Community

Westlake Community

Westlake Community

Government Action

City starts rehilbilitation on the south side of Westlake Park.

You must now displace the homeless encampment.

Community Action

You are volunteering at he local food pantry tonight.

Go to the nearest food pantry.

Illegal Action

You were caught in the Westlake Park tunnels buying meth.

Run 10 blocks away.

Westlake Community

Westlake Community

Westlake Community

Life Action

You missed the Westlake /MacArthur Park and thought there was another stop in Westlake.

You are now in Downtown.

Life Action

You and your family need dinner.

You must go to the nearest Target.

Life Action

A palm tree fell on your vending cart.

Buy a new cart 4 blocks away.

Westlake Community

Westlake Community

Westlake Community

Community Action

The students at the elementary school plant trees to help supply food in this food desert.

Go visit Westlake Elementary School.

Illegal Action

You are street vending on Alvarado but don't have the proper permits.

Your cart is now confiscated and now being held at the police station.

Government Action

The city installs electronic garbage cans but no longer cleans the streets regularly.

Take 1 step forward and 3 steps back.

Westlake Community

Westlake Community

Westlake Community

Illegal Action

There was a man killed by a shooting at a busy bus stop.

Wait for the next bus.

Development Action

Developers from Syracuse are redeveloping your elementary school.

Find the nearest school to transfer to.

Community Action

You refused to barter when selling your goods today.

No action can be taken.

Westlake Chance

Westlake Chance

Westlake Chance

You sell 20 screwdrivers in one day.

Collect 100 dollars.

You are a member of the Columbia Lil Cycos and are shaking down street vendors.

Collect 20 dollars from each player.

The Levitt Pavillion hosts your favorite play.

It's free!

Westlake Chance

Westlake Chance

Westlake Chance

A tourist mistakenly ends up in Westlake instead of Hollywood. In anger they kick over your cart.

You lost 30 dollars.

A palm tree falls on your competitor's vending cart. You get all their customers.

Gain 60 dollars.

You go out to dinner at Yerim Korean BBQ.

Pay 20 dollars for your dinner.

Westlake Chance

Westlake Chance

Westlake Chance

A large protest is happening in the park. The police fence off 7th Street.

You lose 50 dollars that day.

New developers come in and the apartments with a view of downtown have very high rent.

Pay 90 dollars.

Your great-grandfather leaves you a historic mansion in his inheritance.

You get any property you like.

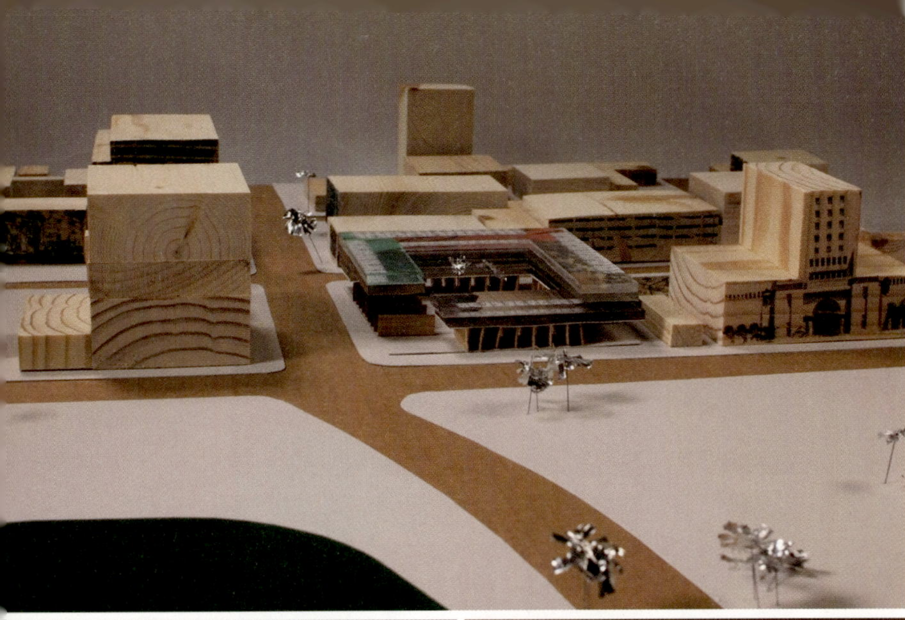

COLLECT $200 WHEN YOU PASS GO

PROJECT BY
ADRIENN VIRAG
SARAH BANNAN

Ituated in Westlake Los Angeles, the project Collect $200 Hundred Dollars when you Pass Go claims a block of the city to establish a collaborative and entrepreneurial environment for young people. It also aims to claim a safe and clean public space bounded by strong, warm, materiality. Pass Go aims to house flexible functions in permanent materiality.
Super Serious interviewed locals in order to get a sense of living and working in the area. The secretary at the Carondelet said her street feels safe, but she avoids venturing too far. Alex from Travel Inn said,"Clean up the homeless, put more bins. Streets are littered with paper and everything. I'm picking up trash all the time. The guests are all saying there is too much trash." About MacArthur Park he said, "I don't go there so much. I don't want to be confronted by homeless people, and it smells like urine." Maria, a Subway Employee said, "It's busy, but there's a lot of violence. Drugged-up people during the day are violent with everyone. But I feel safe because everyone knows each other, and looks out for each other."

이것은 로스앤젤레스 웨스트레이크에 위치하고 지역 젊은이들을 위한 협력적이고 기업가적인 환경을 구축하기 위해 도시의 한 블록을 통과할 때 200$를 모으는 프로젝트이다. 강하고 따뜻하며 물질적인 것으로 경계를 이루는, 안전하고 깨끗한 공공 공간을 주장하고자 한다. Pass Go는 영구적인 물질에 유연한 기능을 받아들이는 것이 목표이다.
우리는 그 지역에서 일하고 사는 느낌을 얻고자 현지인들을 인터뷰했다. 카론드렛의 비서는 거리가 안전하다고 느끼면서도 멀리 나가는 일은 피한다고 말했다. 알렉스는 "노숙자들을 이동시키고 쓰레기통을 더 많이 두세요. 이 거리에는 종이와 모든 것들이 널려 있어요. 나는 항상 쓰레기를 주워요. 손님들이 모두 쓰레기가 너무 많다고 하더라고요." 그는 맥아더 공원에 대해 "그다지 자주 가지 않아요. 노숙자들과 마주치고 싶지도 않고, 소변 냄새가 나거든요." 지하철 직원 마리아는 "바쁘지만, 폭력이 많아요. 낮에 마약에 취한 사람들은 공격적이에요. 하지만 모두가 서로 알고 돌봐주기 때문에 안전하다고 느껴요."

DIAGRAM OF STREET CLEANLINESS

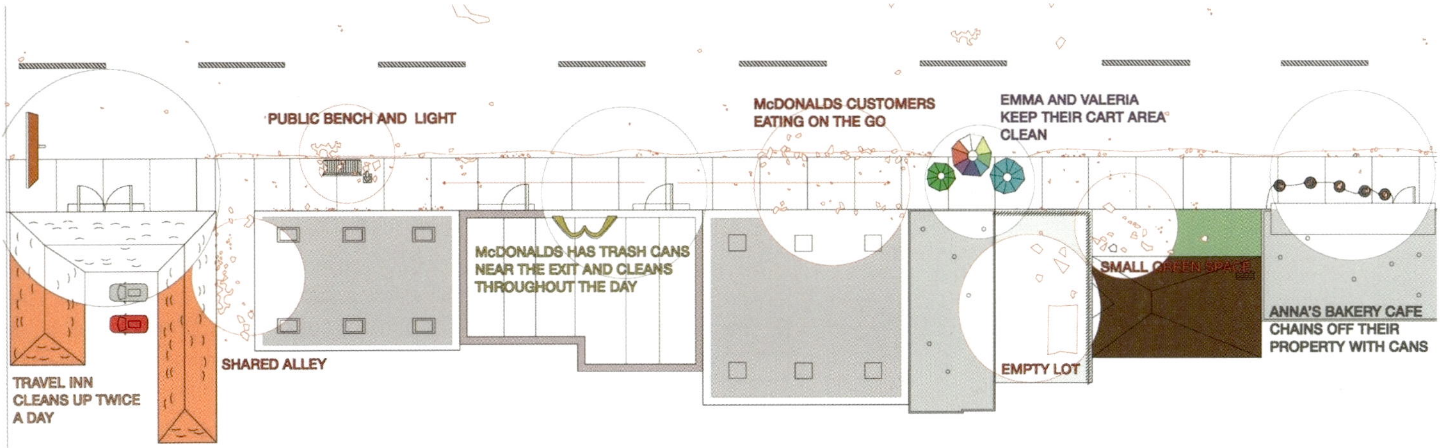

What stood out was how unusable the public space became because of the street culture. As shown in the diagram of street littering, what was evident was that when the street was claimed privately, for example by street vendors, the surrounding area was maintained and became usable. This street vending culture is one of the best things about Westlake. Coming out of the subway, vendors line the street every day, selling everything you can imagine, like a daily flea market. Although these little businesses are very popular, they only claim a narrow portion of the sidewalk, with customers often eating at cramped tables.

Westlake has a large population of young adults when compared to the rest of Los Angeles, with 25% of residents between the ages of 22-34. This is our target group, youth, who have recently become financially independent and are looking to broaden their perspectives and continue to work in or start businesses in the Westlake area. The upper four floors of Pass Go are dedicated to dense communal living, designed as a temporary environment where these young people can meet, learn from each other, and work together. In anticipation of gentrification, the Pass Go perimeter block claims the site for these young individuals as well as the public of Westlake, for as long as possible.

PROGRAM STRATEGIES

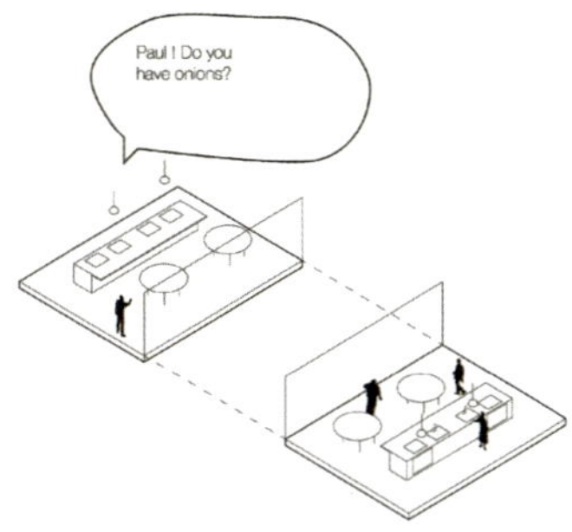

FACING COMMUNITY KITCHEN

CIRCULATION OPENING INTO KITCHEN

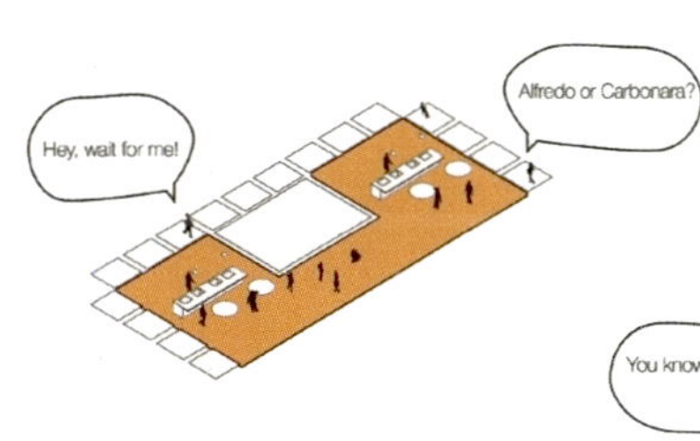

BIG FAMILY FLOOR

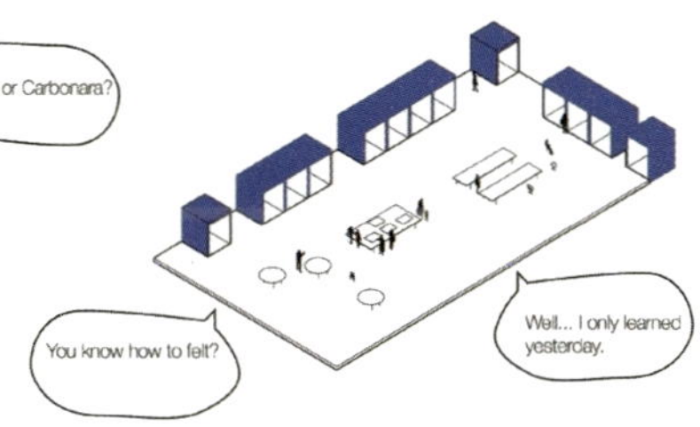

PUBLIC (INSERT COMMERCIAL)

CONVERTIBLE MICROUNITS

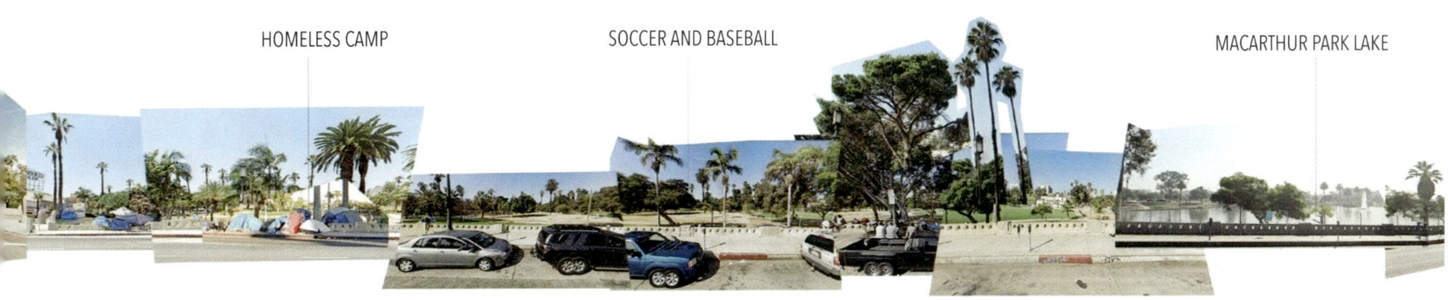

In Pass Go, a primary objective was to reinforce the best parts of Westlake street culture. So, on the ground floor, taking inspiration from the street vendors, are small rentable pods, for our residents, creating a ground floor that's open to the public, but with a porosity that can be controlled. Ideally, there are as many people filling up the facade as possible, while leaving some open for passage. In this situation, there are a lot of eyes everywhere, many people who know each other, and our ground floor becomes safe. However, if the openness ends up feeling unsafe, there remains the option to close the exterior doors, limit the entrances, and vend inwards. The public space circulates all the way up to the rooftop, with workspaces and classrooms to host meetings and workshops. On the roof, is a fully public recreation space, with the goal of putting out eyes over the park, and developing a new image of Westlake, the view overlooking the park and the LA skyline.

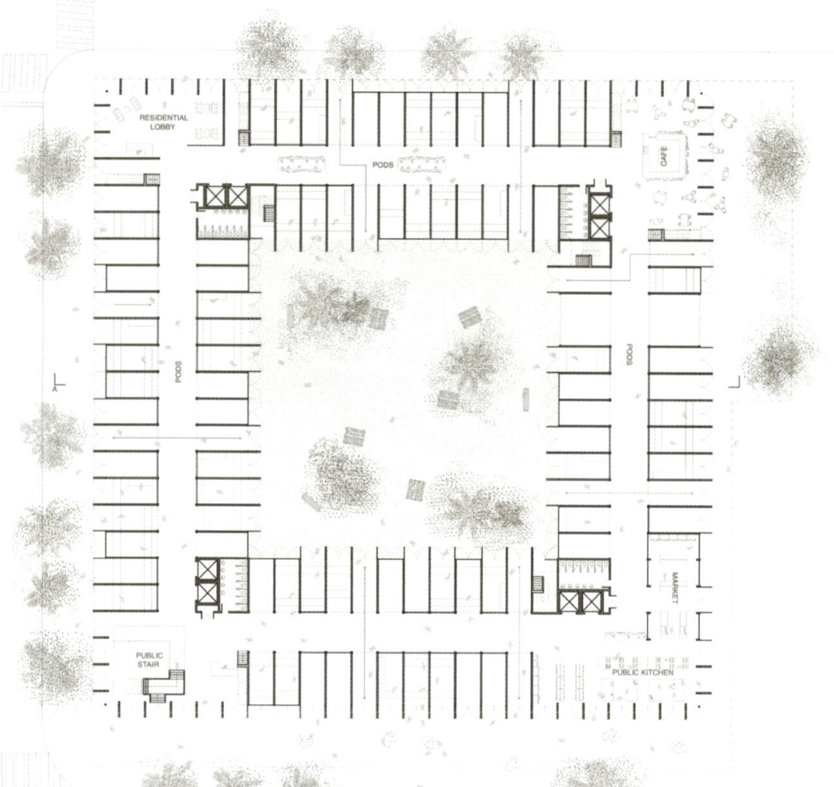

GROUND FLOOR

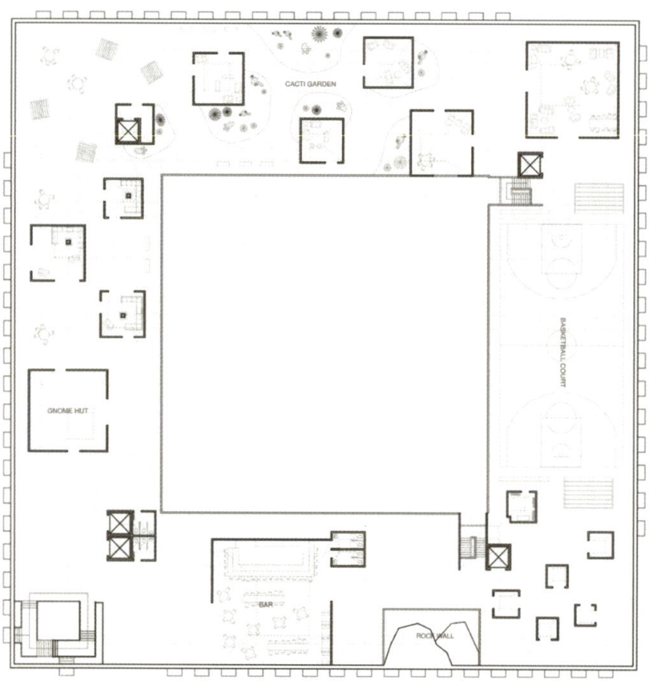

ROOF

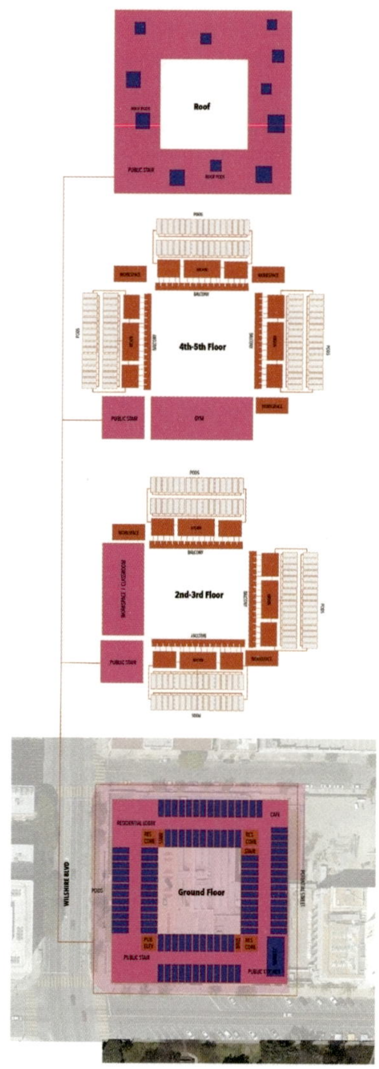

TOPOGRAPHIC PLAN

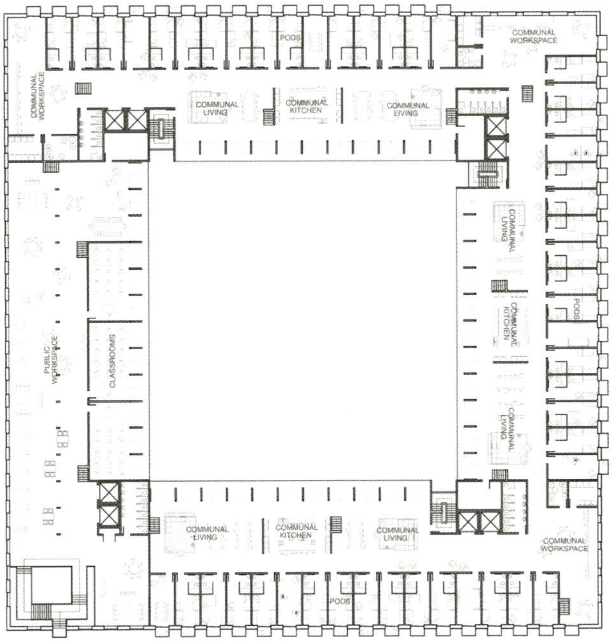

RESIDENTIAL BOTTOM FLOOR

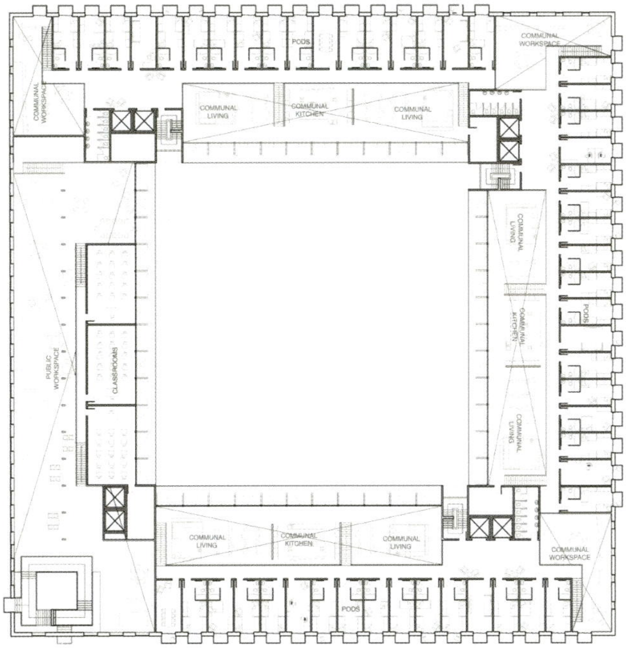

RESIDENTIAL TOP FLOOR

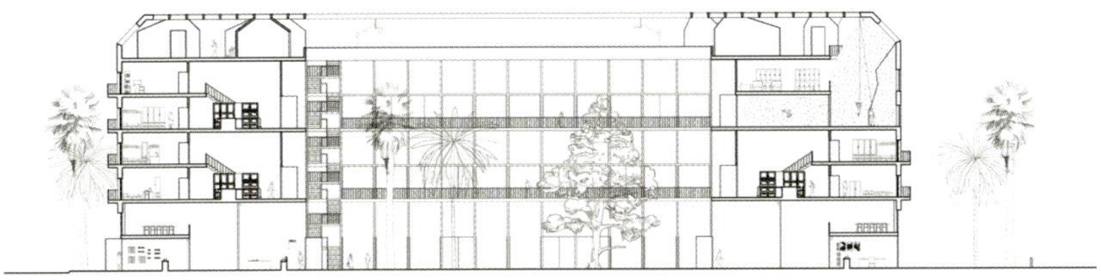

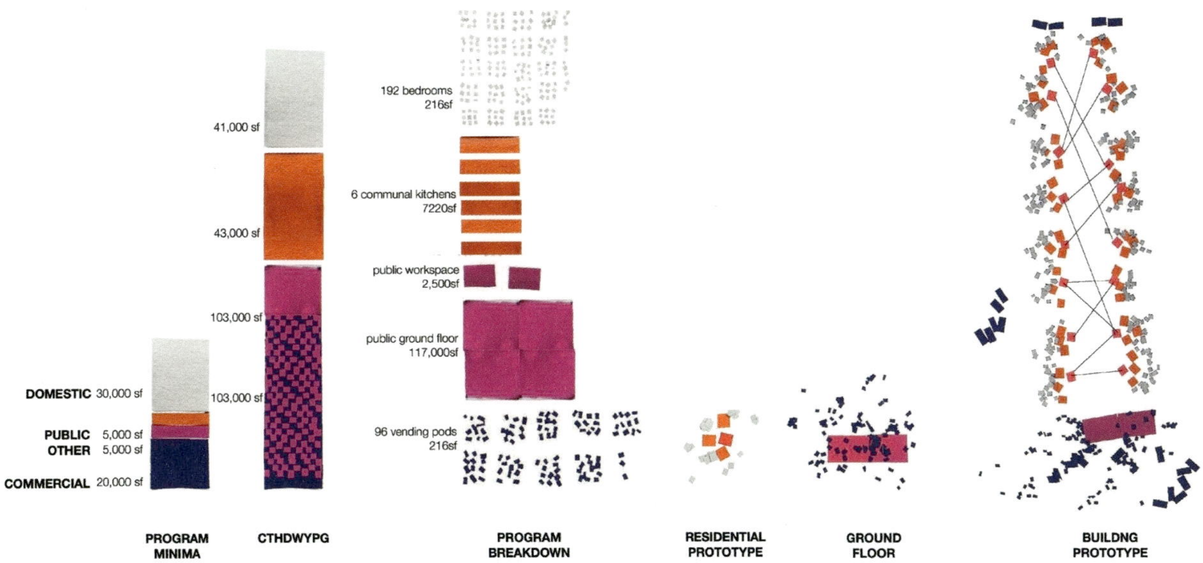

PROGRAM BREAKDOWN

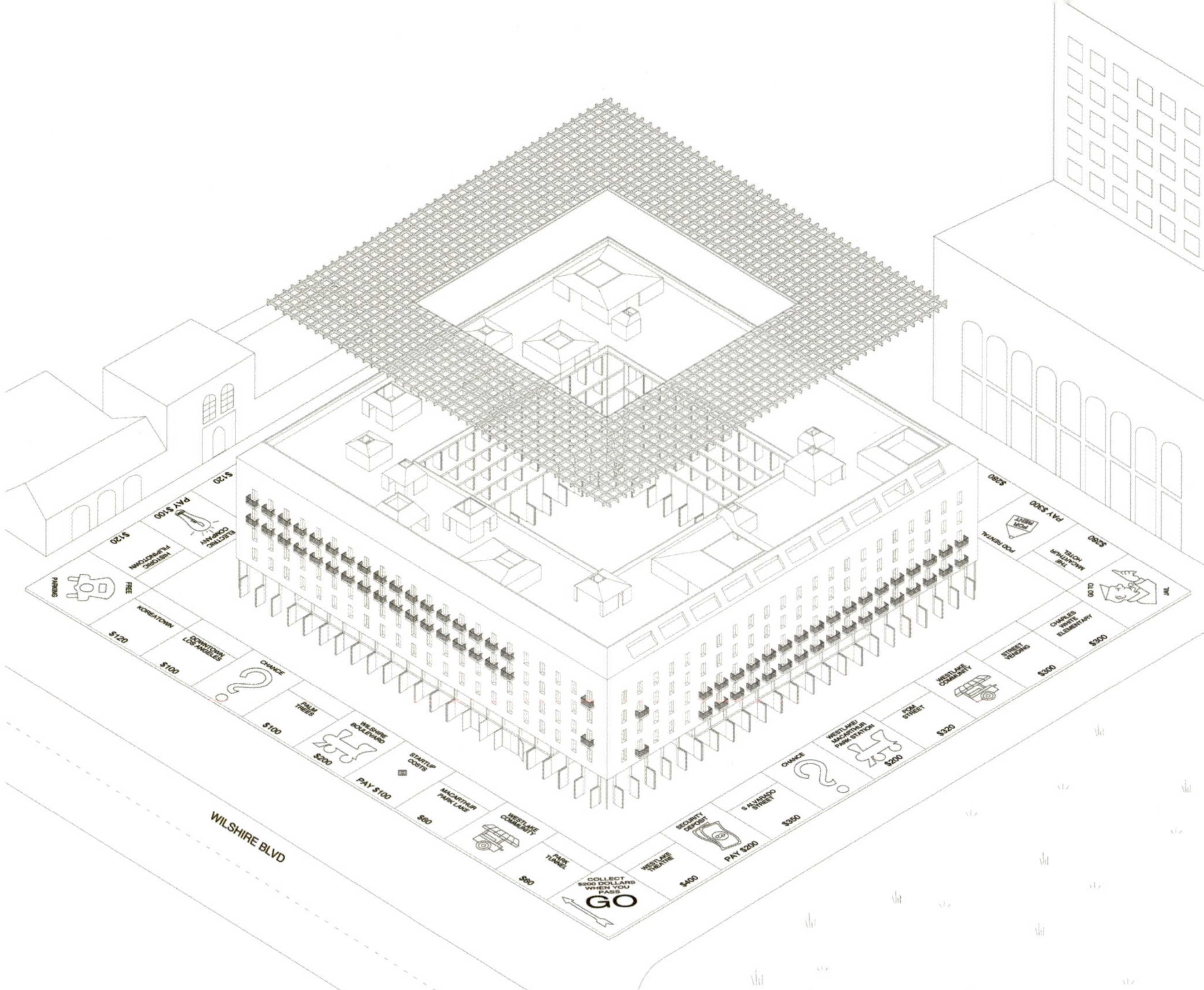

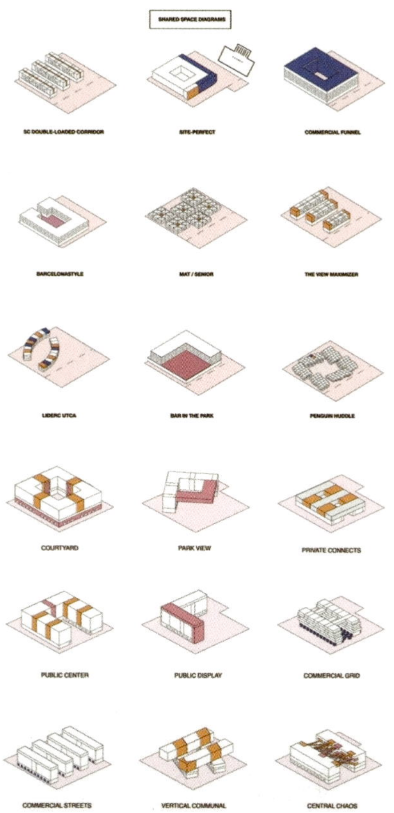

TYPOLOGY STUDIES

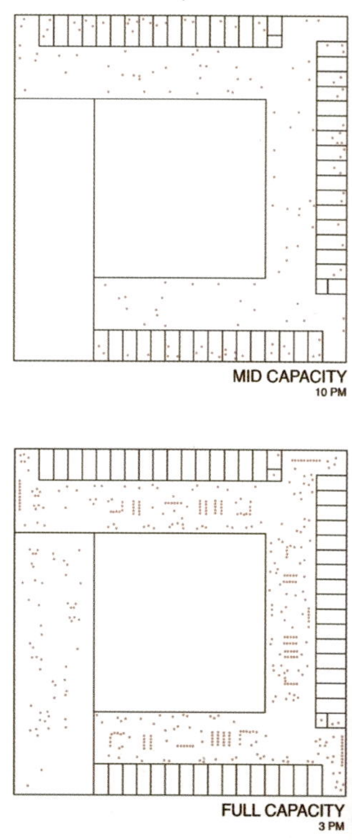

OCCUPANCY STUDIES

FINE FURNITURE

blu

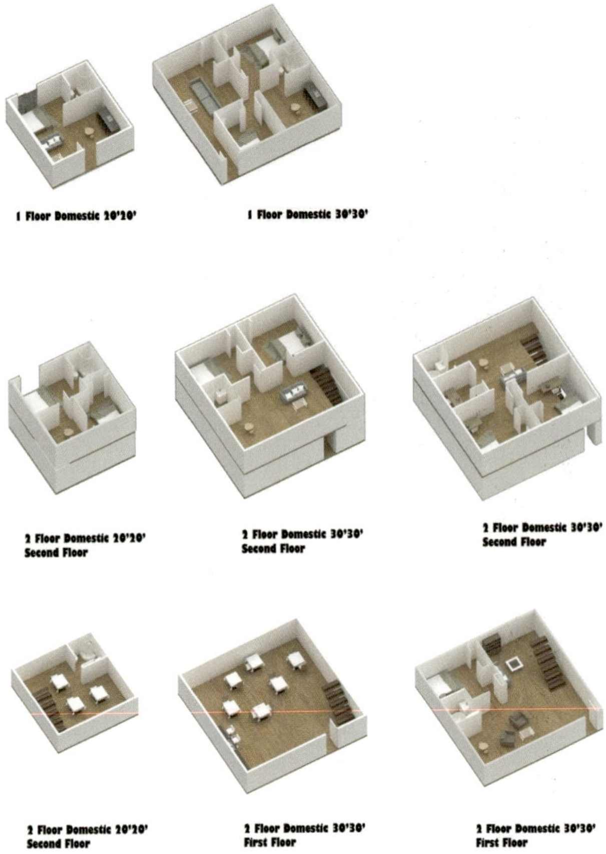

BACK ON OUR FEET

PROJECT BY

DONOVAN HERNANDEZ
RUOHAN JIANG

PROJECT INFORMATION

Location : Westlake, La
Use : Mid-Rise Mix-Use
Site area : 90,000sf
Floor : 5F
Height : 44ft
The number of parking spaces : 152
The number of households : 200

ADAPTABLE LIVING

An exploration into hybrid mat-housing paired with and inspired by Parisian block housing guided the project. Intended to grow and adapt over time, the main structure of Back On Our Feet is two large concrete slab floors that sandwich concrete bearing walls within the individual apartment units. Featuring a unique emphasis on two-story living spaces, the living arrangement encourages a "mix-use" program distribution, with live and work apartments in shared space. This variability added to the wide target audience for individuals looking for anywhere from a single story 400 sq foot unit to a 1300 sq foot multi story unit.

파리의 블록 주거와 짝을 이루고 영감을 받은 하이브리드 매트-주거에 대한 탐구가 이 프로젝트를 이끌었다. 시간이 지남에 따라 성장하고 적응하게 설계된 Back On Our Feet의 주요 구조는 개별 아파트 단위 내에서 콘크리트 내력벽을 사이에 둔 두 개의 대형 콘크리트 슬래브 바닥이다. 2층짜리 주거 공간을 독특하게 강조한 것이 특징인 이 주거 배치는 공동 공간에서 생활 및 작업이 이루어지는 아파트와 함께 "혼합 사용" 프로그램을 장려한다. 이러한 다양성은 단층 400sf 단위부터 다층 1300sf 단위까지 모든 곳을 찾는 광범위한 개인들을 대상으로 한다.

SITE ANALYSIS

Our site needs a new community core that directly benefits those who live and work in the area. There is a large homeless population but a work-stay program can address this by empowering and providing the opportunity to help revive Westlake while also providing a place to call home for those who need it the most. This area is also designated as a food swamp, where residents have to travel more than a mile in some areas to access fresh, healthy, and affordable food which has a multitude of negative effects. Infrastructure for community gardening initiatives can make it possible for the people to have access to good clean food while also playing an important part in producing it. Westlake does not need another housing or gentrification project, it needs architecture that will uplift and empower the people to become the identity of the area again. This project can help Westlake get back on its feet.

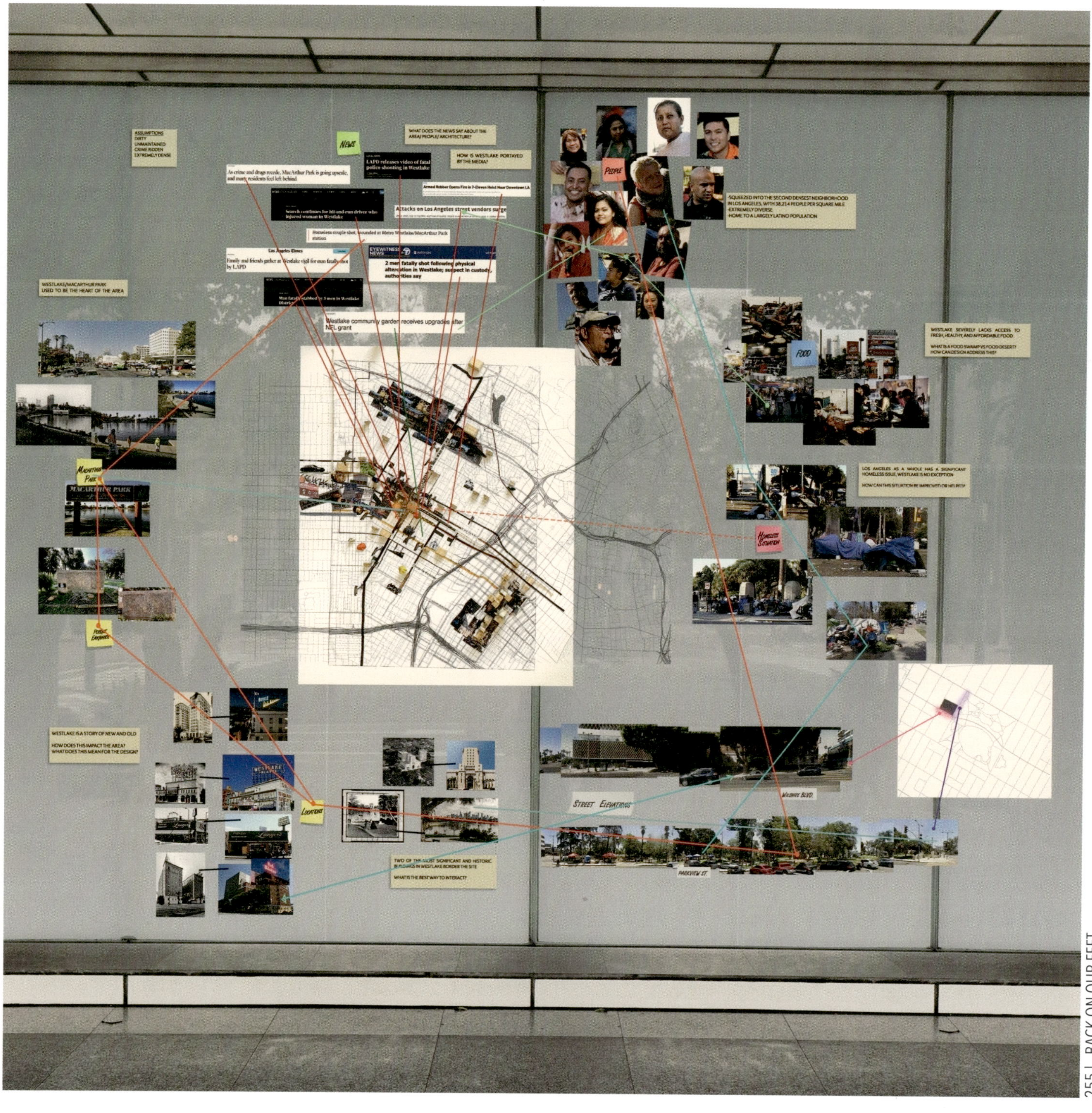

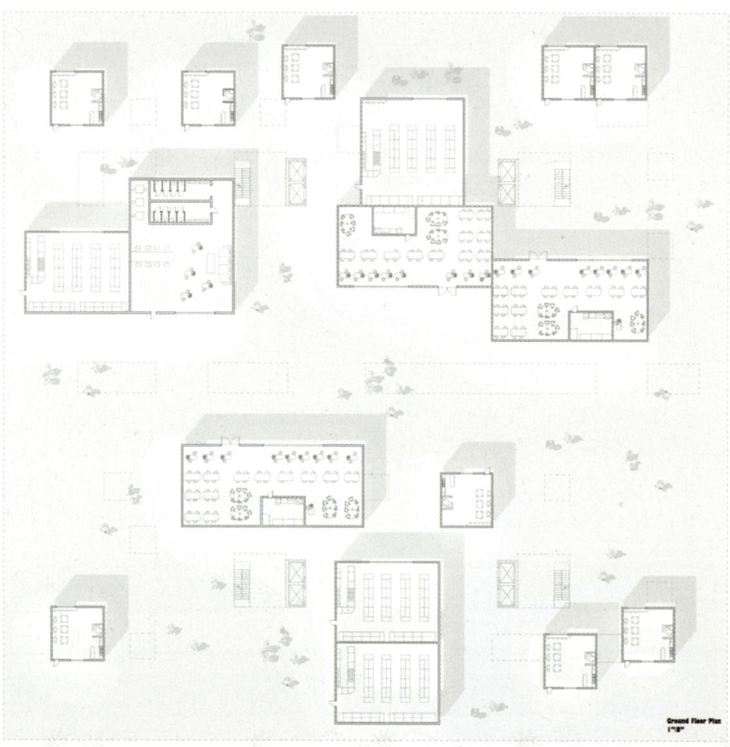
Ground Floor Plan

OVERALL STRATEGY

The influence of Parisian housing and the idea of creating a "wall" to the city, is the inspiration behind the general aggregation of the units. This clear strategy adds significance to the openings created that connect both sides of the project as well as the wall facing MacArthur Park.

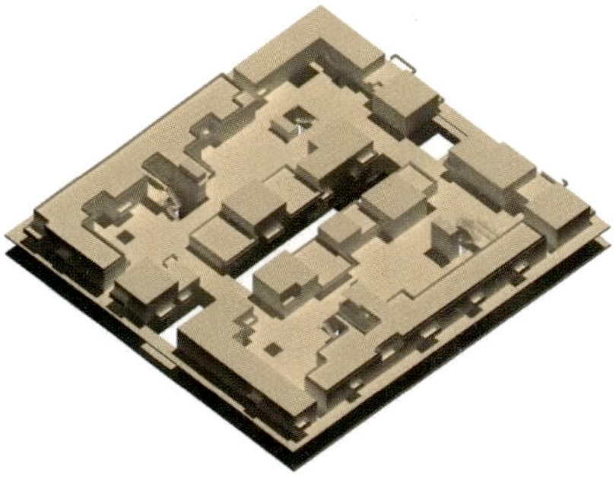

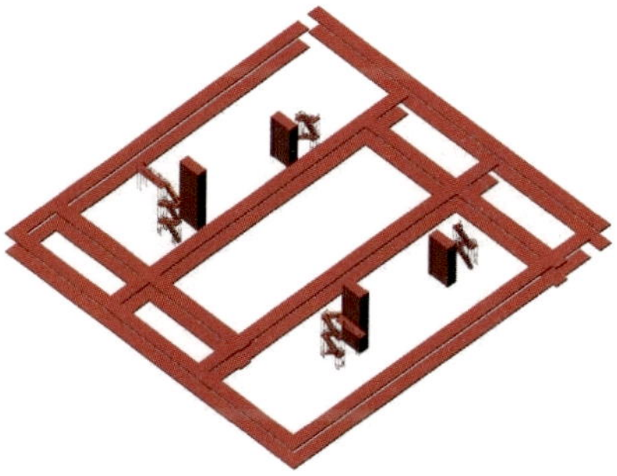

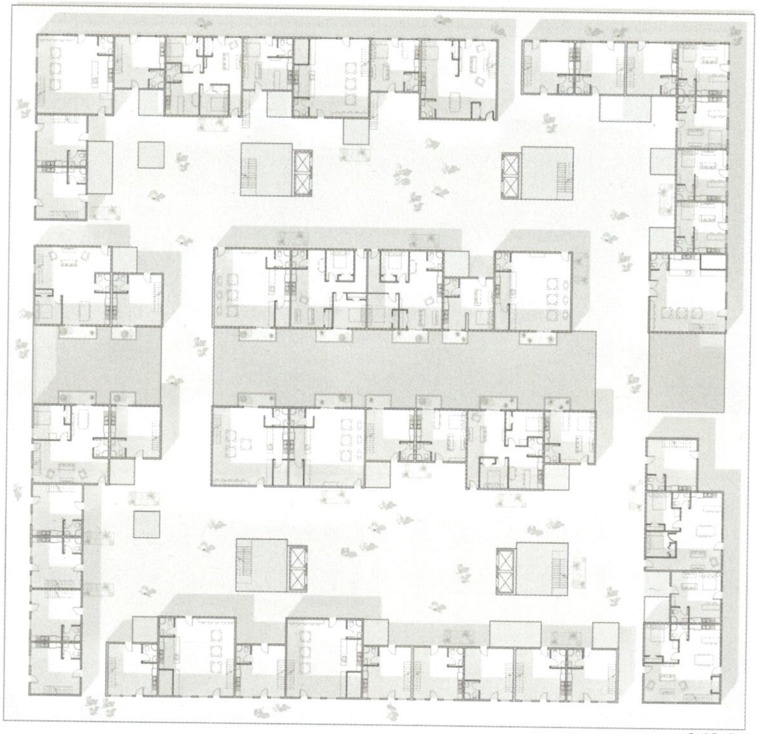

Fourth Floor Plan
1"/8'

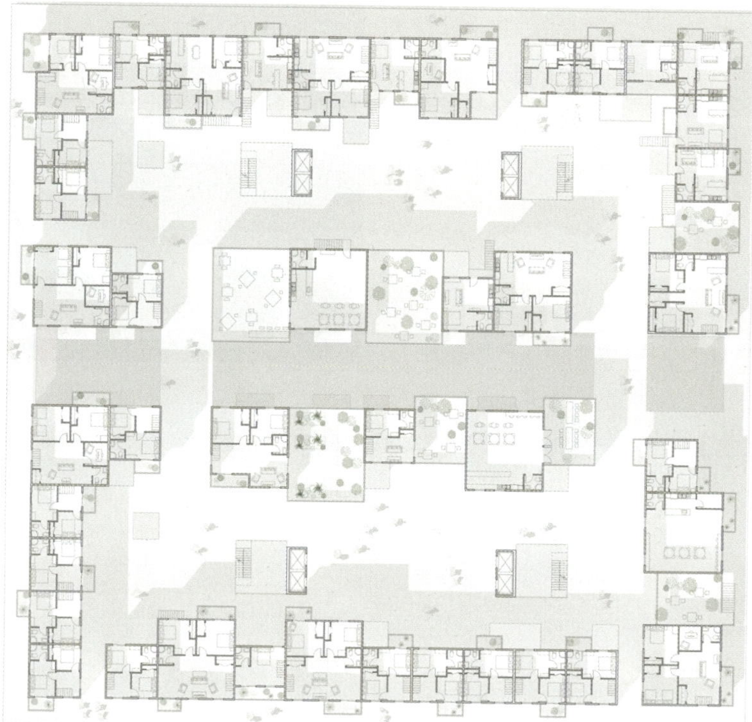

Fifth Floor Plan
1"/8'

With all of the additional horizontal space created with the large floor plates, the opportunity to do more than just provide housing and commercial space in the city presents itself. With the previously mentioned issues of homelessness and lack of access to fresh and healthy food, the project creates space to facilitate the addressing of these issues. With the inclusion of small and mobile agricultural gardens, they allow for community interaction, adaptability, and cultivable space that ordinarily would not exist in such a dense urban context. These gardens will not only become the center of the project but also of the surrounding community and will uplift the people of Westlake together through design.

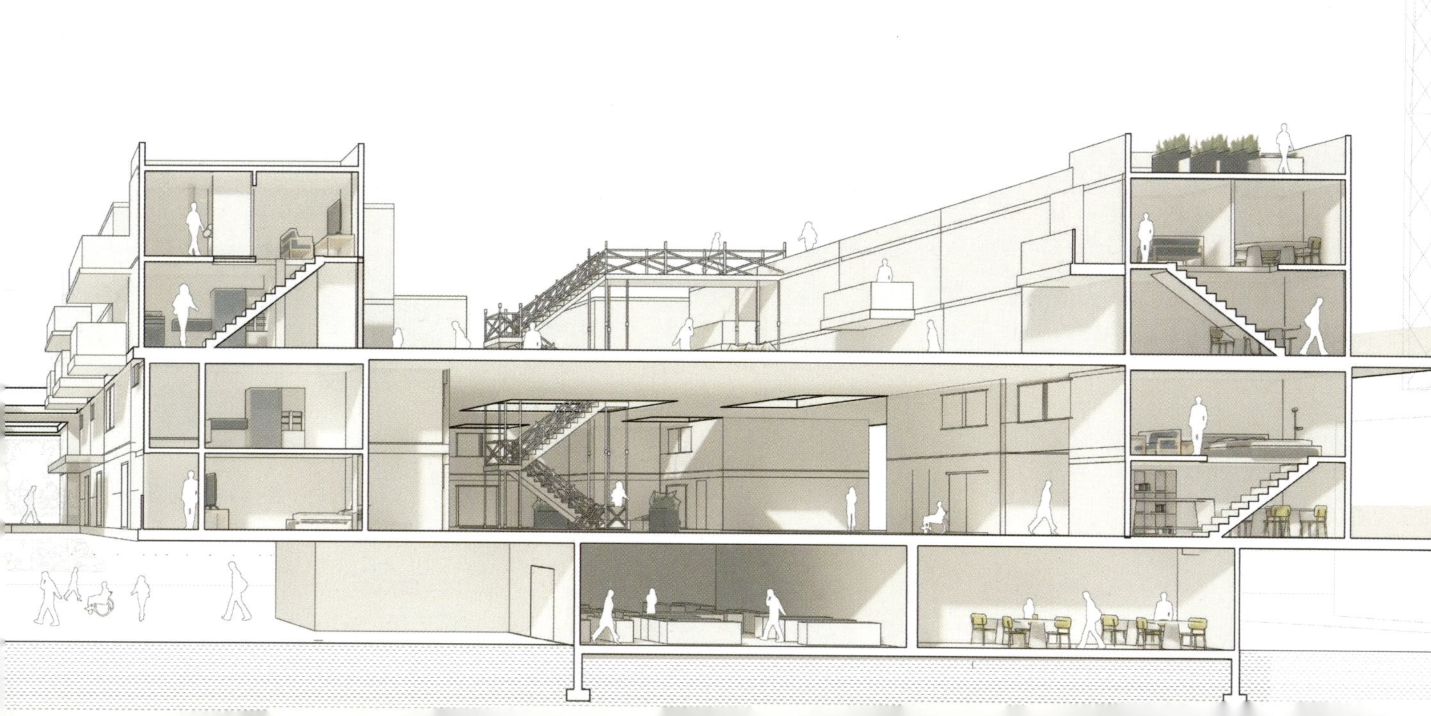

SANTA MONICA, LOS ANGELES COUNTY, CALIFORNIA

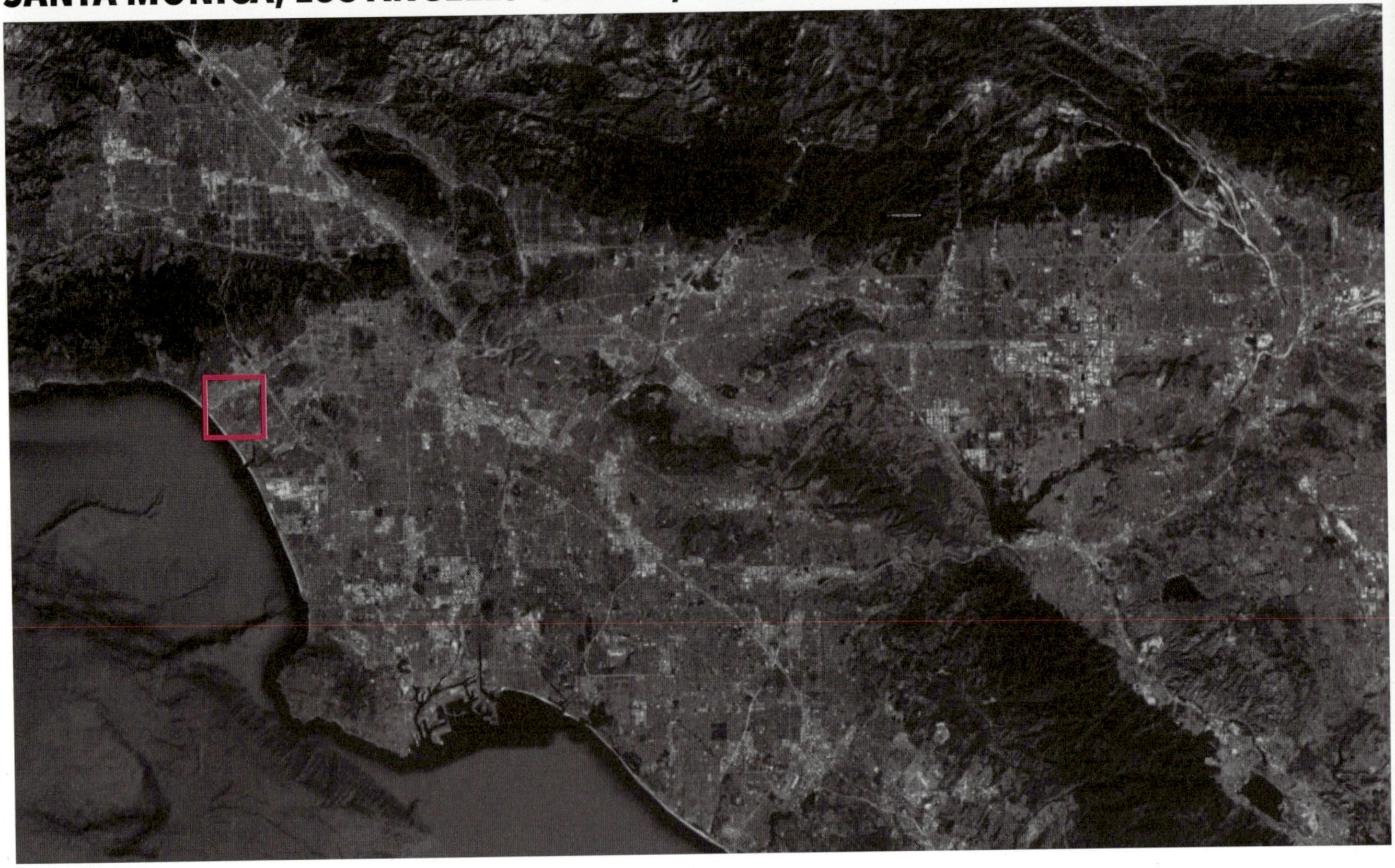

RESEARCH BY
KELLY CHAN
TONY DAI

Santa Monica is a city in Los Angeles County, situated along Santa Monica Bay on California's South Coast. Santa Monica's 2020 U.S. Census population was 93,076. Santa Monica is a popular resort town, owing to its climate, beaches, and hospitality industry. [14] It has a diverse economy, hosting headquarters of companies such as Hulu, Universal Music Group, Lionsgate Films, and The Recording Academy.

산타 모니카는 미국 캘리포니아 남부 해안의 산타모니카 베이를 따라 있는 로스앤젤레스 카운티의 도시이다. 2020년의 미국 인구 조사 결과 산타 모니카의 인구는 93,076명으로 집계되었다. 산타모니카는 기후, 해변, 환대 산업으로 인해 인기 있는 리조트 타운이다. Hulu, Universal Music Group, Lionsgate Films, The Recording Academy 등 회사에서 이곳에 본사를 두는 등 다양한 경제가 자리 잡고 있다.

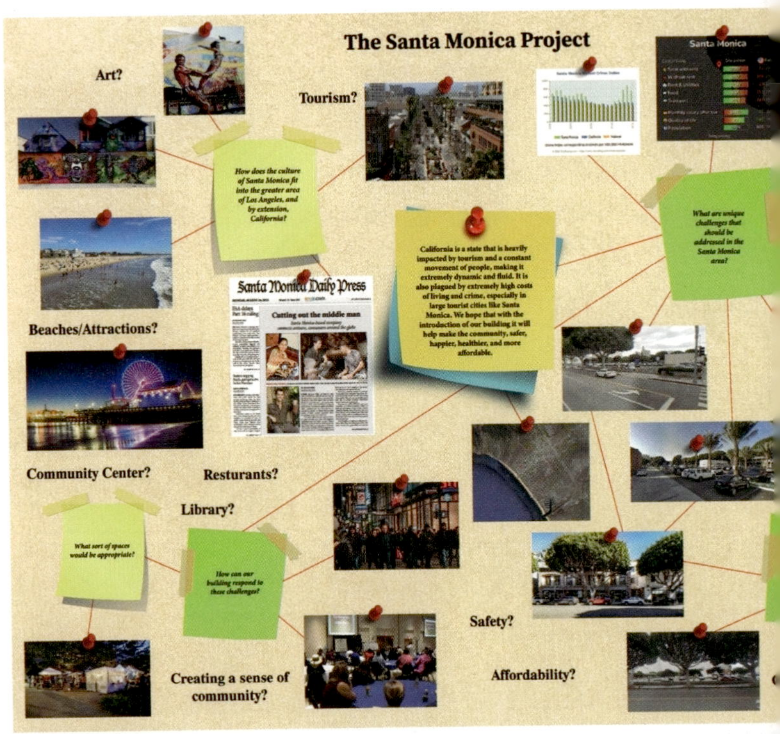

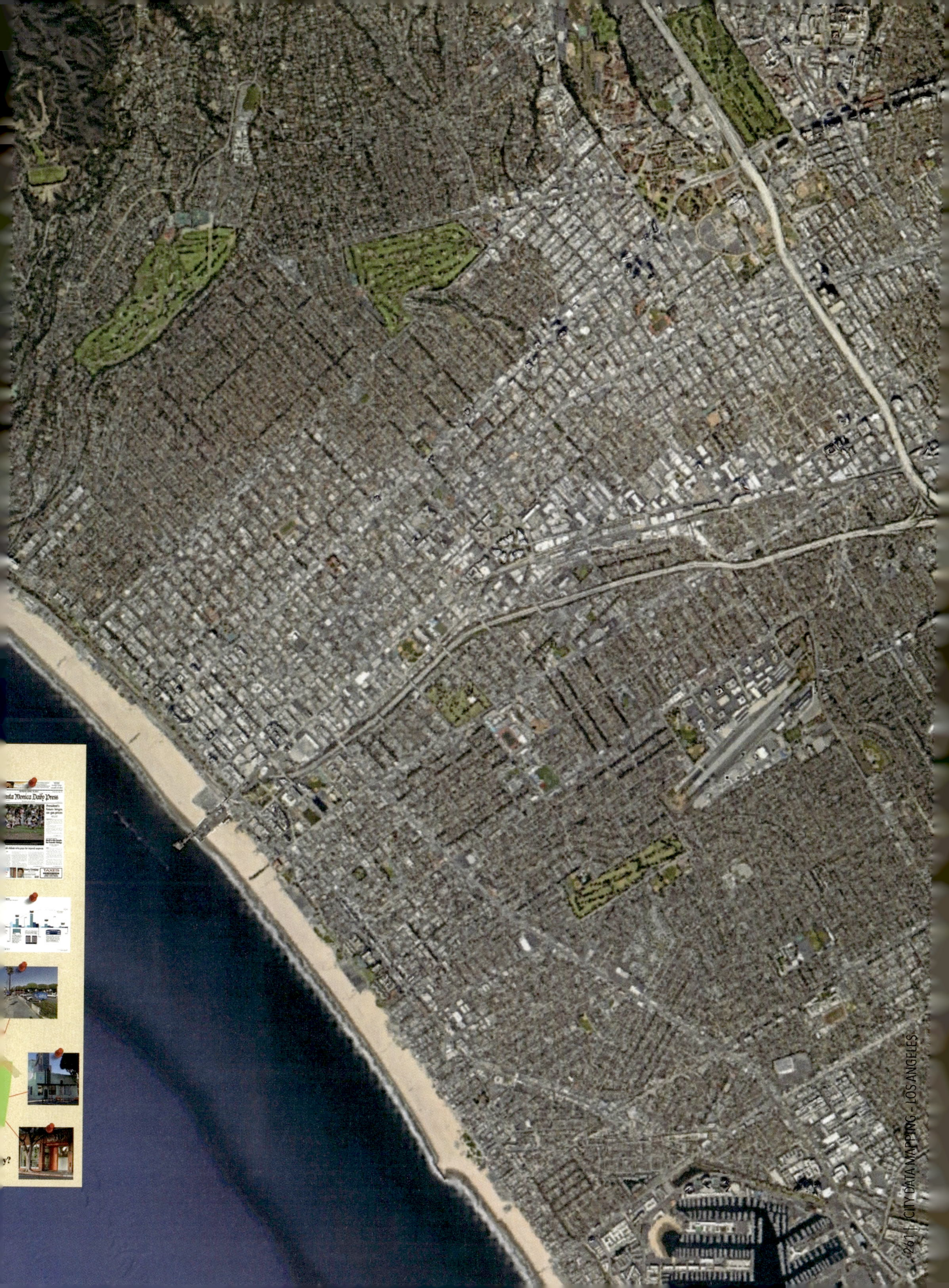

INFRASTRUCTURE

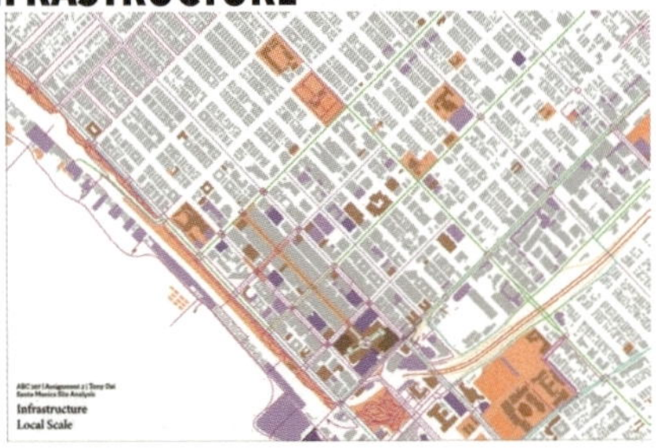

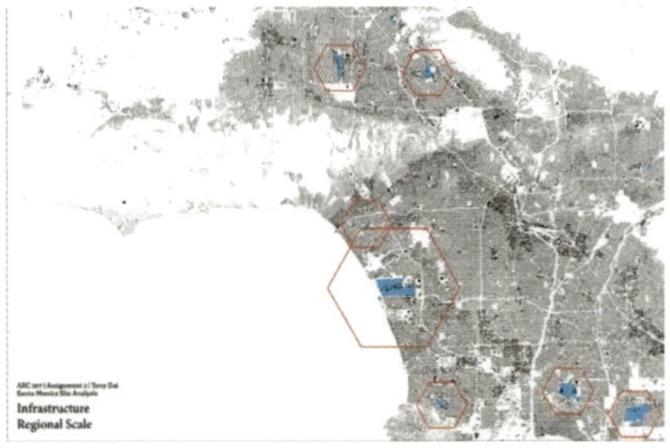

CLIMATE

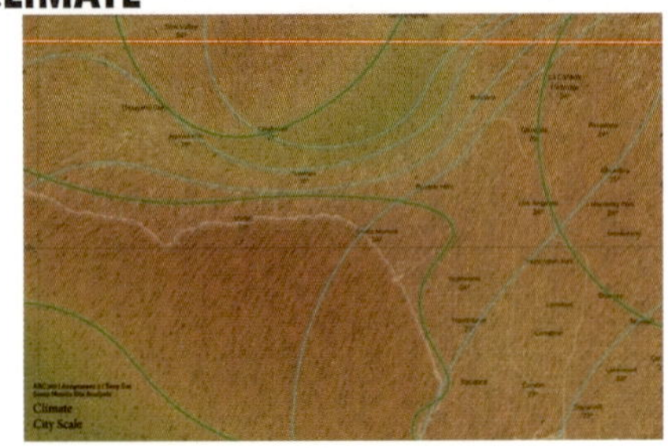

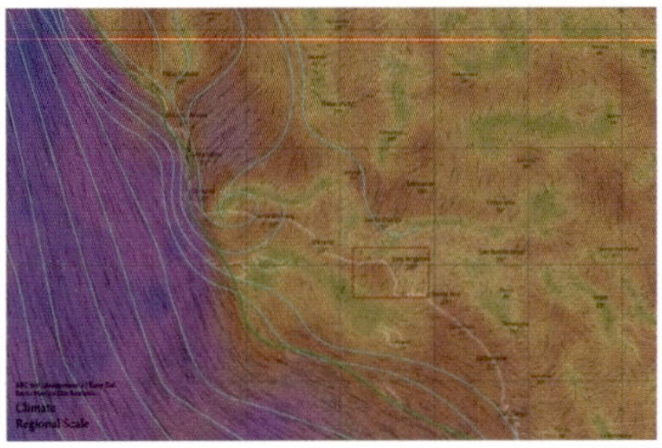

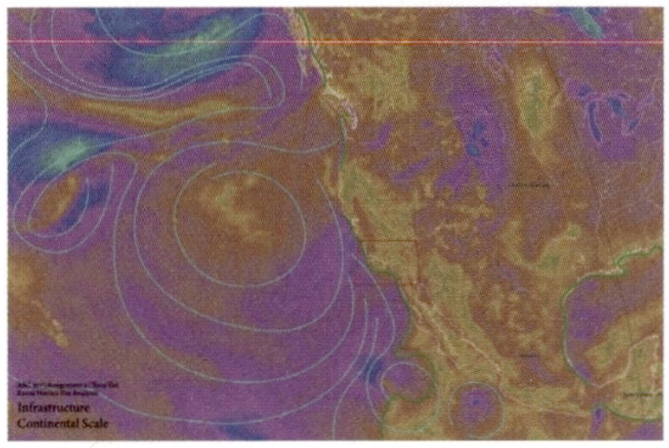

WATER

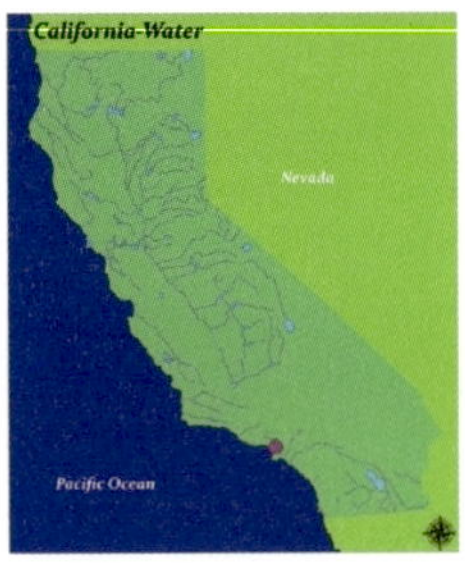

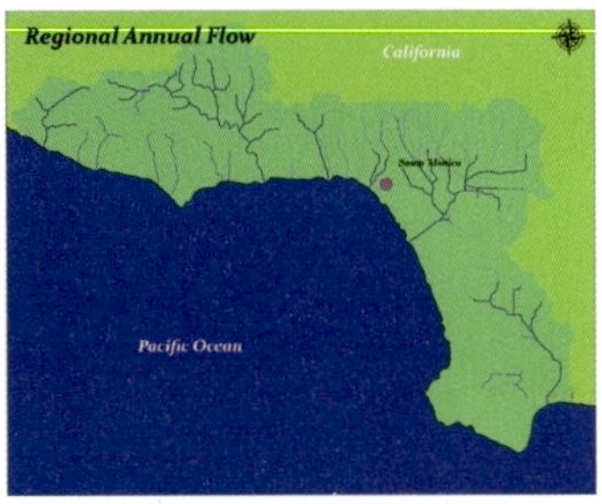

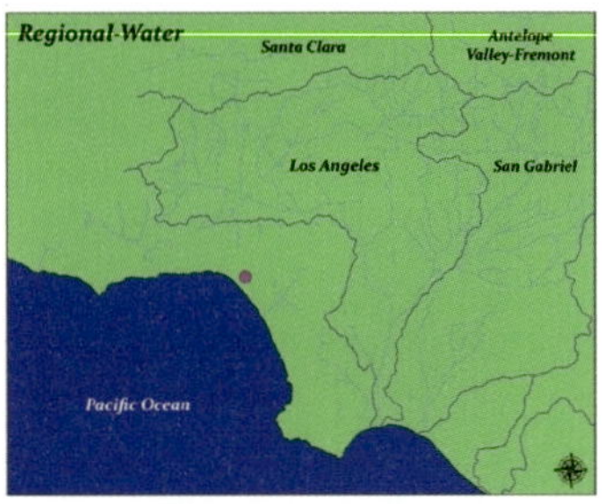

MATERIALITY

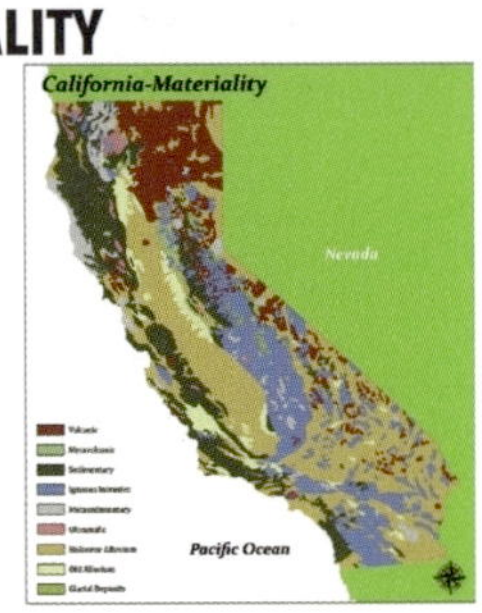

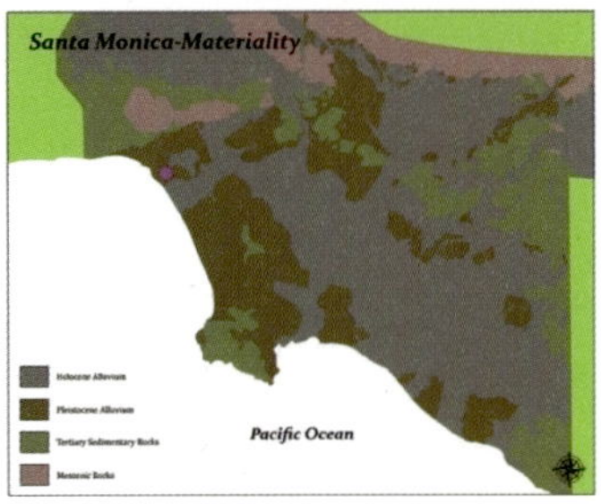

DENSITY

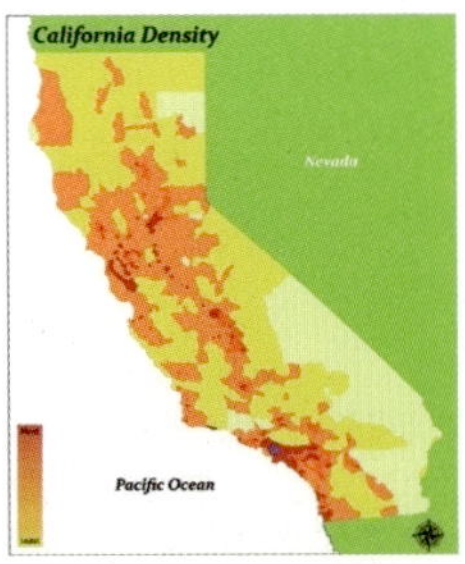

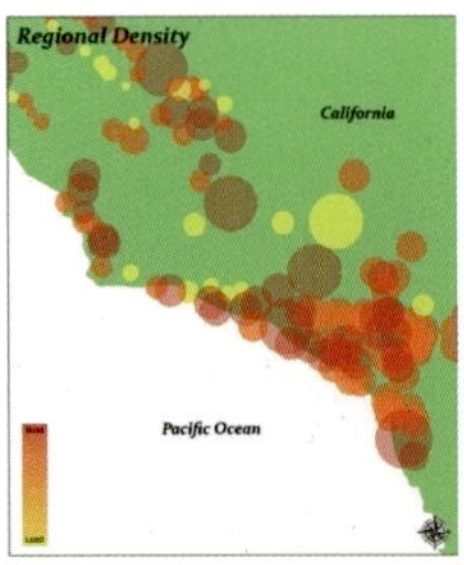

BULDING MORPHOLOGY

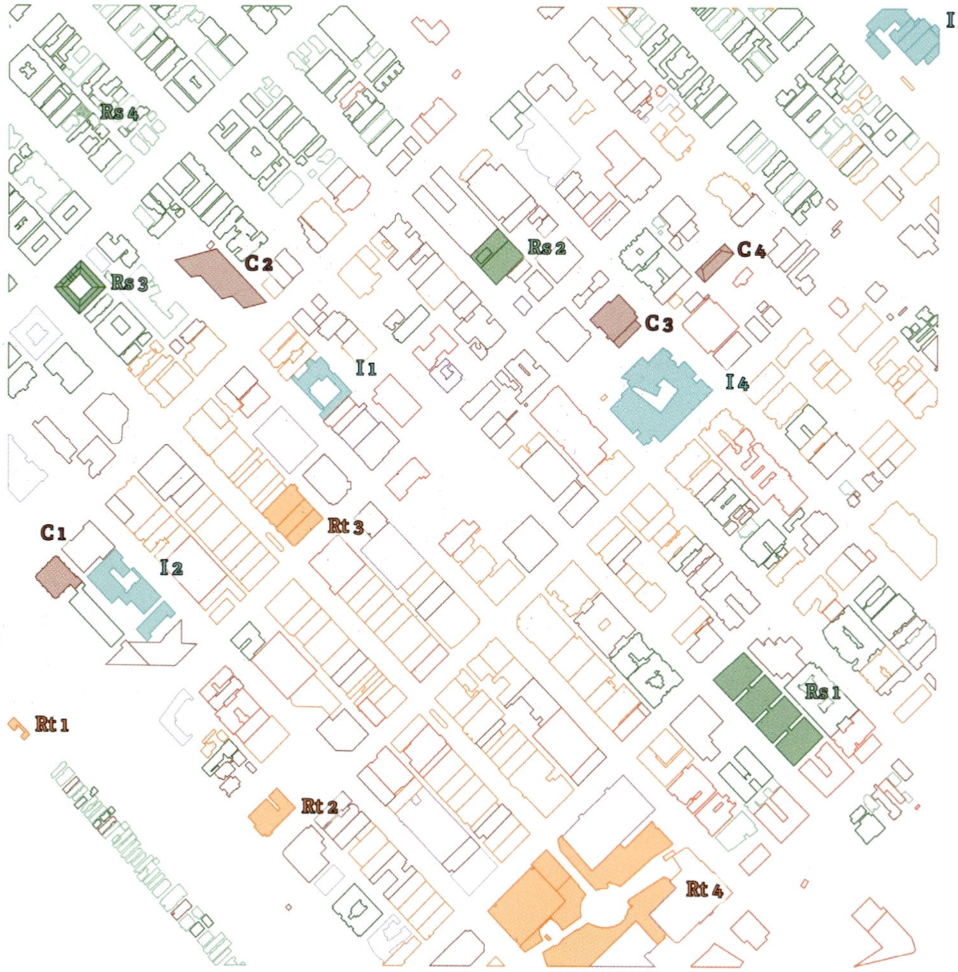

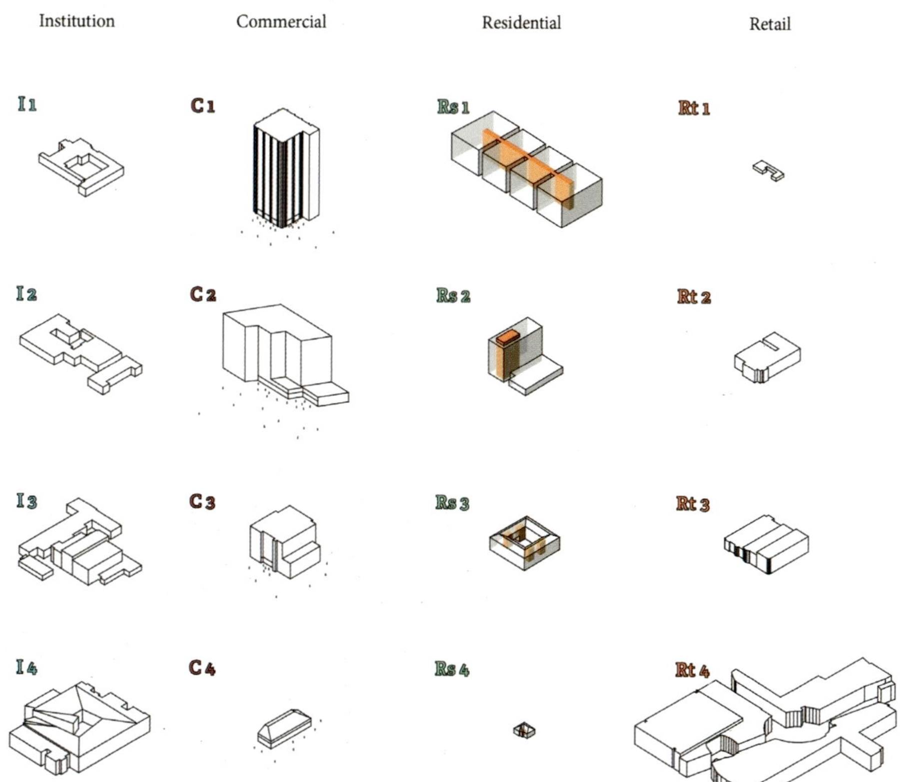

Institutions in the area are limited to libraries, educational institutions and religious institutions. The Santa Monica public Library is an important community gathering place.

Commercial buildings in the are largely comprised of office spaces in dedicated office buildings closer to the coast. As the buildings move inland, office spaces are more frequently adapted from MRMU buildings.

Retail spaces within the area are quite diverse. Retail spaces range from temporary popup locations built with shipping containers in a high traffic area all the way to traditional open air shopping malls such as the Santa Monica Place.

Residential structures also allow a similar pattern as commercial structures where larger apartment buildings are closer to the coast and the smaller single family homes are further inland.

REINSTATING CONNECTIONS

PROJECT BY
KELLY CHAN
TONY DAI

In our project, the main objective is to give the local residents affordable housing units to make the streets safer and promote income equality at our Santa Monica site. The location around our designated site is mainly populated by commercial buildings, which generally shut down at night-time causing the streets to be empty and dangerous. There is a high crime rate and high costs of living throughout California, but especially in heavily populated tourist areas like Santa Monica. The famous pier draws in a lot of visitors, making the general population in the area transitory. Thus, we hope to provide a mixing space for both tourists and locals alike, which would also help relieve tensions between the very polarized groups.

우리 프로젝트의 주된 목표는 지역 주민들에게 저렴한 주택 단위를 제공하면서 거리를 더 안전하게 만들고 산타 모니카 부지의 소득 평등을 촉진하는 것입니다. 대지 주위에는 주로 상업용 건물들이 들어서 있는데, 일반적으로 야간에 문을 닫아 거리가 텅 비고 위험하다. 캘리포니아 전역은 원체 범죄율이 높고 생활비 또한 비싼 편이지만, 특히 산타 모니카처럼 인구가 밀집한 관광지에서는 더욱 그렇다. 유명한 부두는 많은 방문객을 끌어들여 해당 지역의 일반 인구를 일시적으로 만든다. 따라서 우리는 관광객과 현지인 모두에게 혼합 공간을 제공하여 양극화된 그룹 간의 긴장을 완화하는 데도 도움이 되기를 희망한다.

Our residential building area consists of multiple types of units: single units, two-story units, and three-story units. Additionally, here are six different sized units to maximize variety and choice. There are three floors of units, two full floors and one half floor. Furthermore, all the units are uniquely designed. Some have balconies, outdoor hallways, both, or none, and are carefully oriented and organized to provide distinctive experiences for all inhabitants. In this building, we emphasize choice. Perforations on each level help regulate sunlight, which is more than abundant in our Santa Monica location. The residential units are located above a public platform with commercial spaces, along with a large public underground garage.

Below the housing units, the large, warped public platform provides a flat basketball court that can double as a stage or area for popup markets, along with indents for small community gatherings. Situated below that are public commercial spaces. These areas feature lounge spaces and food courts, along with a variety of shops. The public platform dips down to the sidewalk level at certain times to invite any passersby to visit our building.

The commercial and residential spaces are connected by three large egress columns, which is a communal gathering space mainly for residents. All three columns are open on one side for air circulation and sunlight. The space is large enough to accommodate garage sales, picnics, and more on each floor. Below the residential floors, the egress columns also host multiple exterior balconies as a way to integrate the public and the private and is shared by all.

Above the three residential floors is a communal green roof, with trees and a grass floor. A community garden lines the edges of the roof. Residents are able to host local events, meetings, and whatever else they desire. The thick, three feet floor slabs help reduce noise and vibrations, so residents directly below do not have to worry about disturbances.

In total, there are seven inhabitable spaces. The green roof, three residential floors, a public platform, one purely commercial floor, and an underground parking lot that has mixed commercial spaces within. With this variety, there should be more than enough space to enrich and support the local community and any visitors that decide to come.

Site Analysis-Pedestrian Movement

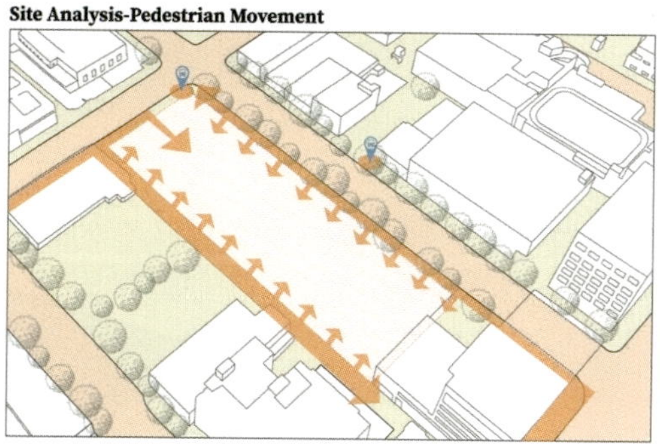

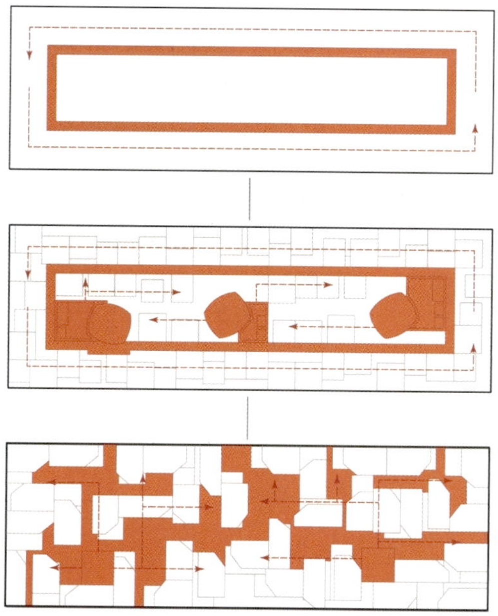

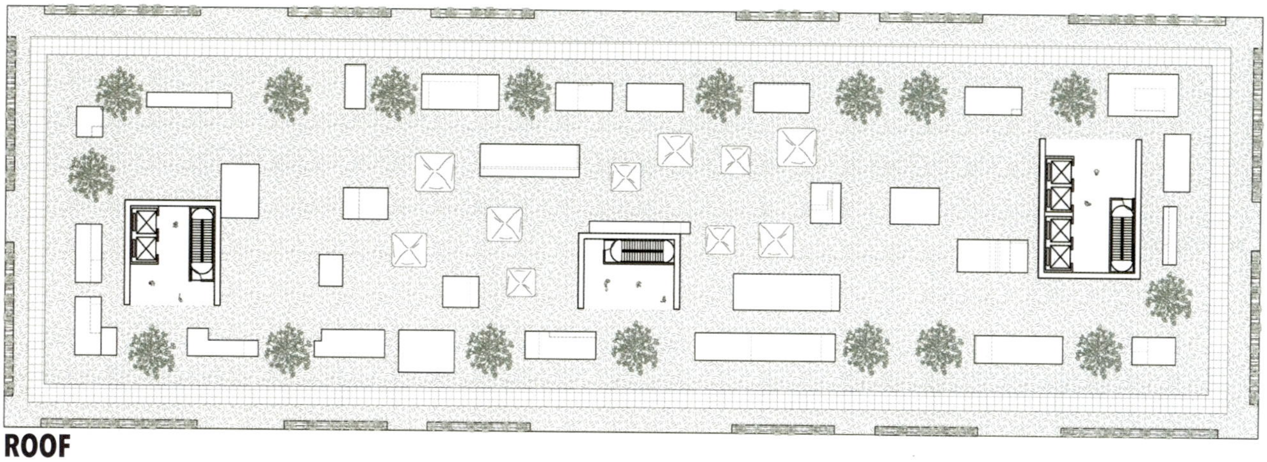

ROOF

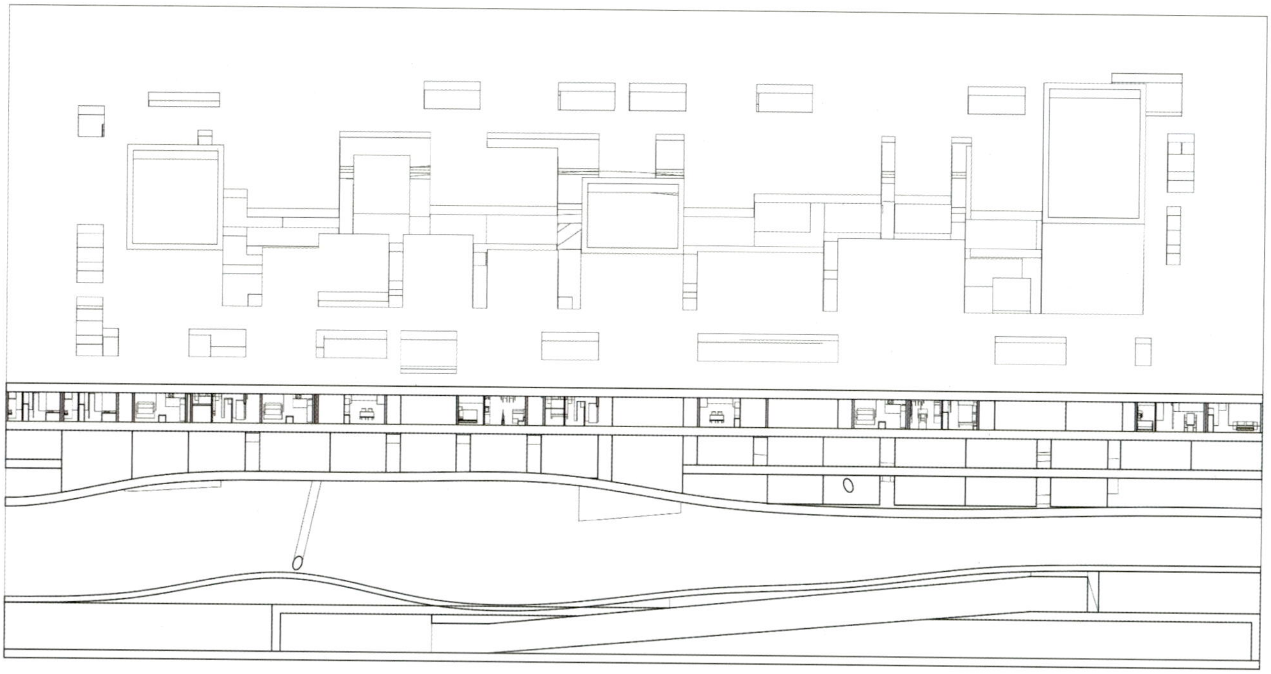

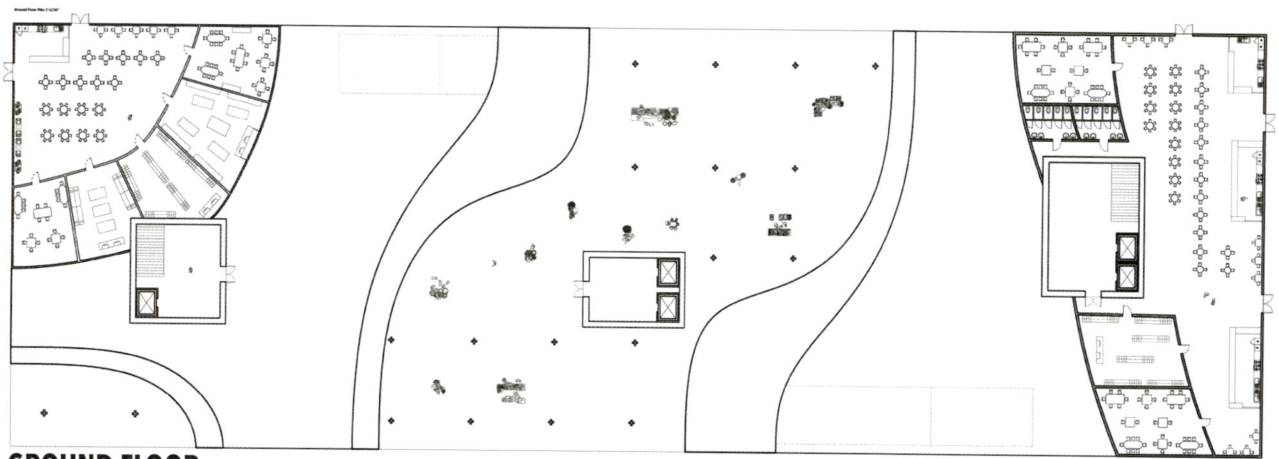

GROUND FLOOR

5TH FLOOR

4TH FLOOR

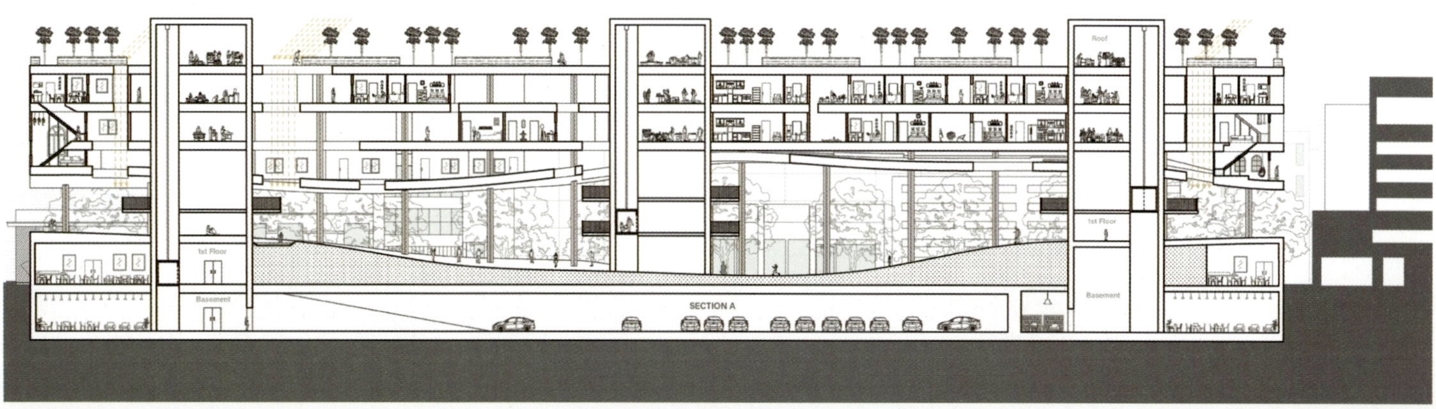

2023년 2월 10일 1쇄 발행

저자 : 진교진 · Bing Bu

펴낸이 / 김기현

발행처 / 도서출판 시공문화사 (***Spacetime***)
등　록 / 1993년 3월 12일
주　소 / (03733) 서울시 서대문구 독립문공원길 13 극동프라자 5층
전　화 / 02)3147-1212 · 2323
팩　스 / 02)3147-2626

www.spacetime.co.kr spacetime@korea.com

인　쇄 / 예림인쇄
제　본 / 보경문화사
제　판 / 에코프린트

ISBN 978-89-5592-466-4

정가 20,000원